A Color Atlas of Post-Harvest Diseases and Disorders of Fruits and Vegetables

Volume 1: General Introduction and Fruits

Anna L. Snowdon Ph. D., D.I.C.

University of Cambridge

CRC Press, Inc.
Boca Raton, Florida

Copyright © A.L. Snowdon, 1990
All rights reserved
First published in 1990 by Wolfe Scientific Publications,
2-16 Torrington Place, London WC1E 7LT, UK

Printed by Grafos, Arte Sobre Papel, Barcelona, Spain
This book is Volume 1 in a two-volume set. Volume 1 covers a general
introduction and fruits. Volume 2 covers vegetables.

This book represents information obtained from authentic and highly
regarded resources. Reprinted material is quoted with permission, and
sources are indicated. A wide variety of references are listed. Every
reasonable effort has been made to give reliable data and information, but the
author and the publisher cannot assume responsibility for the validity of all
materials or for the consequences of their use.

All rights reserved. This book, or any parts thereof, may not be reproduced
in any form without written consent from the publisher.

Direct all inquiries to CRC Press, Inc., 2000 Corporate Blvd., N.W., Boca
Raton, Florida, 33431.

International Standard Book Number 0-8493-7101-5

Contents

Preface	7
Acknowledgements	9
Abbreviations	9

1. NATURE AND CAUSES OF POST-HARVEST DETERIORATION — 11
  Investigation of losses — 12
POST-HARVEST PHYSIOLOGY OF FRUITS AND VEGETABLES — 13
  Structure and function — 13
  Maturation, ripening and senescence — 14
NATURE OF POST-HARVEST DETERIORATION — 16
  Physiological change — 16
  Physical damage — 16
  Chemical injury — 16
  Pathological decay — 16
FACTORS INFLUENCING POST-HARVEST DETERIORATION — 18
  Pre-harvest factors — 18
  Harvesting and handling techniques — 19
  Conditions during storage — 22
  Conditions during transport — 24
ASPECTS OF FORENSIC PLANT PATHOLOGY — 26
  Cargo surveys and the interpretation of evidence — 26
  Identification of diseases and disorders — 29
CLASSIFICATION OF DISORDERS — 30
  Nutritional disorders — 30
  Respiratory disorders — 30
  Temperature disorders — 31
  Miscellaneous disorders — 33
CLASSIFICATION OF DISEASES AND DISEASE ORGANISMS — 34
  Types of disease — 34
  Agents of disease — 39

2. CITRUS FRUITS — 54
Oranges, Mandarins, Grapefruits, Lemons and Limes — 54
Diseases:
  Alternaria rot — 56
  Anthracnose — 58
  Aspergillus black mould rot — 59
  Bacterial canker — 60
  Black pit — 61
  Black spot — 62
  Brown rot — 64
  Cottony rot — 66
  Fusarium rot — 67
  Greasy spot rind blotch — 68
  Green and blue mould rots — 70
  Grey mould rot — 73
  Melanose — 74
  Scab — 76
  Septoria spot — 77
  Sour rot — 78
  Stem-end rots — 80
  Trichoderma rot — 82
  Other diseases — 84
Disorders:
  Chilling injury — 87
  Freezing injury — 88
  Oleocellosis — 89
  Other disorders — 90

3. MISCELLANEOUS TROPICAL AND SUBTROPICAL FRUITS — 92
Avocados — 93
Diseases:
  Anthracnose — 94
  Cercospora spot or blotch — 95
  Dothiorella rot — 96
  Scab — 98
  Stem-end rots — 99
  Other diseases — 100
Disorders:
  Chilling injury — 102
  Other disorders — 103

Bananas — 104
Diseases:
  Anthracnose — 106
  Cigar-end rot — 108
  Crown mould, crown rot and finger-stalk rot — 110
  Finger rot — 112
  Johnston spot or pitting — 113
  Sigatoka — 114
  Other diseases — 116
Disorders:
  Chilling injury — 120
  Other disorders — 121

Guavas	**122**
Litchis	**126**
Mangoes	**128**
Diseases:	
Anthracnose	130
Bacterial black spot	132
Black mould rot	133
Botryodiplodia rot	134
Stem-end rots	135
Other diseases	136
Disorders:	
Chilling injury	138
Other disorders	139
Papayas	**140**
Diseases:	
Anthracnose	141
Black rot	142
Phytophthora rot	144
Rhizopus rot	146
Stem-end rots	147
Other diseases	148
Disorders:	
Chilling injury	151
Passion fruits	**152**
Persimmons	**154**
Pineapples	**155**
Diseases:	
Black rot	156
Fruitlet core rot	158
Other diseases	160
Disorders:	
Chilling injury	162
Pomegranates	**163**
Sour sops and sweet sops	**164**
Other tropical and subtropical fruits	**166**
4. POME FRUITS	**170**
Apples and Pears	**170**
Diseases:	
Alternaria rot	172
Bitter rot	174
Black rot and white rot	176
Blue mould rot	178
Brown rot	180
Core rot and mouldy core	182
Cylindrocarpon rot	184
Gloeosporium rot	186
Grey mould rot	188
Mucor rot	190
Pink mould rot	191
Phacidiopycnis rot	192
Phytophthora rot	194
Powdery mildew	196
Rhizopus rot	197
Scab	198
Side rot	200
Other diseases	201
Disorders:	
Bitter pit of apples	203
Bitter pit or corky spot of pears	204
Breakdown	205
Brown heart	206
Bruising and other injuries	207
Chilling injury or low temperature breakdown	208
Core flush or brown core	209
Freezing injury	210
Jonathan spot and lenticel spot	211
Senescent breakdown	212
Soft scald	213
Superficial scald	214
Water core	215
Other disorders	216
5. STONE FRUITS	**218**
Peaches, Nectarines, Apricots, Plums and Cherries	**218**
Diseases:	
Alternaria rot	220
Anthracnose or bitter rot	221
Bacterial spot	222
Blue mould rot	223
Brown rot	224
Cladosporium rot	226
Grey mould rot	227
Pustular spot	228
Rhizopus rot	230
Rust	232
Other diseases	234
Disorders	236
6. SOFT FRUITS AND BERRY FRUITS	**238**
Strawberries and Raspberries	**238**

Diseases:
- Anthracnose or black spot — 240
- Calyx blotch and stem-end rot — 242
- Grey mould rot — 244
- Leak — 246
- Leather rot — 248
- Other diseases — 250

Blueberries, Cranberries, Currants and Gooseberries — 253

Grapes — 255
Diseases:
- Aspergillus rot — 256
- Blue mould rot — 257
- Grey mould rot — 258
- Rhizopus rot — 260
- Other diseases — 261

Disorders:
- Sulphur dioxide injury — 265
- Other disorders — 266

Kiwifruits — 267
Diseases:
- Grey mould rot — 268
- Other diseases — 269

7. MELONS AND WATERMELONS — 270
- Diseases — 272
- Disorders — 278

Appendices — 279
1. Glossary — 279
2. Lexicon — 283
3. Recommended Storage Conditions — 286

Index — 291

Contents of Volume 2: Vegetables

1. CUCURBITS
2. SOLANACEOUS FRUIT VEGETABLES
3. MISCELLANEOUS FRUIT VEGETABLES
4. LEGUMES
5. BRASSICAS
6. MISCELLANEOUS FLOWER, STEM AND LEAF VEGETABLES
7. BULBS
8. TEMPERATE ROOTS AND TUBERS
9. TROPICAL ROOTS AND TUBERS

Dedicated to the memory of

Frederick A. Buttress (1908–1984)

Librarian in the School of Agriculture (which became the Department of Applied Biology), University of Cambridge. Over a period of 45 years he assembled an incomparable collection of scientific journals from all parts of the world. It was the richness of this collection which provided the inspiration for the book.

Preface

The origins of this book go back more than twenty years, when a London solicitor, working on a cargo claim, sought expert advice in the University of Cambridge School of Agriculture. His case concerned the *Baltic Swift*, a refrigerated ship which had discharged a cargo of carrots in mouldy condition after a voyage from Cyprus to the UK. The lawyer was Mr K.G. Elmslie, of Richards Butler & Co., acting on behalf of cargo receivers. The plant pathologist he consulted was Dr P.H. Lowings, who learned that the case was one of many in which cargoes had out-turned in poor condition, and in which the claimants were fruit and vegetable importers belonging to the National Federation of Fruit & Potato Trades. The court hearing threw light on several aspects of the problem, but cargo claims continued to be a source of concern to shippers, ship owners and receivers alike. As a prospective research student in plant pathology, I was offered the problem as a 3-year doctoral research project, funded by the importers and exporters of Cyprus carrots. I am grateful to all the people who engineered this link, most particularly Peter Lowings, who became my research supervisor, Denis Mead, then Chief Executive of the National Federation, and the late Joseph Emanuel, former Chairman of the Federation Claims Committee.

Subsequently I was invited to take up a research post, funded by the National Federation, in the Department of Applied Biology. The association between university and fruit trade continued for fifteen years, with active encouragement from Sir Joseph Hutchinson and later Sir James Beament (Drapers' Professors of Agriculture), from Leslie Norris of the National Federation, and from many in the fruit trade, especially Wilfred Hockfield, Alick Glass, Tony Price and the late Geoff Jones. I am grateful to all the importers who commissioned me to survey their cargoes, and I am indebted to the many ships' officers, engineers and marine surveyors who taught me about ships.

In writing this book I have attempted in the introductory chapter to convey the essentials of post-harvest biology to those involved in handling, storing and shipping fresh fruits and vegetables. The nature and causes of deterioration are discussed in general terms, and it is emphasised that contributory causes may include a combination of adverse pre-harvest factors, post-harvest handling techniques, and storage/carriage conditions. There follows a series of descriptions of specific diseases and disorders, and it is intended that these be used in two ways. The photographs and text provide the basis for a preliminary identification, enabling the fruit owner or surveyor to assess the nature of the problem. Diagnosis can then be confirmed or modified by a specialist, making use of the microscope drawings and references together with appropriate practical techniques.

The majority of the photographs show samples of produce encountered during surveys of deteriorated consignments. A significant number show specimens received by the Plant Clinic of the Agricultural Development & Advisory Service (Ministry of Agriculture, Fisheries and Food), Cambridge, and I acknowledge with pleasure the help given by David Yarham, Michael Foley and, especially, Barry McKeown. Further material was kindly provided from time to time by Daniel Cutts, Dai Humphreys-Jones, Trevor Wyatt, and several bemused greengrocers. The remaining gaps in my collection were filled by borrowed photographs, for which I thank the owners, acknowledged separately. For the photomicrographs of fungi, I had the benefit of excellent departmental facilities, and I am grateful to Allen Hilton for his expert technical assistance. My initial interest in fungi had been stimulated by my former teachers, Professor J.A. Macdonald, Professor R.K.S. Wood and Dr Bryan Wheeler.

For permission to reproduce drawings from the invaluable *CMI Descriptions of Pathogenic Fungi and Bacteria* I am grateful to Professor D.L. Hawksworth, Director of the CAB International Mycological Institute (formerly the Commonwealth Mycological Institute or CMI). The scheme for the more stylised drawings in Chapter 1 was inspired by George Agrios' excellent *Plant Pathology* (Academic Press, 3rd ed. 1988), but new drawings have been made, based on numerous sources mentioned in the present text. The relevant genera of fungi are arranged in classes according to the 'general purpose classification' outlined in *Plant Pathologist's Pocketbook* (Ed. by A. Johnston & C. Booth, Commonwealth Agricultural Bureaux, 1983). Nomenclature is based (for the most part) on usage at the CAB International Mycological Institute, as defined in the above handbook, in *Ainsworth & Bisby's Dictionary of the Fungi* (Ed. by D.L. Hawksworth, B.C. Sutton & G.C. Ainsworth, CAB, 1983) and in *Guide to Plant Pathogenic Bacteria* by J.F. Bradbury (CAB, 1986), together with revisions as reported in the literature. A few synonyms are given in the text; some others may be traced via the index. Common names of diseases have been used wherever possible; in the absence of a common name it has been necessary to use the generic name of the causal organism, unsatisfactory though this may be. Perhaps readers conversant with particular diseases can devise appropriate descriptive common names, as urged by Paul Holliday in *A Dictionary of Plant Pathology* (Cambridge University Press, 1989), published as this book was about to go to press. For plant names and authorities, reference was made to E.E. Terrell's Checklist of Names for 3,000 Vascular Plants of Economic Importance (*USDA Agriculture Handbook* No. 505). Further information was obtained from *The Oxford Book of Food Plants* (Oxford University Press, 1969), *The Fruits of the Earth* (F. Bianchini, F. Corbetta & M. Pistoia, Cassell 1975), *Evolution of Crop Plants* (Ed. by N.W. Simmonds, Longman, 1976), and J.W. Purseglove's *Tropical Crops: Dicotyledons* and *Tropical Crops: Monocotyledons* (Longman, 1968, 1972). Production data were taken from *FAO Production Yearbook* **41** (Food and Agriculture Organisation, 1988).

The text is elementary, technical terms being largely in parentheses. The aim is to provide the basic facts, with pointers to more detailed sources. No attempt has been made to recommend specific prophylactic chemicals, and local advice should be sought.

The references were chosen with the aim of including, for each main disease and disorder, a wide selection of journals, a range of countries and authors, one or two early classical papers (which are often rich in detail and illustration), together with recent reports describing current methods of control. The literature search was based on the CABI abstracting journals *Review of Plant Pathology* and (to a lesser extent) *Horticultural Abstracts*, which were scanned up till the beginning of 1989. Holliday's *Fungus Diseases of Tropical Crops* (Cambridge University Press, 1980) proved to be an invaluable source of key references up to and including 1978. *Post-Harvest Pathology of Fruits and Vegetables* (Ed. by C. Dennis, Academic Press, 1983) provided detailed reviews relating to temperate crops. Titles of papers in German, Dutch, French, Italian, Spanish and Portuguese have been retained in the original language; the use of English titles within square brackets denotes papers written in other foreign languages, usually with summaries in English.

A very large proportion of the relevant journals were available in the library of the Department of Applied Biology. (Following the closure of the Department in September 1989 the collection will come under the aegis of the Scientific Periodicals Library.) The remainder were consulted in the University Library, the Scientific Periodicals Library, and the libraries of the Botany School, the University Botanic Garden, the Shipowners' Refrigerated Cargo Research Association, Cambridge, the CAB International Mycological Institute, Kew, and the Overseas Development Natural Resources Institute, London (now at Chatham). I thank all the librarians, and am especially indebted to Peter Filby for bringing to my attention numerous additional publications which I might otherwise not have seen.

I thank Awad Hassan Ahmed for help in compiling Appendix 3, which was first published in *The Storage and Transport of Fresh Fruit and Vegetables* (National Institute of Fresh Produce, 1981). I also acknowledge the valuable advice and assistance given by John Turner during the final stages of preparing the text for press.

Many other generous individuals have helped me during the preparation of this book. It is not possible to mention them all, but they include Anne Alvarez, Joe Britton, Marilynne Burbage, Annette Cobb, Roger Cook, Michael Cowley, Desma Goddard, Robert Heap, Lyn Hinton, Chris Hubbert, Andrea Johanson, Greg Johnson, Barbara Leonard, Iain MacSwan, Francis Marousky, Dick Murfitt, Gordon Revell, Frederick Snowdon, Mary Snowdon, Peter Steer, Henry Tribe, Elaine Watson and Gordon Wrigley. Sections of the completed text were checked by Joseph Eckert, Felicity Proctor, Robin Scrine, Richard Sharples, Peter Twiss and Jack Whiteside, and the whole was refereed by Professor F.M.R. Isenberg of Cornell University. I am grateful to all these people for their helpful comments.

Above all I am indebted to Dr Margaret A. Keay for her constructive criticism of the manuscript as it developed. Had it not been for our regular discussions, the book would have been much the poorer, if indeed it had appeared at all. I hope that it will be useful in its present form, and I would welcome suggestions as to how it might be improved or corrected.

Lastly, I should like to thank family and friends for their great forbearance during my temporary withdrawal from normal life.

Anna L. Snowdon
Wolfson College, Cambridge, 1989

ACKNOWLEDGEMENTS

Copyright relating to borrowed illustrations rests with the following individuals and/or institutions:

AFRC Institute of Horticultural Research, East Malling **375, 378, 380**; Ivan W. Buddenhagen, formerly at the University of Hawaii, Honolulu **285**; Daniel F. Cutts, Geest Organisation **243, 244, 251**; Michael A. Ellis, Ohio State University **422, 423, 440** (**440** previously published in *Compendium of Grape Diseases*, American Phytopathological Society 1988); C.A. Goodman, South African Plant Improvement Organisation, Stellenbosch **388, 389**; David Jones and James Mitchell, Macaulay Land Use Research Institute, Aberdeen **70** (previously published in *ARC Research Review* 3, 79–81, Agricultural Research Council 1977); Arthur Kelman and Gary Gaard, University of Wisconsin-Madison **26** (previously published in *Phytopathology* 70, 1110–1115, American Phytopathological Society 1974); J.M. Kotzé, University of Pretoria **168, 269, 274, 278**; Lim Tong-Kwee, Universiti Pertanian Malaysia **261**; Wayne T. Nishijima, University of Hawaii, Hilo **284**; Alan M. Paton, University of Aberdeen **25**; D. Prusky, Volcani Center, Bet Dagan, Israel **275**; Queensland Department of Primary Industries **270** (Anthony Cooke), **276** (Ian F. Muirhead), **277** (Scott Ledger), **299, 305** (B.L. Oxenham), **311, 312, 402** (the five last-mentioned previously published in *A Handbook of Plant Diseases in Colour. Volume 1 Fruit and Vegetables*, Queensland Department of Primary Industries 1982); William F. Reeder, United States Department of Agriculture, Miami, Florida **218, 219**; Mario Salerno, University of Bari, Italy **166, 191**; M. Sepiah, Malaysian Agricultural Research and Development Institute, Kuala Lumpur **260, 279**; Peter J. Shipton, Shipton (Crop Consultancy) Ltd, Aberdeen **413, 414**; David Sugar, Oregon State University **367**; Turner B. Sutton, North Carolina State University **360** (previously published in APS Slide Set 24, No. 6, American Phytopathological Society 1980); UK Ministry of Agriculture, Fisheries and Food (Crown Copyright) **410** (John Morrison), **411** (**410** previously published in *The Fruit Grower*, January 1989); Stuart J. Wale, School of Agriculture, Aberdeen **20**; David G.A. Walkey, AFRC Institute of Horticultural Research, Wellesbourne **23, 24**; J.O. Whiteside, University of Florida **165, 176**; H.J. van Zyl, Fruit and Fruit Technology Research Institute, Stellenbosch **370** (previously published in *Deciduous Fruit Grower* **28**, 115–117).

ABBREVIATIONS

cf.	(confer) compare
cm	centimetre(s)
cv.	cultivar
ed.	edited, edition
e.g.	(exempli gratia) for example
f.sp.	forma specialis (sub-division of a species of fungus)
illus.	illustrated
mm	millimetre(s)
pH	hydrogen ion concentration
pp	pages
ppm	parts per million
pv.	pathovar (sub-division of a species of bacterium) (below the level of subspecies)
q.v.	(quod vide) which see
RH	relative humidity
sp.	species (singular)
spp.	species (plural)
ssp.	subspecies
syn.	synonym
var.	variety

°C	°Celsius (Centigrade)
°F	°Fahrenheit
degrees Celsius	– a temperature range (as distinct from a specific temperature)
CMI Descr.	*CMI Descriptions of Pathogenic Fungi and Bacteria* Published by CAB International Mycological Institute in sets of 10, four sets per year. Each description gives morphological details of the pathogen (including spore measurements) together with information on hosts, diseases, geographical distribution, physiologic specialization, transmission, and relevant literature.
CMI Map	*CMI Distribution Maps of Plant Diseases* Published by CAB International Mycological Institute and updated as necessary. Each map shows the world distribution of the pathogen, together with references to the relevant literature.

CHAPTER 1
NATURE AND CAUSES OF POST-HARVEST DETERIORATION

In 1926 a government laboratory was set up adjacent to the wholesale fruit and vegetable market in Covent Garden, London. This was to enable a team of plant pathologists to identify diseases and disorders in imported fruit (1). 'Previously', it was explained, 'no attempt at scientific identification was made, and the descriptions "wasty", "wet and wasty" and "frozen" were used to cover a wide range of diseased conditions. In the absence of accurate diagnosis it was inevitable that there should be a good deal of misunderstanding amongst growers, shipping companies, and salesmen as to the actual causes of wastage, and that little progress should be made towards its prevention'. That was written in 1930. After much useful research and advisory work, the laboratory was closed down in the early 1970s, and the catch-all term 'wasty' re-entered the English language.

In the United States of America the first major survey of post-harvest losses was carried out by two plant pathologists who examined truckloads of vegetables arriving in the markets of New York and Chicago. They identified a wide range of diseases and disorders (3). In 1919 they pointed out that 'Production pathology deals with the diseases which affect the growing plant; market pathology with the diseases economically important in fruits and vegetables after harvest, whether in the field, in the packing house, in transit, in storage, or in the markets. Since market pathology affects directly many business interests it presents economic and legal problems quite unlike those of production pathology'.

For many years insufficient regard was given to post-harvest losses. The emphasis had always been on crop production and the breeding of better varieties. Much effort was devoted to maximising yield by the use of these new cultivars (cultivated varieties) in conjunction with improved irrigation and high inputs of fertilizers, pesticides, fungicides and herbicides. The achievements in many parts of the world were undeniable. In 1965, however, the United Nations Advisory Committee on the Application of Science and Technology to Development made a strong recommendation that more attention be paid to the prevention of food losses by improving methods of storage and preservation of harvested crops. Initially, most work had been on grain storage, research having received an impetus from the needs in Europe during the Second World War. Work on the more perishable food crops followed, and attempts were made to evaluate the magnitude of losses in horticultural produce (2). It was acknowledged that overall losses are often substantial but difficult to quantify, and that the prevailing state of the market (supply and demand for a particular commodity) has a decisive influence on the extent of both physical and monetary losses. Furthermore, deterioration is cumulative, occurring at all stages between harvest and final consumption (8). By the time the produce has been harvested, cleaned, graded, packed and shipped, its value will have increased many times, and therefore the more serious is loss through deterioration (3).

In the 1970s there were several international conferences on food losses, and it was reiterated that increased food could be made available more effectively by protecting it after harvest than by pouring yet more effort into production (5,9). It is important to appreciate that the world food problem can only be solved by political change, starting with land reform and the redistribution of resources. However, the study of post-harvest biology can make a very significant contribution (6,10). Not only grains but some fruit and vegetable crops may constitute the staple diet, for example 'cooking bananas' in East Africa, breadfruit in the Pacific Islands, yams in West Africa, cassava and potatoes in South America (5). Leguminous crops (peas, beans and lentils) are important sources of protein, and many fruits and vegetables provide essential vitamins, minerals and dietary fibre, as well as adding interest and variety to the diet (4,7). Furthermore, horticultural produce grown for export can be a valuable source of revenue (6). Thus, for many reasons, there is a clear need to investigate post-harvest losses. Indeed, whenever a consignment of fresh produce is bought and sold, problems arise if deterioration has taken place (3).

1 BARKER J. (1930) Wastage in imported fruit; its nature, extent and prevention. *Special Report of the DSIR Food Investigation Board* No. 38, 62 pp illus.

2 COURSEY D.G. & PROCTOR F.J. (1975) Towards the quantification of post-harvest loss in horticultural produce. *Acta Horticulturae* No. 49, 55–66.

3 LINK G.K.K. & GARDNER M.W. (1919) Market pathology and market diseases of vegetables. *Phytopathology* **9**, 497–520.
4 LIU M.S. & MA P.C. (1983) Post-harvest problems of vegetables and fruits in the tropics. *Asian Vegetable Research and Development Center, Taiwan, 10th Anniversary Monograph Series, Publication* No. 83-197, 14 pp.
5 PARISER E.R. (Chairman) (1978) *Postharvest food losses in developing countries*. Washington, DC: National Academy of Sciences, 202 pp.
6 SINGER D.D. (1980) Post-harvest food losses – world over-view. *Progress in Food and Nutrition Science* **4**(3/4), 3–9.
7 SWANSON B.G. (Ed.) (1979) Symposium: Nutritional quality of fruits and vegetables. *HortScience* **14**, 243–258.
8 WRIGHT W.R. & BILLETER B.A. (1975) Marketing losses of selected fruits and vegetables at wholesale, retail and consumer levels in the Chicago area. *Marketing Research Report of the United States Department of Agriculture* No. 1017, 21 pp.
9 ZAEHRINGER M.V. & EARLY J.O. (Eds) (1976) *Proceedings of the National Food Loss Conference*. University of Idaho, USA, 142 pp.
10 ZENTMYER G.A., THURSTON H.D. & NIEDERHAUSER J.S. (1983) International co-operation in agricultural research; the basis for feeding more people and feeding people more. In *Challenging problems in plant health* (Ed. by T. Kommedahl & P.H. Williams), pp. 501–515. St Paul, Minnesota: American Phytopathological Society.

Investigation of losses

The first step is to identify the nature of the deterioration, and to find out whether or not an organism is involved, for example a bacterium, a fungus (mould) or an insect. The second step is to investigate possible contributory causes, for example the weather during the growing season, the post-harvest handling techniques, the methods of transport and storage. Examination of all these influential factors helps in the decision about appropriate control measures which can reduce losses in the future (6). In the shorter term it assists in the settlement of claims, where more than one party may share responsibility for the loss.

The task of diagnosis does not necessarily call for a highly trained specialist (but see page 29). In the USA there appears to have been considerable expertise amongst agricultural marketing inspectors as early as the 1940s, in that specific fungal and bacterial diseases were recorded on the 'dumping certificates' issued when consignments were condemned as unfit for sale (7). A large-scale project is now underway (4), the aim being to scan tens of thousands of US Department of Agriculture in-spection certificates and, with the aid of computers, collate data on the nature of deterioration. The first commodity to be studied was potatoes, and it was found that the relative incidence of different diseases had altered over the years, perhaps as a result of changes in cultural practices or post-harvest handling techniques. Post-harvest losses have recently been investigated in many countries, by a variety of methods (1,3,5,6,8,9,10).

The first detailed global report on perishables was the 1981 Report of the Food and Agriculture Organisation/United Nations Environment Programme Expert Consultation on the Reduction of Food Losses in Perishables of Plant Origin (2). In tropical countries deterioration tends to be especially rapid because of prevailing high temperature and, in many places, high humidity (10). Problems may be compounded by lack of money to improve communications, hygiene and facilities (10). Technical innovations may reduce the incidence of deterioration or merely change its nature (6).

There remains a need to devote more resources to research and also to the training of personnel engaged in the handling, carriage and storage of perishable produce (2). In all countries, including the most 'advanced', significant losses will continue to occur wherever non-technical personnel remain unaware of elementary principles (5,8). With the exception of instances of local over-production and glut, post-harvest losses could be substantially reduced by the correct application of existing knowledge (10).

1 AMEZQUITA R. & GRA J. la (1979) A method-ological approach to identifying and reducing postharvest food losses. *Inter-American Institute of Agricultural Sciences, Dominican Republic, Miscellaneous Publication* No. 219, 84 pp.
2 ANON. (1981) Food loss prevention in perishable crops. *FAO Agricultural Services Bulletin* No. 43, 72 pp.
3 ASLAM M. & KHAN A.H. (1983) *Post-harvest loss reduction in fruits and vegetables: a review with special reference to Pakistan*. Islamabad: Pakistan Academy of Science, 89 pp illus.
4 CAPPELLINI R.A., CEPONIS M.J. & LIGHTNER G.W. (1988) Disorders in apricot and papaya shipments to the New York Market, 1972–1985. *Plant Disease* **72**, 366–368.

5 HARVEY J.M. (1978) Reduction of losses in fresh market fruits and vegetables. *Annual Review of Phytopathology* **16**, 321–341.
6 MUKAI M.K. (1987) Postharvest research in a developing country: a view from Brazil. *HortScience* **22**, 7–9.
7 RAMSEY G.B., SMITH M.A. & HEIBERG B.C. (1947) Fruit and vegetable losses on the Chicago market as indicated by dumping certificates. *Plant Disease Reporter* **31**, 387–390.
8 RIPPON L.E. (1980) Wastage of post-harvest fruit and its control. *CSIRO Food Research Quarterly* **40**, 1–12 illus.
9 SNOWDON A.L. (1979) Diseases and disorders of imported fruits and vegetables. *Annals of Applied Biology* **91**, 404–409.
10 TINDALL H.D. & PROCTOR F.J. (1980) Loss prevention of horticultural crops in the tropics. *Progress in Food and Nutrition Science* **4**(3/4), 25–39.

POST-HARVEST PHYSIOLOGY OF FRUITS AND VEGETABLES

Structure and function

Plant material is composed of cells, which continue to function after harvest. In other words, all fresh fruits and vegetables are living products. Whether they have been plucked from a tree, cut from a vine, or dug up out of the ground, plant organs continue to respire. The process of respiration is a series of biochemical reactions of the utmost complexity but, in essence, it may be expressed as the intake of oxygen, the breakdown of carbohydrates (e.g. starch, sugars), and the evolution of carbon dioxide, water vapour and a considerable quantity of energy in the form of heat (7,9).

The rate of respiration in plant cells is determined in the first instance by the basic function of the organ involved. Horticultural produce comprises all manner of plant parts (4,5,6). Physiologically immature structures such as young leaves (e.g. spinach, watercress), shoots (asparagus, bean shoots), buds (globe artichokes, brussels sprouts), inflorescences (calabrese, sprouting broccoli), certain young fruits (okra, sweetcorn) and fleshy seeds (green beans and peas) have extremely high rates of respiration (2,10). This is because they are still undergoing rapid growth, i.e. cell division; indeed, harvested asparagus spears are capable of such significant elongation that, when they are packed for market, it is necessary to leave a 'head space' in the pack.

Much less physiological activity is shown by mature leaves (cabbage, head lettuce) and leaf stalks (celery, rhubarb); they have comparatively low rates of respiration. The same may be said of mature bulbs (onions, garlic), roots (carrots, sweet potatoes) and tubers (potatoes, yams) (1,8). These are storage organs which may be in a state of dormancy, being specialised structures with the biological function of surviving until the next growing season (2,7). If regrowth (sprouting and rooting) begins during storage, there is a corresponding increase in respiration and heat output. If, in the first place, they are harvested immature (salad onions, spring-crop carrots, 'new' potatoes) then they exhibit moderate to high respiration rates (8). Commodities differ from one another in this respect by several orders of magnitude; for example, under similar conditions, the rate of respiration of sprouting broccoli is about 10 times greater than that of cabbage (8). This means that a tonne of broccoli produces 10 times as much heat as a tonne of cabbages. Clearly this has important implications for post-harvest handling.

Respiration rate is also influenced by the ambient temperature and, within limits, the rate doubles, trebles or even quadruples for each 10 degree Celsius rise in temperature of the produce (1). The significance of respiration is that it determines the perishability of the product (5). Commodities with intrinsically high respiration rates have short storage lives even under optimal conditions, while other commodities store well if kept cool, but poorly if held under warm conditions (8,10).

Another relevant physiological process characteristic of plants is transpiration, which is the loss of water by evaporation from the plant surface (9). When the plant is in the ground, moisture lost from the leaves can be replaced by water drawn up through the roots from the soil. Once harvested, the plant organ is deprived of the means of replenishment, and is prone to weight loss, wilting and shrivelling (2). Substantial weight loss can occur before the appearance of visible symptoms. There are a few instances in which water is re-imbibed (e.g. carrots in washing tanks) but, in the main, prevention of moisture loss is a major concern in post-harvest handling of horticultural produce (3,8).

The rate of moisture loss from plant material is

determined by the nature of the surface tissues. Water is lost slowly through the skin of a watermelon (which has a waxy surface) and comparatively rapidly from carrot roots (which possess irregular discontinuous outer layers). Transpiration is rapid in immature potato tubers, especially if they have suffered abrasion; moisture loss decreases if the wounds heal and if the skin is allowed to set. In cabbages with a loose and open structure, wilting can be extremely rapid; in those with compact heads the outermost leaves lose water but protect the densely packed leaves beneath. The outermost scales of onion bulbs become papery during maturation and post-harvest drying, and act as a barrier to moisture loss from the inner fleshy scales (2).

1 BERG L. van den & LENTZ C.P. (1972) Respiratory heat production of vegetables under refrigerated storage conditions. *Journal of the American Society for Horticultural Science* **97**, 431–432.
2 BURTON W.G. (1982) *Post-harvest physiology of food crops*. London: Longman Group, 339 pp.
3 FOCKENS F.H. & MEFFERT H.F.T. (1972) Biophysical properties of horticultural products as related to loss of moisture during cooling down. *Journal of the Science of Food and Agriculture* **23**, 285–298.
4 NICHOLSON B.E., HARRISON S.G., MASEFIELD G.B. & WALLIS M. (1969) *The Oxford book of food plants*. Oxford University Press, 206 pp illus.
5 PANTASTICO E.B. (1975) *Postharvest physiology, handling and utilization of tropical and subtropical fruits and vegetables*. Westport, Conn.: AVI Publishing Co., 560 pp illus.
6 PURSEGLOVE J.W. (1972) *Tropical crops. Monocotyledons. Vols 1 and 2*. London: Longman Group, 607 pp illus.
7 RHODES M.J.C. (1980) The physiological basis for the conservation of food crops. *Progress in Food and Nutrition Science* **4**(3/4), 11–20.
8 ROBINSON J.E., BROWNE K.M. & BURTON W.G. (1975) Storage characteristics of some vegetables and soft fruits. *Annals of Applied Biology* **81**, 399–408.
9 WEICHMANN J. (Ed.) (1987) *Postharvest physiology of vegetables*. New York: Marcel Dekker, 616 pp.
10 WILLS R.H.H., McGLASSON W.B., GRAHAM D., LEE T.H. & HALL E.G. (1989) *Postharvest: an introduction to the physiology and handling of fruit and vegetables*. 3rd ed. Oxford: Blackwell Scientific Publications, 176 pp illus.

Maturation, ripening and senescence

It may be seen from earlier statements that horticultural maturity (readiness for harvest) is not the same as physiological maturity, many commodities being harvested at an early stage in their development (9). The appropriate maturity for harvesting depends on intended use. Thus peas (various cultivars) can be harvested as very young tender pods, to be eaten whole ('mange-tout'); or they may be allowed to grow a little more, for consumption of the developing fleshy seeds ('petits pois'); or a higher yield may be obtained, of slightly larger seeds ('green peas'); or the pods can be left on the plant until fully mature, i.e. until the seeds become dry.

Courgettes and cucumbers are considered ready to eat when only half-grown, and they do not have a noticeable sugar content. Despite being, in the botanical sense, young fruits, they are thought of as vegetables in the culinary sense because they are eaten as a savoury (8). (Conversely, rhubarb, consisting of a vegetative plant part, the leaf stalk, is nevertheless loosely called a 'fruit' simply because it is customary to eat it with added sugar). The wide range of familiar fruits which are appreciated for their sweetness, flavour and aroma appear to comprise two basic types (2). On the one hand, fruits such as citrus, grapes, cherries and pomegranates have, in common with vegetables, a respiration rate which at a given temperature remains roughly constant (or even falls slightly) during the post-harvest period. Having no starch reserves, these fruits are not capable of significant changes after harvest, and hence must be allowed to remain on the plant until they have attained satisfactory eating quality. (Sour grapes do not become sweeter after harvest.) The second group of fruits includes apples, pears, plums, bananas, mangoes and many others. They are characterised by a sudden surge in respiration rate (usually after harvest) known as the climacteric rise (3,5). This is triggered by ethylene gas, produced in minute quantities by the fruits themselves (1,6). Over a period of several days (at moderate temperature) a characteristic skin colour becomes apparent, the flesh softens, starch is converted into sugar, and flavour and aroma develop (1,2). This is the stage of prime eating quality or 'ripeness'. Subsequently, as the rate of respiration declines, a process of senescence sets in, off-flavours develop, and the dying tissues become increasingly susceptible to attack by decay organisms (3,7).

In some climacteric fruits (e.g. avocados) the climacteric rise is inhibited until the fruit has been detached from the tree (3). In others (e.g. apples, papayas) the entire ripening process can occur whilst the fruit is still on the tree, enabling it to be harvested 'tree-ripe' and eaten immediately. In practice, the usual aim is to harvest just before the climacteric, in order to allow sufficient time for transport and distribution to the place of consumption. Assessment of maturity for harvest is often a difficult problem. If harvested immature, fruits may not have the capacity to ripen normally, while if harvested at an advanced stage (though still appearing green), they may enter the climacteric rise prematurely, during the journey to a distant market.

Although the term 'ripe' is applied loosely to any commodity, denoting that it is ready to eat, it is often used in a narrower sense, though still with various connotations (9). A climacteric fruit which is harvested when fully mature can subsequently ripen to optimal eating quality. Most can ripen on the tree. A non-climacteric fruit harvested when fully mature is immediately edible, and does not undergo a comparable ripening process, either on or off the plant; however, the term 'ripe' is frequently used to denote maturity in this context (9). Classification of fruits into two basic types is perhaps rather simplistic; some, especially the aggregate fruits, are not so easily categorised. A Japanese study (4) distinguished a third type (the 'late peak type') in which there is a rise in respiration at some stage *after* the fruit has ripened. Examples include the strawberry, peach and persimmon.

Vegetables (excluding some 'fruit vegetables' such as tomatoes) begin to decline as soon as they are harvested, the rate depending partly on their intrinsic storage potential and partly on the environmental conditions. The most characteristic aspect of senescence in green vegetables is the loss of chlorophyll and resultant yellowing of tissue (8). This occurs rapidly in commodities such as parsley and watercress, especially if held at ambient temperature. Another effect is abscission of plant parts, exemplified by the falling off of the 'wrapper leaves' of cauliflower and the calyces of eggplant (1). Internal features of senescence in vegetables may include toughening of the tissues and deterioration in flavour and aroma (7). All these changes are hastened by the presence of ethylene, whether from the vegetables themselves or from an external source (1,6).

In growing plants, the carbohydrates which are broken down during respiration are replaced by products of photosynthesis manufactured by the leaves. In harvested plant parts, only the existing carbohydrates are available and so ultimately there are only two alternatives, cell death from 'natural causes' or colonisation by micro-organisms.

1 BLANPIED G.D. (Ed.) (1985) Symposium: Ethylene in postharvest biology and technology of horticultural crops. *HortScience* **20**, 39–60.
2 FRIEND J. & RHODES M.J.C. (1981) *Recent advances in the biochemistry of fruits and vegetables.* London: Academic Press, 275 pp.
3 HULME A.C. (Ed.) (1971) *The biochemistry of fruits and their products. Vol. 2.* London: Academic Press, 788 pp.
4 IWATA T., OMATA I. & OGATA K. (1969) Relationship between the ripening of harvested fruits and the respiratory pattern. 2. Respiratory pattern of fruits and its classification. *Journal of the Japanese Society for Horticultural Science* **38**, 279–286.
5 LIEBERMAN M. (Ed.) (1983) *Post-harvest physiology and crop preservation.* New York: Plenum Press, 572 pp.
6 ROBERTS J.A. & TUCKER G.A. (Eds) (1985) *Ethylene and plant development.* London: Butterworth Press, 416 pp.
7 SHERMAN M. (Ed.) (1987) Symposium: Postharvest senescence of horticultural products. *HortScience* **22**, 851–868.
8 THOMAS T.H. (1981) Hormonal changes during senescence, ripening and regrowth of stored vegetables. In: *Quality in stored and processed vegetables and fruit* (Ed. by P.W. Goodenough & R.K. Atkin), pp 253–265. London: Academic Press.
9 WATADA A.E., HERNER R.C., KADER A.A., ROMANI R.J. & STABY G.L. (1984) Terminology for the description of developmental stages of horticultural crops. *HortScience* **19**, 20–21.
10 YANG S.F. & HOFFMAN N.E. (1984) Ethylene biosynthesis and its regulation in higher plants. *Annual Review of Plant Physiology* **35**, 155–189.

NATURE OF POST-HARVEST DETERIORATION

Physiological change

It may be seen that, while the ripening of climacteric fruits is a desirable process, most other post-harvest physiological changes signify decline and deterioration. Regrowth (sprouting and rooting) in bulbs, roots and tubers is clearly not a decline but a normal and natural process; nevertheless, from a marketing point of view, it represents a deterioration in quality and value (2).

Besides these normal post-harvest changes there are a number of abnormal changes which occur only under certain circumstances. For example, in some commodities, low temperatures (whether before or after harvest) may disrupt the complex sequence of biochemical reactions taking place in plant tissue, and cause irreversible damage known as 'chilling injury'. There are many other examples (page 32) (1). In general, tissue which has suffered physiological disruption is predisposed to microbial decay (5).

Physical damage

This is a very important aspect of post-harvest deterioration and is the primary cause of many losses (5). Various types of injury can be sustained before, during or after harvest. Causal agents include the weather, insects, birds, rodents and farm implements. Harvesting machines generally cause more damage than human pickers, but at all stages of harvesting and handling there is generally an insufficient awareness of the consequences of rough treatment. Impact bruising occurs when produce is dropped on to a hard surface, whether before, during or after packing. Injury is not immediately apparent. Later the bruise may show externally (e.g. apples) or it may be evident only on peeling (e.g. potatoes). Compression bruising may result from the over-stacking of bulk produce in store or from the over-filling of packages. A common example of the latter is grapefruit. Vibration damage can occur in under-filled packs, and is especially likely during long-distance road transport in poorly sprung vehicles.

The implications of injury are various. Most types of damage cause an increase in respiration rate and hence a greater heat output; other adverse effects include increased moisture loss and enhanced production of ethylene. Cuts and abrasions heal promptly under suitable conditions, but the appearance of the produce may nevertheless be marred (2). This is important in view of consumer demand for an unblemished product. Internal bruising leads to wastage since affected flesh is discarded, and a high incidence of this defect can result in rejection of consignments intended for processing (2). A further significance of physical damage is that it very often facilitates invasion by micro-organisms, resulting in a progressive decay which may affect the entire organ (5).

Chemical injury

Adverse effects from chemicals may result from field application of insecticides, fungicides, herbicides, growth regulators or nutritional sprays. Alternatively, damage may be caused by post-harvest treatments. The chemical itself may be present as a visible residue (2), or it may induce a reaction in the produce itself. Onions can be seriously disfigured by the action of water on the brown papery scales; staining occurs, in the form of grey blotches or, at its most severe, a jet-black discoloration resembling scorching by fire. Another example is bleaching in grapes (q.v.) subjected to excessive concentrations of sulphur dioxide (a fungistat).

Pathological decay

Infection by fungi and bacteria may occur during the growing season, at harvest time, during handling, storage, transport and marketing, or even after purchase by the consumer (6). Plants, or plant parts, by virtue of their particular genetic make-up, are resistant to attack by the majority of micro-organisms; disease is the exception rather than the rule, and occurs only when a number of conditions are fulfilled (10). Some micro-organisms are capable of causing disease in a wide range of unrelated plant species, while others are restricted to a single 'host' (8).

Disease development may be divided into two stages – infection, followed by the manifestation of symptoms, either directly or after a period of time (6). The sequence of events occurring during disease development has been likened to a battle between the micro-organism and the plant, with the environment as referee to the contest (9). Complex interactions determine whether or not particular micro-organisms succeed in attacking plant tissues and how much damage they do (7,10).

Bacteria gain entry through wounds or natural openings (such as stomata, lenticels or hydathodes) and multiply in the spaces between plant cells (8). Entry via wounds or natural openings is also characteristic of many fungi (3). Certain species of fungi, however, are capable of direct penetration of the intact cuticle, the waxy outermost layer possessed by leaves, stems and fruits (8). Breach of this barrier is often facilitated by a special procedure following germination of the mould spore on the plant surface; the fungus produces a swelling (appressorium) from the underside of which a thin strand grows through the cuticle and into or between the plant cells (6,9). Penetration is achieved by mechanical pressure and, more importantly, by an array of enzymes specific to the fungus involved. Enzymes and toxic substances produced by fungi and bacteria may also be capable of breaking down the constituents of plant cells and killing them (4,10).

The plant tissue, on the other hand, has several lines of defence. If physical injury has been sustained, an active process of wound-healing may ensue, during which corky cells are formed as a means of protection (8). If (or even before) a fungus or bacterium gains entry, its growth may be inhibited by plant substances which are either present already or else produced in response to injury or infection (10). The plant tissue may in any case be too acid to permit survival and development of the micro-organism; this explains why most of the common bacteria are of greater importance as causes of decay in vegetables than in the more acid fruits (3,6). Further, the water content of the plant cells may be too high or too low to permit infection. Some of these factors change with time and, if the micro-organism remains viable (either on or in the plant tissue), then invasion and complete colonisation may eventually take place (9). Many fruits are resistant to fungal attack when unripe; the infection process is halted almost as soon as it has begun, but the fungus remains alive, entering a 'quiescent' or 'latent' phase (6). The process of ripening is accompanied by a weakening of cell walls and a decline in ability to synthesise anti-fungal substances, until eventually the fruit is no longer able to resist the advance of the fungus.

The development of decay results in increased respiration of the host tissue; heat production and ethylene evolution are enhanced (7). Some moulds themselves produce ethylene, and this in turn predisposes healthy tissue to infection. Such interactions between plant tissue and micro-organism are influenced to a very great extent by environmental factors, notably temperature, humidity and composition of the atmosphere (3,5,6). These factors determine whether, when and how the symptoms of disease are expressed (10).

In the present text it is proposed to restrict the term 'disease' to decay or disfigurement caused by an infectious micro-organism. Diseases may be loosely classified according to symptoms (page 34) and disease organisms are classified by the form and habit of their spores (fungi) or by their biochemical and other characteristics (bacteria). Disorders can occur in the absence of disease organisms, and the term 'disorder' is here used to denote either a physical or a physiological state which is abnormal or inappropriate in the marketing context.

1 ANON. (1985) Plant physiological disorders. *Reference Book of the Ministry of Agriculture, Fisheries and Food, UK* No. 223, 83 pp illus.
2 ARTHEY V.D. (1975) *Quality of horticultural products.* London: Butterworth Press, 228 pp.
3 BURCHILL R.T. & MAUDE R.B. (1986) Microbial deterioration in stored fresh fruit and vegetables. *Outlook on Agriculture* **15**, 160–166 illus.
4 COLLMER A. & KEEN N.T. (1986) The role of pectic enzymes in plant pathogenesis. *Annual Review of Phytopathology* **24**, 383–409.
5 COURSEY D.G. & BOOTH R.H. (1972) The post-harvest phytopathology of perishable tropical produce. *Review of Plant Pathology* **51**, 751–765.
6 DENNIS C. (Ed.) (1983) *Post-harvest pathology of fruits and vegetables.* London: Academic Press, 264 pp.
7 SCHIFFMANN-NADEL M., MICHAELY H., ZAUBERMAN G. & CHET I. (1985) Physiological changes occurring in picked climacteric fruit infected with different pathogenic fungi. *Phytopathologische Zeitschrift* **113**, 277–284.
8 WALKER J.C. (1969) *Plant pathology.* New York: McGraw-Hill, 819 pp illus.
9 WHEELER H. (1975) *Plant pathogenesis.* Heidelberg, West Germany: Springer-Verlag, 106 pp.
10 WOOD R.K.S. (1967) *Physiological plant pathology.* Oxford: Blackwell Scientific Publications, 570 pp.

FACTORS INFLUENCING POST-HARVEST DETERIORATION

Pre-harvest factors

CULTIVARS Different cultivars vary greatly in their susceptibility to field diseases, and one of the aims of plant breeding and 'genetic engineering' is to incorporate resistance genes in new varieties of crop plants (2). It is also desirable to breed for post-harvest quality; for example, melons with a thick skin and raspberries with a firm texture are better able to withstand the rigours of harvesting and handling. Brussels sprout cultivars differ in their susceptibility to internal browning, a physiological disorder, and yams show variation in susceptibility to chilling injury. Thus the choice of cultivar can have an important bearing on ultimate storage life.

PLANTING MATERIAL The next factor is the state of health of the planting material, whether true seeds, seed tubers, sets or cuttings. Many fungi and bacteria persist in or on planting material, and can cause disease in the field and after harvest. Onion seed is often infected with grey mould and, unless subjected to a pre-planting fungicide treatment (3), may produce a crop of bulbs which develop neck rot in store. Plant health regulations exist to limit the movement of disease organisms within and between countries, for example the ring rot bacterium which affects potatoes. The importance of obtaining clean or treated seed cannot be overemphasised, and seed certification schemes have been established for many crops (3).

WEATHER Once the crop is planted, a vast range of interacting factors determines not only yield but also storage potential. Unfavourable high or low temperatures during growth may not cause visible injury at the time, but can result in physiological disorder after harvest (4,7). Wind, rain and hail can cause physical damage. As well as affecting crops directly, the weather has a profound influence on disease organisms. So much is known about the characteristics of the potato late blight fungus that an analysis of meteorological data permits accurate forecasting of the likely onset of disease, and timely preventive measures can be taken.

CROP HUSBANDRY Influential cultural practices (depending on crop) include density of planting, orientation of rows, shading, tillage, irrigation, fertiliser regime, weed control and mulching (6,7). Nutritional sprays affect the incidence of certain post-harvest disorders (such as apple bitter pit, q.v.) (7) and the use of nematicides, acaricides, insecticides and fungicides influences the level of infestation and disease (10). An important aspect of chemical control is the ability of organisms to produce new strains which are tolerant of the chemical used (9). This problem can be averted (or at least postponed) by the use of a range of chemically unrelated compounds with differing modes of action (9,10). The application of various growth-regulating chemicals (7) can confer disease resistance, achieve uniform harvest maturity (fruits) or suppress post-harvest sprouting (bulbs and tubers).

Cultural practices between crops are also important, for example pruning of fruit trees and destruction of crop debris can have a decisive effect on the survival of disease organisms (6). Other methods of killing harmful organisms include flooding the land (perhaps for several weeks) or, in hot countries, covering the soil with a clear plastic tarpaulin so that the sun heats the soil to a lethal temperature. This technique is known as 'solarization' (8). In cooler countries steam sterilization or fumigation can be used to cleanse the soil, although the cost of such treatment tends to limit it to high-value greenhouse crops or to seed-beds. Finally, the sequence of rotation can influence the health of the next crop (6).

For a given disease or disorder it is unlikely that one method of control will be sufficient or appropriate. In view of the complex interactions between crop, soil, weeds, pests, disease organisms and changing environmental conditions (interactions which are still being elucidated), there is now increasing emphasis on integrated control (2). The aim of Integrated Pest Management (IPM, in which 'pest' encompasses 'disease' and 'weed') is to co-ordinate control measures in the most efficient way (9). It may include the strategy of 'biological control', in which one organism (harmless) is used to kill another (1). It may include monitoring of the weather coupled with computer forecasting of disease; this can provide a basis for rational use of pesticides, for the sake of economy and also to minimise residues on the harvested crop. The choice of control methods is dictated by circumstances, for example in hot countries it may not be feasible for workers to wear the protective clothing necessary for safe application of certain chemicals, and less toxic compounds are often more appropriate as well as cheaper (5).

1 COOK R.J. (1985) Biological control of plant pathogens: theory to application. *Phytopathology* **75**, 25–29.
2 HARRIS M.K. & FREDERIKSEN R.A. (1984) Concepts and methods regarding host plant resistance to arthropods and pathogens. *Annual Review of Phytopathology* **22**, 247–272.

3 JEFFS K.A. (Ed.) (1986) *Seed treatment.* 2nd ed. Thornton Heath, UK: British Crop Protection Council, 332 pp.
4 KADER A.A., LYONS J.M. & MORRIS L.L. (1974) Post-harvest responses of vegetables to pre-harvest field temperature. *HortScience* **9**, 523–527.
5 MABBETT T. (1984) Copper fungicides offer versatility and economy for the tropics. *World Crops* **36**, 86–90, 103.
6 PALTI J. (1981) *Cultural practices and infectious crop diseases.* Heidelberg, West Germany: Springer-Verlag, 243 pp.
7 SHARPLES R.O. (1984) The influence of pre-harvest conditions on the quality of stored fruit. *Acta Horticulturae* No. 157, 93–104.
8 STAPLETON J.J. & DeVAY J.E. (1986) Soil solarization: a non-chemical approach for management of plant pathogens and pests. *Crop Protection* **5**, 190–198.
9 STAUB T. & SOZZI D. (1984) Fungicide resistance: a continuing challenge. *Plant Disease* **68**, 1026–1031.
10 WORTHING C.R. & WALKER S.B. (Eds.) (1987) *The pesticide manual. A world compendium.* 8th ed. Thornton Heath, UK: British Crop Protection Council, 1081 pp.

Harvesting and handling techniques

HARVESTING The choice of harvesting date is of great significance, especially for fruits to be stored for extended periods. Maturity indices for various crops include number of days elapsed since full bloom; colour, size, shape and flesh firmness; content of starch, sugar, juice or oil. Despite the use of these and other criteria, estimation of maturity is difficult for some commodities, and prediction of optimal harvest date is often imprecise. For many crops this is an important area of research (4). Even the theoretical optimum may be inappropriate if the weather is unfavourable; harvesting shortly after rain is liable to increase the risk of infection by disease organisms. Thus the choice of date may be a compromise. With regard to harvest schedules, it can be advantageous (for most crops) to avoid mid-day and afternoon picking, and to prefer the cool of the night or early morning. The availability of personnel is a crucial factor, harvesting by hand being the predominant method for fruits and vegetables destined for the fresh market (4). Human pickers can select for maturity and, with proper training and close-trimmed finger nails, can keep damage to a minimum. Machines harvest the entire crop at one 'pass' and, despite ingenious design, can cause substantial damage even when used correctly (1,4). Mechanical harvesting is therefore confined to the more 'robust' commodities (e.g. carrots, potatoes) and to crops which are intended for immediate processing (e.g. raspberries, peas).

CURING In countries with a suitable climate, harvested bulbs, roots and tubers are sometimes left in the field for a few days (protected from direct sunlight). This practice is to facilitate the process of 'curing' (4) which involves drying and coloration in bulb onions, wound-healing in potatoes and tropical root crops. In temperate countries curing is more likely to take place in store. Not all roots are cured. Carrots, for example, are usually transferred as promptly as possible to high-humidity cold storage in an effort to maintain turgidity. Likewise, fruits and leafy vegetables should be protected from exposure to sun and wind during and after harvest, so as to minimise moisture loss and rise in respiration rate (7).

HANDLING The handling of harvested produce varies widely according to commodity and circumstances. The fragility of strawberries, for example, means that it is preferable to harvest directly into 'punnets' (or plastic baskets) in which the fruit will remain until it reaches the consumer (4). Field-packing of bananas for export (page 104) is a similar technique, although a more usual method is to haul banana bunches by cable-way through the plantation to the packing station (4). The aim is always to minimise opportunity for bruising. Other commodities are tipped from harvesting bags or buckets into field boxes for transport to the packhouse. Care or lack of it at each stage determines the amount of injury sustained. Other influential factors include mode of operation of machinery (either in field or packhouse), height of drop, and presence of sharp corners or projections on boxes and machinery (1). Packhouse layout can have an important bearing on the likelihood of infection by disease organisms. For example, studies on citrus (page 55) have shown that minor alterations in design can achieve significant reduction in the level of infection. Wounds are especially likely to become infected if there is poor hygiene, exemplified by the debris left in field boxes or the cull-piles of rotting produce in the vicinity of some packing stations. Another common error is the failure to renew washing water, which can become heavily charged with fungal spores and bacteria; chlorine (in the form of sodium hypochlorite or bleach) rapidly loses its disinfectant properties in the presence of soil particles or organic debris (2).

TREATMENTS For produce requiring fungicide treatment, prompt application is essential, since delay enables fungi to establish themselves beyond the reach of the chemical (2). If citrus fruit is to be 'degreened' (page 55), it is advisable to apply a fungicide drench immediately after harvest. Growth-regulating chemicals are widely used on harvested produce. While ethylene hastens ripening and senescence, auxins and gibberellins retard these changes. Each can be useful in the right context. For example, auxin is applied to citrus to maintain vitality of the 'buttons', which would otherwise be adversely affected by the ethylene used for promoting coloration of the peel. Hot water treatments can be effective against diseases and pests of tropical fruits, but only if temperature and time of immersion is closely controlled (9). Some commodities (e.g. apples) benefit from a calcium drench, which helps to protect against physiological disorders and fungal decay (4). A number of products have been formulated to modify physiological processes. Semi-permeable coatings can reduce respiration rate (5). Artificial waxes inhibit moisture loss, may also modify respiration, and enhance the appearance of some commodities. Fruits such as citrus may be enclosed in individual plastic film bags which are then heat-shrunk; this technique is known as 'seal-packaging' and, in conjunction with fungicide treatment, can extend storage life (9).

Although the chemical industry is more concerned to develop fungicides for use in the field (vast quantities being sprayed on the major crops), about 20 organic compounds have been extensively evaluated as post-harvest treatments during the last 30 years (2). Sodium *ortho*-phenylphenate (SOPP) and *sec*-butylamine are effective against several wound-invading fungi, while diphenyl (biphenyl) has long been used to inhibit spore-formation in the penicillium moulds and thus minimise 'soiling' in citrus. The most far-reaching development was the introduction of the benzimidazole group in the late 1960s. Benomyl (Benlate) and thiabendazole (TBZ) can provide excellent control of a range of fungi. If used before harvest, their 'systemic action' means that they are translocated within the plant and can give long-lasting protection even against quiescent infections. If applied after harvest, benomyl in particular can penetrate the tissue (and also the stem 'button' of citrus fruits). However, intensive use of these compounds has often led to the proliferation of tolerant strains of fungi, efficacy being lost within one or two seasons. It is now recommended that a range of chemically unrelated compounds be used at different stages (e.g. before and after harvest, before and after storage for re-packing) so as not to select out insensitive strains (2). Newer compounds such as imazalil and prochloraz are superior in that the development of such strains is intrinsically less likely. Also, in common with iprodione, they can provide substantial control of alternaria rot, which is unaffected by most fungicides. Etaconazole, propiconazole and guazatine are among the first compounds effective against geotrichum sour rot. Useful compounds with specific activity are dichloran (dicloran) against rhizopus rot, metalaxyl and fosetyl aluminium against phytophthora rot.

Some fungicides may be applied in the vapour phase; fumigation is appropriate for grapes (q.v.), strawberries and potatoes, which might suffer deleterious effects if wetted. For most other fruits and vegetables, fungicides are applied in water solution or incorporated in a heated wax. Adjustment of pH (acidity/alkalinity) is sometimes important for effective fungicide action (2).

With the exception of chlorinated washing water, bactericides are not in common use on harvested produce, and there remains a need for a compound effective against bacterial soft rot of potatoes, for example, and against some vegetable diseases which are not controlled by refrigeration. In parallel with the development of anti-microbial chemicals there has been an increasing consumer demand for produce which is unblemished but at the same time free from chemical residues. Hence the current emphasis on alternative pest and disease control strategies such as biological control and modified atmosphere techniques (q.v.).

Irradiation of produce with γ-rays is an alternative (or additional) way of killing micro-organisms (9). It is also effective against insect pests (4) and has the further advantage of inhibiting sprouting in potatoes and onions. On the other hand, in some fruits irradiation impairs texture, flavour or nutritional value, though a measure of success has been achieved with strawberries, mangoes and especially papayas. Irradiation leaves no 'residue'; indeed it is impossible to detect whether or not produce has been irradiated. Although a well-established practice in other areas of the food industry, irradiation has not been widely adopted as a treatment for fresh fruits and vegetables, and remains a controversial issue (9).

COOLING Cooling of produce (either before or after packing) is for many commodities the most critical procedure, and plays a large part in determining ultimate storage life (4). Low temperature results in a lowered respiration rate, a reduced moisture loss, and an inhibition of the growth of decay organisms. It is important to distinguish between cooling and cold storage, and for this reason the term 'pre-cooling' is used. The amount of refrigeration required for this initial cooling process is many times greater than that required to maintain the produce at a constant low temperature (10). In general, pre-cooling should be both prompt (as soon as possible after harvest) and rapid, the aim being to remove the 'field heat' from the produce and to attain a flesh or pulp temperature appropriate to the commodity. The more quickly the temperature can be reduced, the less moisture will be lost by evaporation (4). An important proviso is that at no time should the temperature be allowed to fall below the critical temperature for the commodity (the critical value being the freezing point or, for chill-sensitive commodities, the temperature below which chilling injury is likely). Immediate cooling is beneficial for the vast majority of commodities, exceptions being melons and certain stone fruits.

There are several methods of extracting heat from the produce (4,8). Hydro-cooling involves contact with cooled water (by spraying, drenching or immersion), and is suitable for commodities such as peaches, melons, asparagus, sweetcorn, celery and carrots. It is important that the produce remain in the hydro-cooler for sufficient time to ensure thorough cooling; chlorination and regular renewal of water are also necessary. More common is air-cooling, in which air is passed over cooling coils and then blown on to the produce. Pressure-cooling, in which special provision is made to force the cold air through the packages, is most effective, and very short cooling times can be achieved. For commodities requiring a temperature around 0°C and a humidity close to saturation, it is appropriate to use an air/water counterflow system (e.g. the 'ice-bank cooler'). Vacuum cooling, inside a specially constructed chamber, is suitable for leafy vegetables such as lettuce, which have a large surface area relative to mass. Reduced pressure causes evaporation of moisture from the tissue, and this has a cooling effect. The uniformity of moisture loss means that significant evaporation can occur without the onset of wilting. The process can be further refined by generating a fine mist in the chamber. A covering of crushed ice may sometimes be put in packages of broccoli and celery and a few other commodities which are not harmed by such contact (4,8).

PACKAGING Packaging is a complex subject (6). The purpose of packaging is to contain, to protect and to inform. A compromise must be made between strength, suitability, appearance and cost and, for a given commodity, different types of package are appropriate for local or national marketing and for export by sea or air. Sacks or nets are generally used for items such as bulbs, roots and tubers, while crates, cartons and trays are used for fruits and many vegetables. An appropriate package is one which takes into account the 'major hazard' to which the commodity is subject. Thus a tray for fruits should be free of sharp edges and protruding nails or staples, since physical injury will lead to infection and decay. A carton for broccoli should be well-provided with ventilation holes if self-heating and senescence are to be avoided; alternatively, if 'top-ice' is used, the material of the carton should be water-repellent. Good quality cartons are essential for all types of produce transported by ship or stored for any length of time; problems have been encountered with cartons which appear strong initially but which lose their rigidity after absorbing moisture from the atmosphere (6). The ideal pack will have, in addition to brand-name, origin, weight etc., information about commodity requirements, for example optimal temperatures for storage and (for certain fruits) subsequent ripening, together with appropriate warnings such as sensitivity to ethylene. Storage information is especially important for those commodities (such as avocados, citrus) for which the optimal temperature depends on cultivar and growing conditions, with which only the exporter may be familiar.

In recent years there has been an increasing move towards 'unitisation', for example the stacking of individual packs on to a wooden or plastic pallet (1200 x 1000 mm). Each palletload is overwrapped or strapped for stability and is handled as a unit (8). Palletisation shortens loading time and reduces damage. Thorough pre-cooling is essential for most commodities. It may be advisable to arrange the packages so as to leave a central 'chimney' to permit airflow through the palletload. A similar technique is used in the construction of wooden 'bins', having an integral pallet base and sometimes used to carry onions or apples which are subsequently to be re-selected and packed at destination.

Much work has gone into the design of 'pre-packs' for supermarket display. Sometimes the produce is prepacked before shipment, for example carrots in small polyethylene bags which are then enclosed in a master-bag for transportation. More usually, however, pre-packing takes place shortly before distribution of produce to retail outlets. Pre-packs range from nets (for citrus fruits, onions) and perforated polyethylene bags (apples, potatoes) to small polystyrene trays overwrapped with transparent film such as polyvinyl chloride (3). The choice of film (and its thickness) is important, since different materials vary in permeability to oxygen, carbon dioxide and water vapour; as the produce respires it modifies the atmosphere within the package, and this can prolong storage life for several days (page 23). Ripening of apples and tomatoes can be successfully delayed by this method, and research is in progress to test other films for use on the more rapidly respiring commodities such as Brussels sprouts and cauliflowers (3).

1 DRAKE S.R. (Ed.) (1983) Symposium: The influence of mechanical harvesting on the quality of horticultural crops. *HortScience* **18**, 405–423.
2 ECKERT J.W. (1983) Control of post-harvest diseases with antimicrobial agents. In *Post-harvest physiology and crop preservation* (Ed. by M. Lieberman) pp. 265–285. New York: Plenum Press.
3 GEESON J.D., BROWNE K.M., MADDISON K., SHEPHERD J. & GUARALDI F. (1985) Modified atmosphere packaging to extend the shelf-life of tomatoes. *Journal of Food Technology* **20**, 339–349.
4 KADER A.A., KASMIRE R.F., MITCHELL F.G., REID M.S., SOMMER N.F. & THOMPSON J.F. (1985) Postharvest technology of horticultural crops. *University of California (Division of Agriculture and Natural Resources) Special Publication* No. 3311, 192 pp illus.
5 KESTER J.J. & FENNEMA O.R. (1986) Edible films and coatings – a review. *Food Technology* **40**(12), 47–59.
6 PELEG K. (1985) *Produce handling, packaging and distribution*. Westport, Conn.: AVI Publishing Co., 625 pp.
7 RICKARD J.E. & COURSEY D.G. (1979) The value of shading perishable produce after harvest. *Appropriate Technology* **6**, 18–19.
8 RYALL A.L. & LIPTON W.J. (1979) *Handling, transportation and storage of fruits and vegetables. Vol. 1. Vegetables and melons*. 2nd ed. Westport, Conn.: AVI Publishing Co., 588 pp.
9 SHEWFELT R.L. (1986) Postharvest treatment for extending the shelf life of fruits and vegetables. *Food Technology* **40**(5), 70–80, 89.
10 WADE N.L. (1984) Estimation of the refrigeration capacity required to cool horticultural produce. *International Journal of Refrigeration* **7**, 358–366.

Conditions during storage

Appropriate storage can minimise moisture loss, slow down respiration rate and inhibit the development of micro-organisms. In other words, wilting, re-growth, ripening, senescence and decay can be postponed. Commodities vary widely in their requirements and in their tolerance of non-optimal conditions (1,5,9).

COMMON STORAGE   Storage at ambient temperature may be quite satisfactory for onions and potatoes, for example (5). In many countries it is the sole option for all commodities. The simplest type of storage involves shading from direct sunlight, in order to prevent undue temperature rise or excessive moisture loss. Tropical roots and tubers may be stored in pits or else in covered structures above ground. The traditional 'yam barn' in the humid zone of West Africa consists of closely spaced poles of live wood which quickly produce leafy branches; the yam tubers are tied between the poles with string, and benefit from the airy shaded environment. Onions (provided they are fairly free from disease organisms) store well at quite high temperatures, sprouting being inhibited above about 25°C. Low temperatures are also suitable; in temperate countries onions, and also potatoes, may be kept for several months through the winter, by ventilating the store with ambient air whenever external conditions are favourable (simple thermostats serve to shut off the outside air supply during warmer daytime hours and in the event of frost, and maintain in-store air circulation whenever necessary). Ventilated storage may also be used for carrots and beets, provided that a humidifier is installed to prevent excessive dehydration of the crop (1).

It is sometimes possible to take advantage of naturally-occurring or man-made caves, in which there is minimal temperature fluctuation and a cool environment even in summer. Cave storage is common in parts of Turkey, Japan and China. In the UK, caves in the limestone cliffs of Dover are well-placed to serve the port, and are in frequent use for the temporary storage of imported potatoes. An important factor in all types of storage is the arrangement of the produce so that all parts have access to air (1,9). Various methods are appropriate, according to whether the produce is loose or packaged, and according to the properties of the store (5).

REFRIGERATED STORAGE   In conventional refrigerated storage air is cooled by directing it over pipes through which circulates a refrigerant such as ammonia, brine or 'freon' (the name given to a group of chlorofluorocarbon compounds or CFCs). The greater the surface area of the cooling coils, the more nearly can the refrigerant temperature approximate to the desired air temperature, and hence (since condensation of moisture is minimised) the higher the humidity attainable in the store (5,8). Mechanical humidifiers are sometimes used to add moisture to the store atmosphere. Alternative means of achieving high humidity include the 'ice-bank store' and the 'jacketed store' (1). Salad onions, leafy vegetables and certain root crops benefit from a humidity close to saturation, provided that the temperature is close to 0°C. (At higher temperatures a saturated atmosphere is conducive to mould growth). Slightly lower relative humidities (90 to 95%) are suitable for most fruit crops. Substantially lower relative humidities (70 to 80%) are desirable for bulb onions, garlic and ginger, and are attainable only in certain stores. Optimal storage temperatures vary according to commodity and cultivar, and detailed recommendations are given in Appendix 3).

The rates of growth of fungi and bacteria are markedly influenced by the storage environment (10). In general, the lower the temperature and humidity, the slower the growth of micro-organisms and the smaller the likelihood of new infections. However, the minimum temperature for growth varies according to species. A few are inactive below 8° to 10°C, most grow slowly at 5°C; some can make significant progress even at 0°C (10). The environment also affects the susceptibility of plant tissue to infection and decay, and relationships are not always simple (8). For example, some fusarium rots of potato proceed more rapidly at 12° and at 25°C than they do at intermediate temperatures, evidently because of the variable resistance of the tuber under different conditions. Grey mould rot of carrots is less severe in a saturated atmosphere than it is at 90% relative humidity (8).

SUPPLEMENTARY TECHNIQUES Although refrigeration is the principal method of prolonging storage life of horticultural produce, there are several supplementary techniques (5,10). Modified atmosphere (MA) storage relies on the fact that an increased concentration of carbon dioxide and a reduced concentration of oxygen will retard respiration rate and may also inhibit disease development (10). A modified atmosphere may be achieved by enclosing the produce in semi-permeable film packaging (page 21). Success has been achieved with several commodities, including tomatoes and bananas. It may be advisable to include an ethylene-absorbent, especially if refrigeration is unavailable (7).

Controlled atmosphere (CA) storage involves the monitoring and control of oxygen and carbon dioxide together with temperature and relative humidity so as to provide constant optimal conditions (2,3). CA storage has proved very effective in extending the storage life of apples in particular (1,5). Other commodities which benefit from CA storage include pears, avocados and cabbages (2,3). Ethylene removal helps to delay the onset of physiological disorders and senescence, and is now considered to be an important aspect of CA storage (7).

Low pressure (hypobaric) storage has the effect of reducing respiration rate and dissipating evolved ethylene (6). Leafy vegetables retain their green colour and fruit ripening is retarded. The apparatus is costly, however, and the technique has been applied on a very limited scale (10).

Not only storage environment is important, but also the conditions under which the produce is held temporarily or offered for sale. In the open air (but preferably under shade), misting with a fine spray of water provides the benefit of evaporative cooling and is of great help in maintaining freshness, especially of leafy vegetables (4). Indoors, produce may be displayed loose or in pre-packs; shelf display lights can be as detrimental as the sun, and can cause warming of produce in general and greening of potatoes (5). Date-stamping of pre-packs, adopted by many supermarkets, is helpful in ensuring proper rotation of stock.

1 ANON. (1979) *Refrigerated storage of fruit and vegetables*. London: HMSO, 148 pp illus.
2 BLANKENSHIP S.M. (Ed.) (1985) Controlled atmospheres for storage and transport of perishable agricultural commodities. (Proceedings of the 4th National Controlled Atmosphere Research Conference.) *North Carolina State University Horticultural Report* No. 126, 511 pp.
3 BLANPIED G.D. (Chairman) (1987) Symposium: Factors that influence commodity response to controlled atmosphere storage. *HortScience* **22**, 761–794.
4 COHEN R.A. & HICKS J.R. (1985) Effect of roadside market display method on weight loss of selected vegetables. *HortScience* **20**, 697–698.
5 HARDENBURG R.E., WATADA A.E. & WANG C.Y. (1986) The commercial storage of fruits, vegetables, and florist and nursery stocks. *Agriculture Handbook of the United States Department of Agriculture* No. 66, 130 pp.
6 JAMIESON W. (1980) Use of hypobaric conditions for refrigerated storage of meats, fruits, and vegetables. *Food Technology* **34**(3), 64–71.

7 KNEE M., PROCTOR F.J. & DOVER C.J. (1985) The technology of ethylene control: use and removal in post-harvest handling of horticultural commodities. *Annals of Applied Biology* **107**, 581–595.
8 LIPTON W.J. (Ed.) (1978) Symposium: Relative humidity – physical realities and horticultural implications. *HortScience* **13**, 549–574.
9 MUNOZ-DELGADO J.A. (Ed.) (1979) Recommendations for chilled storage of perishable produce. *Annexe au Bulletin de l'Institut International du Froid*, 148 pp.
10 SOMMER N.F. (1985) Role of controlled environments in suppression of postharvest diseases. *Canadian Journal of Plant Pathology* **7**, 331–339.

See also:

LIDSTER P.D., HILDEBRAND PD, BÉRARD L.S. & PORRITT S.W. (1988) Commercial storage of fruits and vegetables. *Agriculture Canada Publication* No. 1532/E, 88 pp.

Conditions during transport

MODES OF TRANSPORT   Horticultural produce may be carried by road, rail, sea or air, sometimes for many thousands of miles. The principles of ventilation, refrigeration and modified atmospheres applying to transportation are similar to those for storage, but there are additional factors to be taken into consideration. Each journey comprises a number of stages, and considerable expertise is required for proper co-ordination of the chain of events (5,6,10). Furthermore, the environment through which the truck, train, ship or aeroplane passes is always changing, and it may be necessary to make continual adjustments to take account of this. A store on land is often designed for a specific commodity, and the personnel soon gain experience of particular requirements. A ship, on the other hand, may carry frozen meat on one voyage and chilled carrots the next, or, during a single trip, the crew may have to deal with commodities as diverse as cartons of canned goods, blocks of zinc, bales of cotton, bags of fishmeal and bundles of balsa wood, as well as onions and melons. In some shipping companies training is inadequate and there may be no cargo manual on board (1,4,9). Instructions given by the shipper (exporter) are sometimes vague, ambiguous, misguided or non-existent. Nevertheless, a considerable body of expertise exists, and the vast majority of cargoes shipped around the world, by whatever means, are delivered in good condition.

High-value perishables can be sent by air, either in wide-bodied freighters or in the cargo holds of passenger aircraft. Produce should be pre-cooled and suitably packed (5,6). The advantage of a brief transit time is sometimes negated by delays in loading, take-off and intermediate connections.

Land transport has undergone several changes in recent years (6). In the USA rail transport used to predominate, and ice was a common refrigerant, placed either in bunkers or, for certain vegetable commodities, in direct contact with the load (and known as 'top ice'). Mechanical refrigeration (or, if appropriate, heating) is now general, and a greater proportion of produce is carried by road (3). In Europe, on the other hand, there is still an important rail service using ice-bunkers which are replenished at various locations along the main routes. Road vehicles are either ventilated or mechanically refrigerated.

In many countries the choice of transport method now includes the intermodal container, first introduced in the 1960s (1,7). Nearly all types are of standard dimensions, either 20 ft or 40 ft in length, and capable of being carried by road, rail and sea (5). Those which have no means of ventilation are unsuitable for the carriage of perishables. Ventilated containers have openings to the exterior and possess a fan, requiring electrical power (2). They are successfully used for onions, even on month-long voyages, and for potatoes and melons during short trips. Refrigerated containers are insulated and are of two main types; integral containers possess a built-in cooling system powered by an electric generator, while port-hole containers derive refrigerating power from a clip-on unit or from a connection with the vessel or port facility (1,2,7). Some containers are designed to cope with the receipt of 'warm' cargoes and, if the packages are properly stowed, are capable of removing 'field heat' from the load. Most, however, are intended only to maintain a given temperature, and it is essential to pre-cool the goods thoroughly before putting them into the container (2).

STOWAGE   The method of stowage is of crucial importance and, whether in a container or in a conventional hold, the pattern adopted must take account of design of airflow as well as stability of stow (9). The principle of ensuring that air goes through the stow rather than around the extremities is an important distinction between live cargoes and frozen cargoes (3). When frozen goods are carried, the sole object is to prevent ingress of external heat, whereas with living cargoes their own heat has also to be dissipated from the middle of the stow by the circulating air (1,2). The exact procedure depends on factors such as commodity, initial temperature, nature of packaging and particulars of the space to be filled (4,10); the following are general principles. In ventilated containers (using ambient air) it is important to ensure that air can pass beneath the stow as well as through it (5). In ventilated holds, where both inlet and exhaust vents are often at the top of the chamber, bags (of potatoes, onions etc) must be stowed in blocks, separated by air channels which are maintained by horizontally-placed strips or planks of wood (dunnage). In the absence of deck gratings it is advantageous to stow the cargo on pallets or on crossed dunnage wood in order to raise it off the deck and allow under-stow ventilation.

The same principles may be applied in a refrigerated hold possessing a horizontal airflow system (whether longitudinal or athwartships). This design of vessel was appropriate for bananas when they were shipped 'on the stem', but with packages there is a tendency for the air (taking the line of least resistance) to pass over the top of the stow. Therefore, appropriate air channels may have to be constructed so that air can penetrate the cargo (8). Most (but not all) modern refrigerated vessels have a vertical airflow system, with underdeck delivery and powerful fans (1). Here the cargo is best stowed 'solid' without visible air channels, air being forced upwards through the interstices between the packages, so reaching all parts of the stow. (To ensure that air can pass, cartons should be stacked 'in register' rather than in brick-wall fashion). A solid stow is appropriate in refrigerated containers designed with 'bottom air delivery' (5). On the other hand, 'top air delivery' (sometimes via a plastic chute) requires elaborate stowage arrangements, such as 'modified bonded block' and 'pigeon-hole' load patterns (5). As the names suggest, these create substantial gaps in the stow to assist in air distribution (3).

FRESH AIR VENTILATION A refrigerated space is essentially a closed system, the same air being continually recirculated. Airflow rates are usually expressed in terms of the cubic capacity or volume of the chamber. Modern reefer (refrigerated) ships and containers have fans capable of achieving a recirculation rate of the order of 60 to 100 volumes per hour. For living cargoes it is necessary to allow slight ingress of fresh air from outside (to provide oxygen) and corresponding elimination of respiratory gases (carbon dioxide and ethylene). Requirements vary considerably, depending on commodity and carriage temperature but, in the absence of specific instructions, there is a marked tendency for ships' personnel to err on the side of excessive fresh air ventilation of refrigerated chambers (8). Fresh air systems may be capable of 0.5 to 3 air changes per hour (some reefer containers, designed for the carriage of flower bulbs, have up to 10). This means that, with vents fully open continuously, a volume of fresh air up to 3 times (or even 10 times) the volume of the empty chamber is introduced every hour. The implications of such a massive influx of air are often not fully appreciated by cargo owners. This introduced air is usually warm and moist. The extra heat load may jeopardise cargo-cooling and is also wasteful of fuel (4). The excess moisture condenses out on the cooling coils; if refrigerant temperature is below 0°C the condensed water freezes on the coils, necessitating frequent defrosting and resulting in a further reduction in refrigeration efficiency.

In view of these drawbacks, it is preferable to limit fresh air ventilation, especially during the critical period of initial cooling of warm produce. It is desirable to keep the fresh air vents closed for the first 24 hours or so, in order to give priority to rapid cooling (8). Ideally, the concentration of carbon dioxide in the chamber should be monitored, and ventilation carried out only if the level approaches a prescribed maximum. In the absence of a carbon dioxide meter, ventilation can be continuous and slight (vents partially open) or else intermittent (vents fully open for a short period every 12 or 24 hours). With ethylene-sensitive cargoes such as bananas, it may be considered prudent to allow partial ventilation from the outset. There remains a need to define more closely the ventilation requirements of different commodities, and the ventilation capacities of many ships.

One further point must be mentioned with regard to refrigerated holds. Unless the ship has a special conversion facility (modifications to the main air trunking), it is quite unjustified to suggest that a refrigerated space, used with fans on but coolers off, is equivalent to 'ventilated space' for carriage at ambient temperature. Since the bulk of air is recirculated, the heat of respiration of the cargo is not carried away, and serious self-heating can result.

MONITORING   In most refrigerated ships it is customary to record (for each hold) the air delivery temperature, air return temperature and several hatch (hold) temperatures (1,4). In refrigerated containers only air delivery and air return temperatures are measured, though prudent shippers include an independent temperature recorder with each consignment. While temperature can easily be measured, humidity is difficult to determine and still more difficult to control (2). Humidity is influenced not only by the fresh air ventilation regime but also by the surface area of the cooling coils, the volume of air circulated by the fans, the heat flow through the chamber walls, and the nature of the commodity and its packaging (2). Optimal temperatures and relative humidities are given in Appendix 3, but in practice such ideal conditions (e.g. for onions) may not be attainable during transport. Modified atmospheres are used for the container carriage of commodities such as lettuce and avocados, and controlled atmosphere carriage is being developed for long distance transport of apples.

COMPATIBILITY   An important aspect of the carriage of horticultural products is that of compatibility (1,2,3,4,5,8). Reasons for incompatibility between two or more commodities include widely differing optimal temperatures or humidities (e.g. strawberries with tomatoes), susceptibility to cross-taint (e.g. onions with melons) and sensitivity to ethylene (e.g. kiwifruit with apples). In view of the fact that ethylene can be deleterious at concentrations as low as 1 part per million (or even lower), segregation of incompatible cargoes must extend to the prevention of mingling of atmospheres, for example from adjacent holds.

During some journeys, especially those involving intermodal containers, there may be a period when refrigeration is interrupted. Such a break in the 'refrigeration chain' can have serious consequences, since produce which is allowed to warm up takes a long time to re-cool (10). In summary, the problems of transportation are considerably more complex than those of storage.

1. ALDERS A.W.C. (1987) *Marine refrigeration manual.* The Hague: Cip-Gegevens Koninklijke Bibliotheek, 245 pp illus.
2. ANON. (1985) *The transport of perishable foodstuffs.* Cambridge, UK: Shipowners' Refrigerated Cargo Research Association, 50 pp.
3. ASHBY B.H., HINSCH R.T., RISSE L.A., KINDYA W.G., CRAIG W.L. & TURCZYN M.T. (1987) Protecting perishable foods during transport by truck. *Agriculture Handbook of the United States Department of Agriculture* No. 669, 94 pp illus.
4. HALES K.C. (Ed.) (1973) The carriage of refrigerated cargoes. *Annexe au Bulletin de l'Institut International du Froid 1973*, 125 pp.
5. NICHOLAS C.J. (1985) Export handbook for U.S. agricultural products. *Agriculture Handbook of the United States Department of Agriculture* No. 593, 154 pp illus.
6. RYALL A.L. & PENTZER W.T. (1982) *Handling, transportation, and storage of fruits and vegetables. Vol. 2. Fruits and tree nuts.* 2nd ed. Westport, Conn.: AVI Publishing Co., 610 pp illus.
7. SCRINE G.R. (1981) Marine refrigeration and container ships. *Proceedings of the Institution of Mechanical Engineers* **195**, 37–52 illus.
8. SNOWDON A.L. & AHMED A.H.M. (1981) *The storage and transport of fresh fruit and vegetables.* London: National Institute of Fresh Produce, 32 pp.
9. THOMAS O.O., AGNEW J.A. & COLE K.L. (1983). *Thomas' Stowage. The properties and stowage of cargoes.* Glasgow, UK: Brown, Son & Ferguson, 369 pp illus.
10. WORTHINGTON-SMITH P.S. (1985) Importance of the cold chain. Quality and cost considerations. *Deciduous Fruit Grower* **35**, 95–97.

ASPECTS OF FORENSIC PLANT PATHOLOGY

Cargo surveys and the interpretation of evidence

If a consignment arrives in poor condition, the cargo receiver needs to know the causes of deterioration. This may enable him to lodge a claim against another party and in due course to recover his losses, either by mutual agreement or through litigation (hence the term 'forensic plant pathology'). The supplier, carrier and insurer of the consignment therefore also need to know the reasons for deterioration. Furthermore, a knowledge of causes means that steps can be taken to reduce the likelihood of similar losses in the future.

SURVEYS A full diagnosis is possible only if sufficient evidence is available (1,10). While it may be quite easy to identify a disease organism in a small sample of produce, this may be of limited value if the context remains unknown (3,5). On the other hand, if it is possible to carry out a full survey and examination of the consignment, then further conclusions can be drawn (4). The sooner a survey can take place, the more likely it is that important details can be noted. It is especially valuable to be able to examine the cargo while it is still in the ship's hold, container or vehicle, since the pattern of damage is often very significant. Thus, for example, if potatoes with symptoms of freezing injury are found scattered at random in a load, this suggests pre-shipment freezing, whereas if affected tubers are found only at the extremities of the stow or near to an air inlet vent, this indicates freezing during transit (8). Prompt examination may mean that the goods are observed in the frozen state, leaving little room for dispute.

Pulp temperatures of the produce, taken with a spear thermometer at intervals during unloading, give an indication of conditions in the stow during the journey and the efficiency or otherwise of the airflow (2). In containers and road vehicles a portable temperature recorder may have been put in with the load at the time of 'stuffing', to provide a continuous independent record of temperature during the journey. The instrument is usually hung on the interior wall towards the rear, and although temperatures within the stow may be substantially different, the thermograph nevertheless gives a useful indication of air temperature and of any fluctuations during the trip. As well as noting the position of such a recorder, it is important to investigate the air circulation system of the container or vehicle, and to find out whether the thermostat is located in the air delivery stream or in the air return. In refrigerated ships note is taken also of the position of hatch thermometers, which are usually in the form of rubber-encased leads. These are designed to be suspended in the cargo for the duration of the voyage and to be coiled up and attached to the deckhead when not in use. It is often found that in practice they are left permanently in the coiled position and so record the temperature of the air above the cargo rather than the temperature within the stow. Such observations have a bearing on the interpretation of the refrigeration log (which shows, among other details, the temperatures recorded during each watch). It is preferable to examine the 'raw log' rather than an abstract which may give only average temperatures. The presence and state of development of certain mould species in the cargo can often provide reliable independent evidence about temperatures during the voyage.

Throughout the discharge of cargo, observations can be made and photographs taken, showing details of stowage, air channels, ventilation/refrigeration systems and patterns of damage. Samples of deteriorated produce are collected for later examination in the laboratory. Interviews with the Master and officers of a ship, and access to the logs, can provide information about the loading schedule, the shipper's carriage instructions, the design and function of the vessel's equipment, and the weather during the voyage. The driver of a road vehicle may also be able to supply useful information. With regard to pre-shipment factors, details of production, harvesting and handling can often be obtained from the importer, who is usually in close communication with his shipper (exporter). Alternatively, a visit can be made to the producer country in order to observe current practices at first hand. General information about the weather can later be amplified by official meteorological data (3).

PATTERNS OF DAMAGE Diagnosis involves a search for patterns of damage. There may be a pattern in the individual fruit or vegetable, for example pitting confined to the stem-end region and shoulders of oranges is more suggestive of hail damage while attached to the tree than it is of a post-harvest disorder. Suppositions must remain tentative, however, until more information is forthcoming. First thoughts are not always correct. For example, if grapefruits are found to have symptoms of chilling injury (q.v.) on one side of the fruit only, and there is no pattern of damage seeming to relate this with aspect or position in the carton (e.g. in relation to vent holes), then it might be presumed that chilling had occurred in the grove. Some citrus-growing areas are subject to frosty winds, and perhaps only the outward-facing (or alternatively the windward) side of the fruit might be affected. There is an alternative explanation, however. The outward-facing side of any grapefruit has different physiological characteristics from the side which may be shaded by foliage during fruit development. Similarly, a grapefruit which develops deep within the tree canopy differs from a fruit which is totally exposed (6). Either light or heat from the sun may be influential, and moisture loss seems to be relevant. Normally these inherent differences are not discernible and are of no consequence, but if grapefruits are subjected to chilling temperatures *after harvest*, the fruit tissue previously exposed to the sun shows a greater susceptibility to chilling injury (6). Thus, although the pattern of damage may suggest pre-harvest chilling, the reality may well be chilling during transport or storage.

Most explanations of patterns of damage are much more straightforward. For example, in 'tray-pack' cartons of apples, it might be noticed that fruits in the top tray are severely disfigured by brown marks and stalk punctures, and that the incidence of such injuries decreases abruptly in the lower trays. Such a pattern suggests that the injury occurred after packing rather than before, and that the brown marks are vibration damage, probably a result of the apples being too loosely packed; information about a transcontinental road journey completed in record time might suggest over-enthusiastic driving.

Patterns of damage within a consignment are also important indicators of pre-shipment factors on the one hand and carriage factors on the other. For example, onions may be found with grey or black stains indicating water damage. If the stained bulbs are randomly distributed in the ship's hold, this suggests rain before shipment. If only the upper surface of the stow is affected, this can be the result of water dripping off the deckhead, moisture having condensed from the humid air introduced into the hold. If the damage is confined to those onions directly beneath the air delivery vents, it is probably the result of sea-spray entering the vents. In a cargo of potatoes, one 'brand' or 'mark' may show a significantly high incidence of bacterial soft rot compared with other brands. This suggests differing pre-shipment conditions (e.g. a particular packing station or an earlier date of packing). However, caution is needed in interpretation, since the high incidence may instead (or also) be related to position in the stow (e.g. sluggish airflow in the hatch coaming). In a cargo of bananas, if the contents of every hold are found to have ripened during the voyage, this could be due to pre-shipment unfitness of the bananas or to faults on board the vessel, and further investigation is necessary. If, on the other hand, the only chambers with ripe bananas are those adjacent to chambers containing citrus fruits, it could reasonably be postulated that ethylene-laden air from mouldy citrus might have entered the banana chambers and triggered the ripening process.

Finally, in a series of shipments of a particular commodity from a particular country, there may be patterns relating to such factors as length of voyage, season of the year or design of ship. An example is provided by the first classic study of a transport disorder, 'brown heart of apples' (q.v.) occurring in Australian apple cargoes shipped to the UK in the 1920s (7). It was established that the cause was 'suffocation' due to accumulation of carbon dioxide produced by respiration of the apples (the vessels, having been built for the carriage of frozen meat, had no provision for the introduction of fresh air into the holds). Nevertheless, some cargoes arrived in good condition, and the criteria for success appeared to be the use of ships with ill-fitting hatch plugs and the occurrence of stormy weather during the voyage. The explanation lay in the fact that some vessels had inherently 'leaky' hatches (with respect to air) and that beneficial air exchange with the outside atmosphere was especially copious during the pitching and tossing of the vessel in heavy weather (7).

Thus it may be seen that, even after diagnosis of the nature of deterioration, further study and experimentation may be needed to elucidate contributory causes.

ANALYSIS OF CAUSES By the use of methods described above, causes of deterioration in fruit and vegetable cargoes have been investigated over a number of years, often in collaboration with marine surveyors (9). Information was obtained either during out-turn surveys (on behalf of cargo receivers, cargo underwriters, ship-owners or charterers) or during the study of claims documents (on behalf of lawyers acting for one of these parties). For most (though by no means all) of these cargoes the available information permitted a diagnosis of the underlying causes of deterioration, which were assigned to one or more of the following ten categories: weather during the growing season; crop husbandry; harvesting and grading; post-harvest treatments; packaging; pre-cooling; design and function of ship, container or truck; stowage; carriage instructions and policy; and length of voyage. Analysis of results obtained over a 15-year period showed that in over half of the cargoes deterioration resulted from more than one cause. In 39% of the cargoes deterioration resulted from adverse pre-shipment factor(s) only, in 42% from adverse shipboard factor(s) only, and in 19% from both pre-shipment and shipboard factors (9).

It was found that while packaging, for example, was rarely at fault, other aspects of post-harvest handling often left much to be desired; there was frequently evidence of rough treatment of produce and inadequate control of disease organisms (before and after harvest). As for shipboard factors, it was found that the means of transport was sometimes intrinsically unsuited to the task, for example general cargo vessels used for onions, and ill-designed refrigerated ships used for unprecooled carrots or bananas. The commonest fault in ship stowage was a failure to appreciate the risk in continuing the stow up into the hatchcoaming. With regard to container stowage there was sometimes a lack of awareness of the special needs of cargo presented at ambient temperature. Shippers chartering vessels did not always understand how the holds functioned, and gave instructions which were difficult to carry out or which had a deleterious effect on their cargoes. Shipboard policies were sometimes misguided, for example deliberate slow cooling of vegetable cargoes. The effect of a delayed voyage depended on commodity, but tended to be especially serious in ventilated cargoes (9).

These findings relate to a very small proportion of cargoes shipped around the world, but nevertheless give some indication of the reasons for losses. A general rule, illustrated by the results, is that deterioration is frequently the result of a combination of adverse factors.

1 GROGAN R.G. (1981) The science and art of plant-disease diagnosis. *Annual Review of Phytopathology* **19**, 333–351.
2 HARVEY J.M. (1981) Optimum environments for the transport of fresh fruits and vegetables. *International Journal of Refrigeration* **4**, 293–298.
3 ISENBERG F.M.R. (1984) Some problems in maintaining quality of fresh vegetables moving in national and world market systems. *Acta Horticulturae* No. 157, 169–176.
4 LOWINGS P.H. (1980) Factors leading to cargo loss – the diagnostic approach. *Progress in Food and Nutrition Science* **4**(3/4), 21–24.
5 MOLINE H.E. (Ed.) (1984) Postharvest pathology of fruits and vegetables: postharvest losses in perishable crops. *University of California Publication* NE-87 (*UC Bulletin* No. 1914), 80 pp.
6 PURVIS A.C. (1980) Influence of canopy depth on susceptibility of 'Marsh' grapefruit to chilling injury. *HortScience* **15**, 731–733 illus.
7 SMITH A.J. (1925) Experiments on the leakage of carbon dioxide gas from 'unventilated' holds of ships. *Special Report of the DSIR Food Investigation Board* No. 24, 25 pp.
8 SMITH W.L. & WILSON J.B. (1978) Market diseases of potatoes. *Agriculture Handbook of the United States Department of Agriculture* No. 479, 99 pp illus.
9 SNOWDON A.L. (1988) A review of the nature and causes of post-harvest deterioration in fruits and vegetables, with especial reference to those in international trade. *Biodeterioration* **7**, 585–602.
10 STREETS R.B. (1972) *The diagnosis of plant diseases. A field and laboratory manual.* Tucson, Arizona: University of Arizona Press, 234 pp.

Identification of diseases and disorders

A preliminary identification may be made by observing the symptoms with the naked eye and under a hand lens, and comparing them with illustrations and descriptions in the present book, in the US Department of Agriculture Handbooks on Market Diseases (for references see pages 55, 171 219, 255) and in other publications (1,10). It may not be possible to confirm the identity of a physiological disorder unless information is available about the history of the consignment. On the other hand, if a disease organism is present it can usually be identified by microscopic examination and by various specialised techniques (2,3,4,6,8). Identification of a virus requires elaborate techniques and equipment, including an electron microscope. On the other hand, most fungi and bacteria can be isolated and grown in pure culture on a sterile medium, which may be solid or liquid. The most usual is a transparent jelly (agar) containing the necessary nutrients. Some are formulated for general use while others are appropriate for particular kinds of fungi and bacteria (2). In developing countries, where proprietary media may be expensive, scarce or unavailable, there are alternative sources of nutrients; for example, coconut water (preferably incorporated in agar) is reported to be an excellent general purpose medium (5). To isolate the organism, small pieces of decayed tissue from the leading edge of the lesion are transferred, by means of sterile techniques, to the medium contained in covered dishes (2). Fungal or bacterial colonies are usually visible within a few days, though it may take some time for certain fungi to mature sufficiently to permit identification (7). In order to confirm beyond doubt that the symptoms observed in the fruit or vegetable are caused by the micro-organism, certain criteria (Koch's postulates) must be fulfilled as follows. The organism from pure culture is inoculated into healthy plant material. If similar symptoms develop, and the same organism can be re-obtained in pure culture, then it is confirmed as the causal agent.

If the appropriate facilities and expertise are not at hand, then it is necessary to consult a specialist. The agronomist, horticulturist or plant pathologist is familiar with diseases and disorders occurring in the growing crop and after harvest. Other specialists include the plant physiologist (for disorders), microbiologist (various micro-organisms), mycologist (fungi), bacteriologist (bacteria) and virologist (viruses). Depending on the country, such specialists may be consulted in government departments (e.g. Ministry of Agriculture), colleges, universities, research stations and institutes, while some may be in private practice.

Identification services for fungi and bacteria are offered by the CAB International Mycological Institute at Kew, near London, and by the Centraalbureau voor Schimmelcultures at Baarn in the Netherlands. These institutes, and others, receive cultures and diseased plant material from all over the world. Cultures submitted for identification should be on firm agar in small glass bottles or in sealed, plugged test tubes (2,9). If live plant material is sent through the post, it should be packed in such a way as to allow aeration. Perforated cardboard is usually appropriate, together with loose internal packaging (not straw, which may itself become mouldy). The use of polyethylene bags is generally inadvisable, because the resulting high humidity tends to lead to superficial mould growth or bacterial soft rot (2). For diagnostic purposes slightly infected material is of more value than severely decayed tissue, the latter often being contaminated with secondary organisms. A small sample is usually sufficient to permit identification of the causal organism. On the other hand, where an opinion is also sought on possible contributory factors, it is important that the consultant receive one or more complete packages (e.g. sack, carton, tray) unopened, or at least with the contents undisturbed, so that the pattern of damage within the pack can be observed (page 27).

1 HALL E.G. & SCOTT K.J. (1977) *Storage and market diseases of fruit.* Melbourne: CSIRO, 52 pp illus.
2 JOHNSTON A. & BOOTH C. (Eds.) (1983) *Plant pathologist's pocketbook.* Farnham Royal, Slough, UK: Commonwealth Agricultural Bureaux, 439 pp.
3 LELLIOTT R.A. & STEAD D.E. (1987) *Methods for the diagnosis of bacterial diseases of plants.* Oxford, UK: Blackwell Scientific Publications, 216 pp.
4 LUND B.M. (1983) Bacterial spoilage. In *Postharvest pathology of fruits and vegetables* (Ed. by C. Dennis), pp 219–257. London: Academic Press.
5 QUIMIO T.H. (1984) Coconut water as a growth medium for edible and plant pathogenic fungi. *Philippine Agriculturist* **67**, 1–8 illus.
6 RIEUF P. (1985) *Clé d'identification des champignons rencontrés sur les plantes maraîchères.* Paris: INRA.
7 ROSSMAN A.Y., PALM M.E. & SPIELMAN L.J. (1987) *A literature guide for the identification of plant pathogenic fungi.* St Paul, Minnesota: American Phytopathological Society, 252 pp.
8 SCHAAD N.W. (Ed.) (1988) *Laboratory guide for the identification of plant pathogenic bacteria.* 2nd ed. St Paul, Minnesota: American Phytopathological Society, 164 pp illus.
9 SUTTON B.C. (1985) The CMI identification service. *Tropical Pest Management* **31**, 217–218.
10 VOCK N.T. (1982) *A handbook of plant diseases in colour. Vol. 1. Fruit and vegetables.* 2nd ed. Brisbane: Queensland Department of Primary Industries.

CLASSIFICATION OF DISORDERS

Nutritional disorders

These are pre-harvest in origin and may develop as a result of an incorrect balance of minerals in the plant. The mineral may be abundant in the soil and freely absorbed by the growing plant, yet inadequately distributed within the plant organs (1,8). For example, bitter pit of apples, tipburn of lettuce and blossom-end rot of tomatoes are associated with a local deficiency of calcium (1). The beneficial role of calcium is probably in strengthening cell walls and membranes so that the tissue is less likely to undergo collapse and browning. Boron deficiency can lead to internal black spot of beet, while blotchy ripening of tomatoes is associated with a lack of potassium. For control of nutritional disorders, fertilisers must be applied according to requirements (determined by soil and plant analyses), and mineral solutions are sometimes sprayed on the plants. Care should be taken to prevent water stress in the crop, since a lack of water reduces mineral uptake and distribution (1). The harvested crop may be given a supplementary mineral treatment, for example a calcium dip or drench to prevent bitter pit and to prolong storage life of apples (1,8).

1 Bitter pit of apple

Respiratory disorders

Disruption of normal respiration can occur if the produce is held in an unsuitable atmosphere. Insufficient oxygen results in anaerobic respiration, leading to 'black heart' of potatoes and 'invasive alcohol poisoning' in apples. Excessive concentrations of carbon dioxide are also harmful; some apple cultivars suffer external injury and others develop 'brown heart'. In vegetables, non-green tissue appears to be especially susceptible to respiratory disorders, e.g. cauliflower curd and the pale midribs of lettuce leaves (4). Unsuitable atmospheres can develop in a controlled atmosphere store, in a ship's hold, or within a package. Injury is especially likely at high temperatures because a high rate of respiration results in rapid modification of the atmosphere surrounding the produce.

2 Black heart of potato

Temperature disorders

Heat injury may occur before harvest, for example sunscorch of tomatoes and waxy breakdown of garlic. High temperatures after harvest are likely to lead to wilting and respiratory disorders.

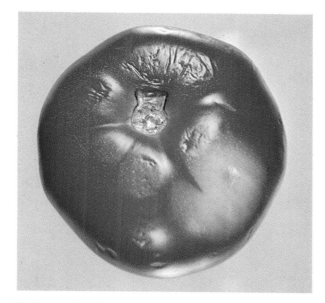

3 Sunscorch of tomato

Freezing injury can occur when the temperature of the tissue falls below its freezing point. Freezing points of fruits and vegetables are slightly below the freezing point of pure water (0°C or 32 F). This is because of the presence of dissolved substances, chiefly sugars, and the sweeter the fruit or vegetable the lower is its freezing point. Thus while lettuce may freeze at −0.2°C, sweet black cherries may not freeze until the temperature is reduced to −3°C (10). Some commodities, for example parsnips and onions, can survive mild freezing; most fruits and vegetables cannot. In severe freezing, ice crystals may protrude from the tissue. Cells are killed, and release fluid on thawing, leaving the fruit or vegetable wilted and often discoloured. Loss of weight on thawing provides the basis for separation by flotation of freeze-damaged citrus fruits (page 88).

4 Severely frozen carrot

5 Carrots after thawing

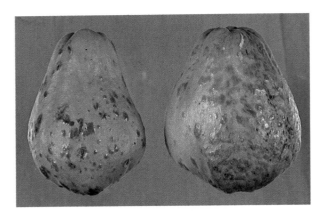

6 Watersoaking of chayote

7 Pitting of grapefruit

Chilling injury is quite distinct from freezing injury, and may occur in certain fruits and vegetables at temperatures between 0° and 15°C (occasionally even higher). Chilling injury may occur in the field or at any stage after harvest (9). Tropical products are especially susceptible, for example bananas, pineapples, sweet potatoes and yams (6). Other susceptible commodities are those with tropical or subtropical ancestors, for example cucumbers and tomatoes. A few temperate crops are slightly susceptible, for example asparagus, cranberries and certain apple cultivars. Symptoms in various commodities include water-soaking, surface pitting, internal discoloration, failure to ripen, accelerated senescence and increased susceptibility to decay (9). The critical temperature, below which injury may occur, depends not only on commodity and cultivar but also on maturity and the conditions under which the crop was grown. Injury results from disruption of normal metabolism, and the severity of damage is determined both by the temperature at which the produce is held and by the duration of exposure (9). Bananas may sustain injury after a few hours at 7°C, whereas cucumbers can be held for a week at 5°C and suffer only slight damage. Injury may not be apparent when the produce is removed from cold storage, but symptoms develop rapidly in warm conditions. Many temperate fruits and vegetables, for example strawberries, grapes, leeks, celery and carrots, are not injured by low temperatures and may with advantage be stored at 0° to 1°C.

The simplest way of avoiding chilling injury after harvest is to ascertain the critical temperature for the commodity in question, and ensure that an appropriate temperature is maintained. For some tropical products, however, the safe temperature is quite high, and storage life is therefore limited (6). Hence there are various strategies (depending on the commodity) for reducing the likelihood of injury at chilling temperatures, for example pre-treatment with ethylene; use of certain post-harvest fungicide treatments; minimisation of moisture loss by means of wax coating or semi-permeable packaging; 'cold conditioning' by stepwise cooling; intermittent warming; and controlled atmosphere storage (9). There are benefits in breeding for reduced susceptibility to chilling injury (6).

Miscellaneous disorders

Some disorders are associated with senescence (5). Mealy breakdown and senescent breakdown of apples can be a consequence of late picking or prolonged storage. Nutritional factors are also influential, and apples treated with calcium show a reduced tendency towards senescent breakdown (8). Other aspects of senescence in fruits and vegetables have been mentioned (pages 15, 20) in descriptions of the normal post-harvest physiology of horticultural produce. Insofar as they render the produce unfit for sale, they can be considered as 'market disorders'. Examples of post-harvest processes which are undesirable or untimely include wilting (3,7), sprouting, rooting, premature ripening and senescence, the latter two being accelerated by the presence of ethylene. Ethylene can also induce specific disorders such as russet spotting of lettuce and bitterness in carrots (5).

Other kinds of defects in freshly harvested or in stored produce include cuts and bruises (page 16); infestation by insects, mites or eelworms (nematodes); contamination by chemicals (e.g. sulphur residues); and the presence of toxins produced by certain fungi (mycotoxins) (2) or by the plant organ itself (e.g. solanine in green potato tubers, and cyanide compounds in cassava roots).

8 Root formation in onions

9 Senescence in broccoli

1 BANGERTH F. (1979) Calcium-related physiological disorders of plants. *Annual Review of Phytopathology* **17**, 97–122.
2 DUPAIGNE P. (1978) Les mycotoxines et les fruits. *Fruits* **33**, 505–519.
3 GRIERSON W. & WARDOWSKI W.F. (1978) Relative humidity effects on the postharvest life of fruits and vegetables. *HortScience* **13**, 570–574.
4 ISENBERG F.M.R. (1979) Controlled atmosphere storage of vegetables. *Horticultural Reviews* **1**, 337–394.
5 KADER A.A. (1985) Ethylene-induced senescence and physiological disorders in harvested horticultural crops. *HortScience* **20**, 54–57.
6 McGLASSON W.B., SCOTT K.J. & MENDOZA D.B. (1979) The refrigerated storage of tropical and subtropical products. *International Journal of Refrigeration* **2**, 199–206.
7 SASTRY S.K., BAIRD C.D. & BUFFINGTON D.E. (1978) Transpiration rates of certain fruits and vegetables. *ASHRAE Transactions* **84**, 237–255.
8 SHARPLES R.O., REID M.S. & TURNER N.A. (1979) The effects of postharvest mineral and lecithin treatments on the storage disorders of apples. *Journal of Horticultural Science* **54**, 299–304.
9 WATADA A.E. (Ed.) (1982) Symposium: Chilling injury of horticultural crops. *HortScience* **17**, 159–186.
10 WHITEMAN T.M. (1957) Freezing points of fruits, vegetables and florist stocks. *Marketing Research Report of the United States Department of Agriculture* No. 196, 10 pp.

CLASSIFICATION OF DISEASES AND DISEASE ORGANISMS

Types of disease

Diseases may be loosely classified according to their signs and symptoms (10). Signs are visible growths of the causal agents, and symptoms the discernible responses produced by the host. In many diseases there is local discoloration and disruption of tissue, with the formation of obvious lesions (1).

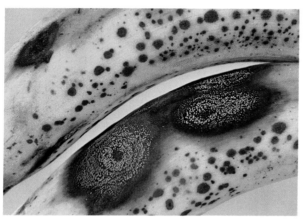

Anthracnose means 'coal-like' and was first used to describe a disease of grapes in which blackening of the tissues was a striking feature (10). It is now used for a group of diseases, characterised by black lesions which become covered with pink spore-masses (2,5). The causal fungi are certain species in the order Melanconiales (1,7). Important hosts include beans, cucurbits, and many tropical and subtropical fruits (8).

10 Banana anthracnose

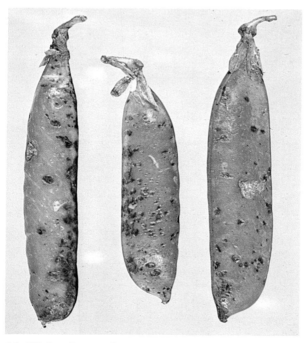

Blight generally signifies a sudden and serious decline. In the field, stem and leaf blight can lead to withering and death of the plant, blossom blight to a failure to set fruit. Examples of blights which may be encountered after harvest include various fungal and bacterial blights (3,6) affecting pods of peas and beans; early and late blights of tomato fruits, potato tubers and celery leaf stalks (1).

11 Blight of pea pods

Blotch denotes an irregular or indefinite lesion, often remaining fairly superficial. Examples are sooty blotch of apples, cercospora blotch of avocados, and purple blotch of onions, caused by various unrelated fungi.

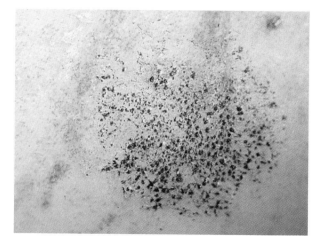

12 Blotch on apple

Canker is a term usually applied to a local malformation of tissue, especially in a tree trunk or branch. Canker diseases may also result in infected stems and fruits, for example bacterial canker of tomatoes (6). In parsnip canker there is gross disfigurement of the root.

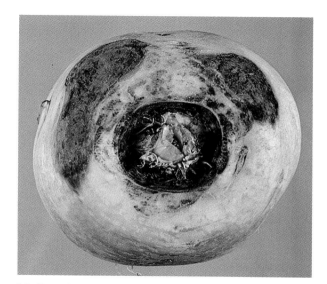

13 Parsnip canker

Dry rot is a type of decay in which the tissue may become dehydrated and eventually stringy or powdery. The development of dry rot (as opposed to soft rot, q.v.) is determined not only by the causal species of fungus or bacterium but also by the environmental conditions (10).

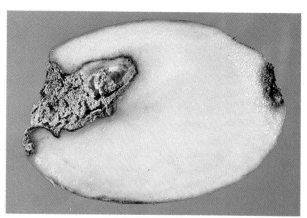

14 Dry rot of potato

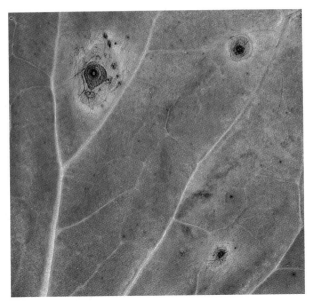

Leaf spot describes a definite, delimited lesion which may, however, develop into a blotch or blight. Many different fungi can cause leaf spots. In cabbage, dark leaf spot and light leaf spot are caused by unrelated fungi. Leaf spot of lettuce may be caused by any of at least three fungi.

15 Leaf spots on cabbage

Mildew is sometimes used loosely to describe any superficial mould growth. Strictly speaking, downy mildews are caused by fungi of the family Peronosporaceae, while powdery mildews are caused by fungi belonging to the Erysiphaceae (4,7,10).

16 Downy mildew of radish

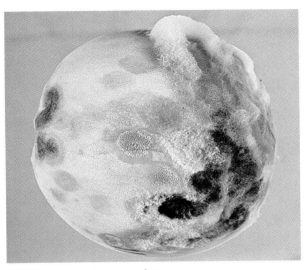

Mould is a term used to describe the strands (usually spore-bearing) produced by all types of fungi. Although the strands are sometimes confined within infected tissue, a humid atmosphere usually permits copious development of visible mould growth on the surface.

17 Various moulds on melon

Rusts are diseases in which there are one or more types of spore-bearing pustules (usually orange, brown or black) produced by fungi belonging to the order Uredinales (5,7,9).

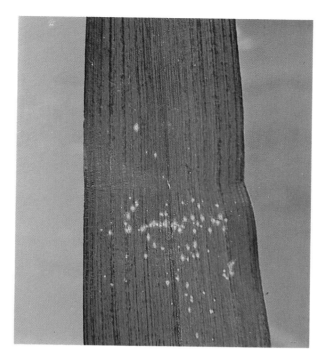

18 Leek rust

Scab describes a prominent corky lesion, of which there are several types. Scab may result from infection by a filamentous bacterium (e.g. common scab of potatoes); a primitive fungus (powdery scab of potatoes); or one of many other types of fungus (apple scab, citrus scab, cucumber scab) (4).

19 Citrus scab

Smuts are diseases caused by fungi of the order Ustilaginales, in which characteristic structures (sori) form in surface or interior tissues. Each sorus contains a mass of dark spores or spore balls (5,9,10).

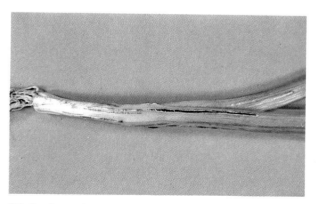

20 Leek smut

21 Soft rot of beans

Soft rot is a type of decay which, in a humid atmosphere, develops into a wet or slimy rot. Soft rots may be caused either by fungi (e.g. *Rhizopus* spp.) or by bacteria (e.g. *Erwinia* spp.) (6,10). Under less humid conditions the same organisms may cause a dry rot (10).

22 Vascular disease of potato

Vascular diseases can have serious effects, since the disease organisms (fungi or bacteria) interfere with water movement within the plant, eventually causing wilting and collapse (4,10). An example is bacterial wilt of potatoes; furthermore, in this disease even apparently healthy plants may give rise to tubers which develop brown rot of the vascular ring (6).

1 AGRIOS G.N. (1988) *Plant Pathology.* 3rd ed. New York: Academic Press, 845 pp illus.
2 ARX J.A. von (1981) *The genera of fungi sporulating in pure culture.* 3rd ed. Vaduz, Liechtenstein: J. Cramer, 424 pp illus.
3 BRADBURY J.F. (1986) *Guide to plant pathogenic bacteria.* Farnham Royal, Slough, UK: CAB International, 332 pp.
4 DICKINSON C.H. & LUCAS J.A. (1982) *Plant pathology and plant pathogens.* 2nd ed. London: Blackwell Scientific Publications, 229 pp.
5 ELLIS M.B. & ELLIS J.P. (1985) *Microfungi on land plants.* London: Croom Helm, 818 pp illus.
6 FAHY P.C. & PERSLEY G.J. (1983) *Plant bacterial diseases: a diagnostic guide.* London: Academic Press, 393 pp.
7 HAWKSWORTH D.L., SUTTON B.C. & AINSWORTH G.C. (1983) *Ainsworth and Bisby's dictionary of the fungi.* 7th ed. Farnham Royal, Slough, UK: Commonwealth Agricultural Bureaux, 445 pp illus.
8 HOLLIDAY P. (1980) *Fungus diseases of tropical crops.* Cambridge University Press, 607 pp.
9 WEBSTER J. (1980) *Introduction to fungi.* 2nd ed. Cambridge University Press, 669 pp illus.
10 WHEELER B.E.J. (1969) *An introduction to plant diseases.* London: John Wiley, 374 pp illus.

See also:

KRANZ J., SCHMUTTERER H. & KOCH W. (Eds) (1977) *Diseases, pests and weeds in tropical crops.* Hamburg: Paul Parey, 666 pp illus.

ROGER L. (1951-54) *Phytopathologie des pays chauds.* Vol. 1, 1126 pp, Vol. 2, 1130 pp, Vol. 3, 895 pp illus. Paris: Paul Lechevalier.

Agents of disease

The word **virus** means poison and, at its simplest, a virus particle is a chemical molecule (composed of a nucleic acid and a protein), too small to be visible through a light microscope. Once inside living cells of plants or animals, viruses can multiply and cause disease by interfering with normal metabolism. They enter plants through injuries (made, for example, by sap-sucking insects or by knives used for propagation) and can cause serious crop losses; symptoms in the field include stunting of the plant, leaf roll, mottling (mosaic) or streaking of leaves, flowers or fruits. The damage is usually obvious before harvest, and abnormal plant parts are left unpicked or are culled during sorting operations. Thus virus diseases are generally not encountered after harvest. Exceptions are tobacco rattle virus of potatoes (causing 'spraing', an internal defect of the tubers) and turnip mosaic virus affecting cabbages, the symptoms of which can develop further in store.

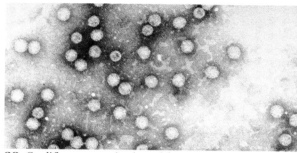

23 Cauliflower mosaic virus (electron microscope view)

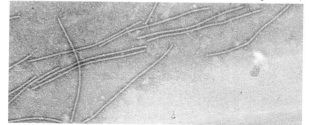

24 Turnip mosaic virus (electron microscope view)

Bacteria are simple micro-organisms visible under the light microscope. Those which attack plants are mostly single-celled and rod-shaped, and they include species of *Erwinia*, *Pseudomonas*, *Xanthomonas*, *Clavibacter* (*Corynebacterium*) and *Bacillus*. They are capable of very rapid reproduction by splitting in two repeatedly. *Streptomyces* is filamentous, and the very thin branching threads give rise to chains of spores. Classification of bacteria is based on a range of criteria, including shape, size, reaction to certain stains, and behaviour on various growth media (8).

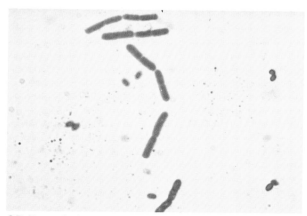

25 Bacteria (*Bacillus* sp.) splitting in two (stained with fuchsin)

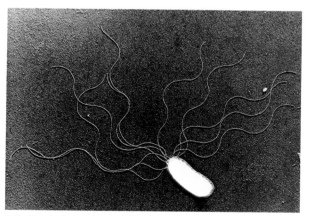

26 Single bacterium (*Pseudomonas marginalis*) with eleven flagella (electron microscope view)

Fungi are organisms which, lacking chlorophyll and conductive tissues, are wholly dependent on plants or animals. Like the bacteria, most fungi perform an essential role in nature, that of helping to decompose dead organic matter which would otherwise litter the earth. Fungi are classified according to their method of reproduction (1,2). Largest are the puffballs and mushrooms, some of which are edible. The fungi which cause diseases in fruits and vegetables are mostly microscopic.

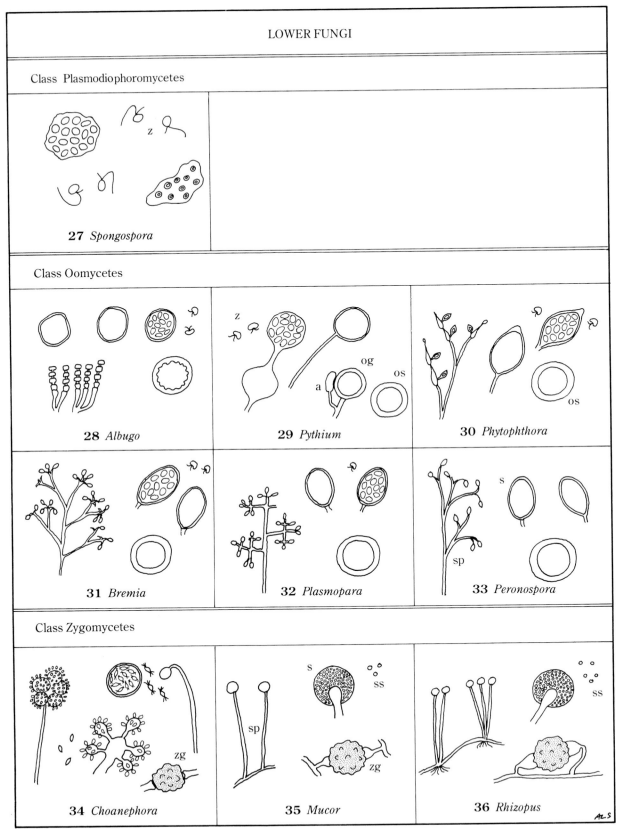

LOWER FUNGI

Class Plasmodiophoromycetes

27 *Spongospora*

Class Oomycetes

28 *Albugo*
29 *Pythium*
30 *Phytophthora*
31 *Bremia*
32 *Plasmopara*
33 *Peronospora*

Class Zygomycetes

34 *Choanephora*
35 *Mucor*
36 *Rhizopus*

a antheridium og oogonium os oospore s sporangium sp sporangiophore ss sporangiospores
z zoospores zg zygospore

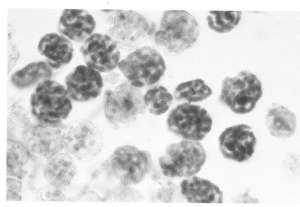

37 *Spongospora* (resting spores)

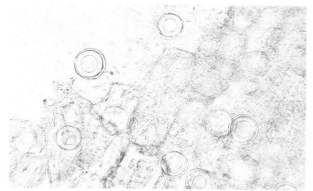

38 *Phytophthora* (oospores)

39 *Peronospora* (sporangia)

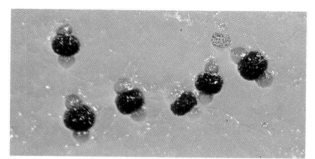

40 *Rhizopus* (zygospores)

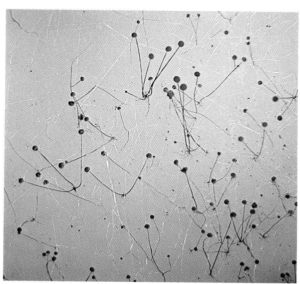

41 *Rhizopus* (sporangia)

The most primitive fungi do not form strands (hyphae, mycelium) but consist of an amorphous **plasmodium** which releases swimming spores (zoospores). An example is *Spongospora subterranea* in the class Plasmodiophoromycetes (2).

'Phycomycete fungi' (an old term, denoting affinity with the algae) (2) possess simple, branching strands without cross-walls (aseptate mycelium). Members of the class Oomycetes reproduce sexually by means of **oospores** formed by the union of an antheridium and an oogonium. They also reproduce asexually by means of sporangia, which may subdivide and release zoospores or which may germinate directly (thereby acting as conidia, q.v.). Important families are the Albuginaceae (which includes *Albugo*), the Pythiaceae (*Pythium* and *Phytophthora*) and the Peronosporaceae or downy mildew fungi (*Plasmopara*, *Peronospora* and *Bremia*). Fungi in the class Zygomycetes reproduce sexually by means of **zygospores** formed by the union of two similar cells. The asexual cycle is characterised by the formation of sporangia which give rise to numerous sporangiospores. Important examples are *Rhizopus*, *Mucor* and *Choanephora*.

ASCOMYCETES

Class Plectomycetes (asci in cleistothecia)

42 *Erysiphe* | **43** *Sphaerotheca* | **44** *Podosphaera* | **45** *Uncinula*

Class Pyrenomycetes (asci in perithecia)

46 *Glomerella* | **47** *Rosellinia* | **48** *Diaporthe* | **49** *Nectria*

50 *Ceratocystis* | **51** *Gnomonia* | **52** *Gibberella*

Class Loculoascomycetes (asci in pseudothecia)

53 *Botryosphaeria* | **54** *Guignardia* | **55** *Didymella* | **56** *Mycosphaerella*

57 *Venturia* | **58** *Leptosphaeria* | **59** *Elsinoe* | **60** *Pleospora*

Class Discomycetes (asci in apothecia)

61 *Pezicula* | **62** *Pyrenopeziza* | **63** *Monilinia* | **64** *Sclerotinia*

ac ascus   as ascospore   ap apothecium   cl cleistothecium   pe perithecium   ps pseudothecium

The higher fungi possess strands with cross-walls (septate mycelium). The various classes of Ascomycetes (1,9) produce sexual spores in sac-like structures called asci, singular ascus, each containing 8 (or sometimes fewer) ascospores. The asci may be borne in a completely closed **cleistothecium**, and this is typical of the Erysiphaceae or powdery mildew fungi, e.g. *Erysiphe*, *Podosphaera*, *Sphaerotheca* and *Uncinula*. When mature, the cleistothecia burst open and spores are discharged into the air. In *Ceratocystis*, *Diaporthe*, *Glomerella* and *Nectria* the asci are borne in a flask-shaped **perithecium** with an opening at the neck. The

65 *Uncinula* (cleistothecia)

66 *Nectria* (perithecia)

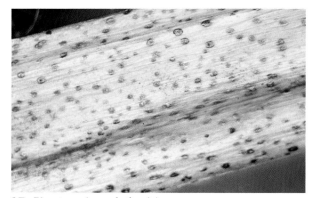

67 *Pleospora* (pseudothecia)

68 *Sclerotinia* (apothecia) (x 2)

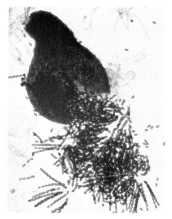

69 Perithecium squashed to show 8-spored asci (x 400)

equivalent in *Elsinoe*, *Guignardia*, *Mycosphaerella*, *Pleospora* and *Venturia* is a cavity-shaped **pseudothecium**, of which several may be formed within a single stromatic layer. In the 'cup fungi', for example *Sclerotinia*, *Monilinia* and *Pyrenopeziza*, asci are produced in a cup- or saucer-shaped **apothecium** from which spores are forcibly ejected.

70 Apothecium discharging ascospores (x 0.5)

BASIDIOMYCETES

Class Hemibasidiomycetes

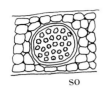
71 *Urocystis*

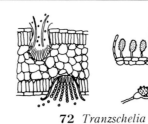
72 *Tranzschelia*

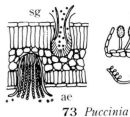
73 *Puccinia*

74 *Angiosorus*

75 *Phragmidium*

76 *Uromyces*

Class Hymenomycetes

77 *Helicobasidium*

78 *Thanatephorus*

79 *Corticium*

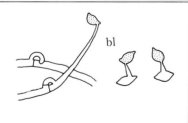
80 *Itersonilia*

DEUTEROMYCETES

Class Agonomycetes

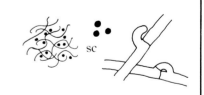
81 *Sclerotium*

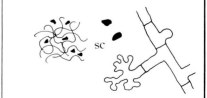
82 *Rhizoctonia*

- ae aecium and aeciospores
- bl ballistospores
- b   basidium and basidiospores
- sc sclerotium
- so sorus
- sg spermagonium (pycnium) and spermatia (pycniospores)
- t   telium and teliospores
- u   uredinium and urediniospores
- us ustilospores

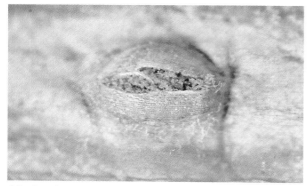

83 *Puccinia* (uredinium)

The various classes of Basidiomycetes (2) comprise fungi which produce sexual spores called basidiospores on a 1- or 4-celled structure called a **basidium**. The rust fungi (Uredinales) have a complex life-cycle in which they produce certain types of spores (urediniospores, teliospores and basidiospores) on one host, and sometimes also other types of spores (pycniospores and aeciospores) on a second host (4). Examples of rusts are *Puccinia* and *Uromyces*. The smut fungi (Ustilaginales) produce only teliospores (ustilospores) and basidiospores, examples being *Urocystis* and *Angiosorus*. In the most advanced types of Basidiomycetes, basidia are produced on a **hymenium**, examples being *Thanatephorus* and *Corticium*. In these fungi, however, the sexual state is uncommon, and they can persist as fungal strands without any spores (sterile mycelium) but with tough resting bodies (sclerotia). These forms have names of their own, so that *Rhizoctonia solani* is the commonly found sterile state of *Thanatephorus cucumeris*, and *Sclerotium rolfsii* the sterile state of *Corticium rolfsii*. The sterile forms are grouped together as Agonomycetes, comprising a class within the Deuteromycetes (q.v.).

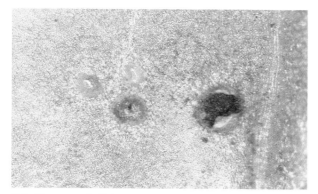

84 *Uromyces* (telia)

85 *Sclerotium* (sclerotia)

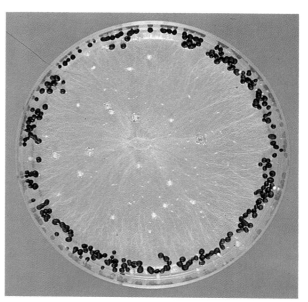

87 *Sclerotium rolfsii* in culture

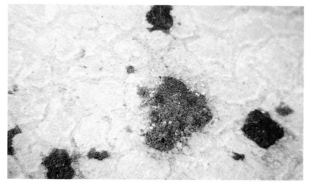

86 *Rhizoctonia* (sclerotia)

DEUTEROMYCETES

Class Hyphomycetes (conidia on hyphae)

Pale or bright aseptate spores

88 *Oidium* **89** *Monilia* **90** *Penicillium*

91 *Geotrichum* **92** *Acremonium* **93** *Trichoderma*

94 *Myrothecium* **95** *Verticillium* **96** *Botrytis*

Pale or bright septate spores

97 *Pyricularia* **98** *Cylindrocarpon* **99** *Fusarium*

100 *Microdochium* **101** *Mycocentrospora* **102** *Trichothecium*

cp conidiophore   c conidium   ma macroconidium   mi microconidium   sd sporodochium

46

103 *Penicillium* (conidiophores)

104 *Botrytis* (conidiophores)

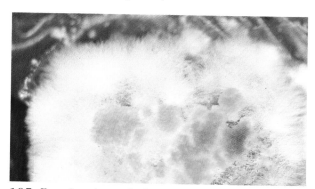

105 *Fusarium* (sporodochia)

106 *Trichothecium* (conidiophores)

Just as some Basidiomycetes may exist in two forms, so also do many Ascomycetes (9). As well as producing sexual spores, most Ascomycetes also produce asexual spores (conidia), indeed the asexual, conidial or imperfect state (anamorph) tends to be much more common than the sexual or perfect state (teleomorph). Furthermore, there are many fungi which appear to exist solely in the asexual state. It is therefore necessary to classify these fungi according to the characteristics of the asexual spore form. These conidial forms, together with fungi which produce sterile mycelium, sclerotia or resting spores (chlamydospores), comprise the Deuteromycetes or Fungi Imperfecti (1).

Conidia may be 1-celled, 2-celled or many-celled, and light or dark in colour. In the Moniliales the spores are borne on specialised hyphae (conidiophores) arising from the mycelium (3,5,6,7). They

107 *Fusarium* (sporodochia)

may arise from a cushion-like structure (sporodochium). The diagrammatic groupings are not always definitive; for example, species of *Aspergillus* may have pale or dark conidia, and *Cladosporium* produces both septate and aseptate conidia (5,6).

DEUTEROMYCETES

Class Hyphomycetes (cont.)

Dark aseptate spores

108 *Chalaropsis*

109 *Thielaviopsis*

110 *Nigrospora*

Dark septate spores

111 *Fusicladium*

112 *Deightoniella*

113 *Stigmina*

114 *Phaeoisariopsis*

115 *Ulocladium*

116 *Stemphylium*

117 *Cercospora*

118 *Curvularia*

119 *Corynespora*

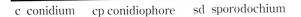

c conidium   cp conidiophore   sd sporodochium

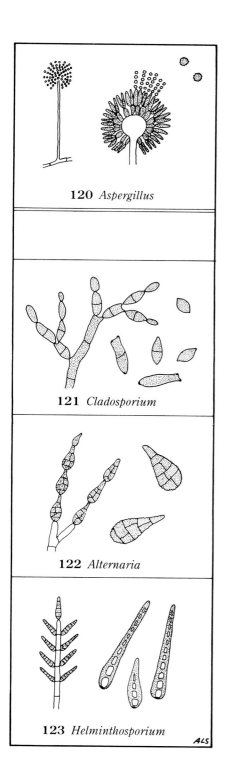

120 *Aspergillus*

121 *Cladosporium*

122 *Alternaria*

123 *Helminthosporium*

124 *Aspergillus* (conidiophores)

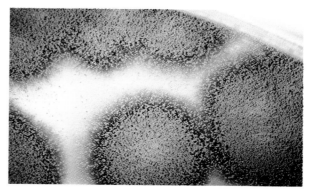

125 *Cladosporium* (conidiophores)

126 *Alternaria* (conidiophores)

127 *Helminthosporium* (conidiophores)

DEUTEROMYCETES

Class Coelomycetes (1) (conidia in acervuli)

Pale or bright aseptate spores

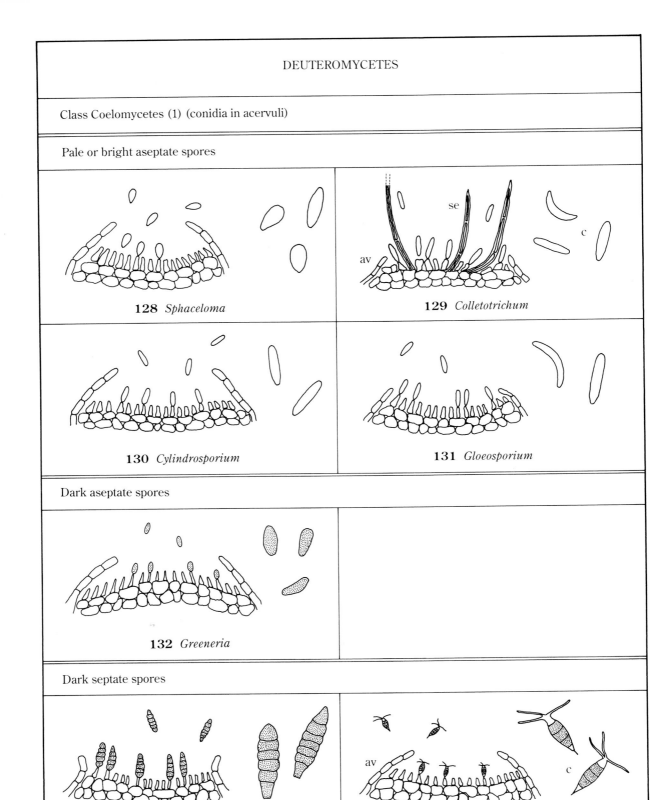

128 *Sphaceloma*
129 *Colletotrichum*
130 *Cylindrosporium*
131 *Gloeosporium*

Dark aseptate spores

132 *Greeneria*

Dark septate spores

133 *Stigmina*
134 *Pestalotiopsis*

av acervulus   c conidium   se seta

135 *Colletotrichum* (acervuli)

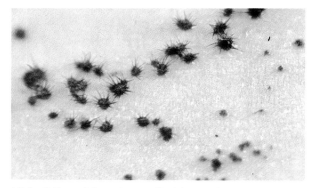

136 *Colletotrichum* (acervuli with setae)

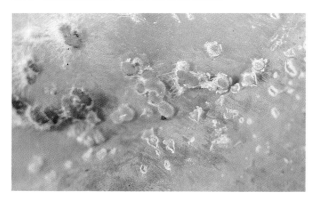

137 *Gloeosporium* (acervuli)

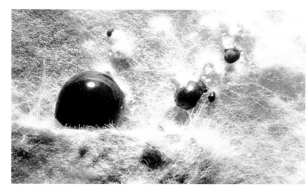

138 *Pestalotiopsis* (acervuli)

In the Melanconiales (1) the spores are borne in an open structure known as an **acervulus**, examples being the anthracnose fungi *Colletotrichum* and *Gloeosporium*. The sporodochia of *Stigmina* have sometimes been regarded as acervuli.

In the Sphaeropsidales (1) spores develop within a flask-shaped structure called a **pycnidium**, somewhat similar in appearance to a perithecium (q.v.) but containing conidia. Examples include *Ascochyta*, *Phoma*, *Phomopsis*, *Septoria* and *Botryodiplodia* (10).

Thus Ascomycetes too may possess two names, for example *Glomerella cingulata* is the sexual stage and *Colletotrichum gloeosporioides* the asexual stage of the same fungus (1). *Colletotrichum musae*, on the other hand, appears to exist solely in the asexual state and so is an Imperfect Fungus.

1 AINSWORTH G.C., SPARROW F.K. & SUSSMAN A.S. (Eds.) (1973) *The fungi: an advanced treatise. Vol. 4A. A taxonomic review with keys: Ascomycetes and Fungi Imperfecti*. New York: Academic Press, 621 pp.
2 AINSWORTH G.C., SPARROW F.K. & SUSSMAN A.S. (Eds.) (1973) *The fungi: an advanced treatise. Vol. 4B. A taxonomic review with keys: Basidiomycetes and Lower Fungi*. New York: Academic Press, 504 pp.
3 CARMICHAEL J.W., KENDRICK W.B., CONNERS I.L. & SIGLER L. (1980) *Genera of Hyphomycetes*. Edmonton: University of Alberta Press, 386 pp illus.
4 CUMMINS G.B. & HIRATSUKA Y. (1983) *Illustrated genera of rust fungi*. St Paul, Minnesota: American Phytopathological Society, 152 pp illus.
5 DOMSCH K.H., GAMS W. & ANDERSON T.-H. (1980) *Compendium of soil fungi. Vols 1 and 2*. New York: Academic Press, 859 pp illus., 405 pp.
6 ELLIS M.B. (1971) *Dematiaceous Hyphomycetes*. Farnham Royal, Slough, UK: Commonwealth Agricultural Bureaux, 608 pp illus.
7 ELLIS M.B. (1976) *More dematiaceous Hyphomycetes*. Farnham Royal, Slough, UK: Commonwealth Agricultural Bureaux, 507 pp illus.
8 KRIEG N.R. & HOLT J.G. (1984) *Bergey's manual of systematic bacteriology. Vol. 1*. Baltimore: Williams & Wilkins, 964 pp.
9 SIVANESAN A. (1984) *The bitunicate Ascomycetes and their anamorphs*. Vaduz, Liechtenstein: J. Cramer, 701 pp illus.
10 SUTTON B.C. (1980) *The Coelomycetes: Fungi Imperfecti with pycnidia, acervuli and stromata*. Farnham Royal, Slough, UK: Commonwealth Agricultural Bureaux, 696 pp illus.

DEUTEROMYCETES

Class Coelomycetes (2) (conidia in pycnidia)

Pale or bright aseptate spores

139 *Dothiorella*
140 *Phoma*
141 *Phyllosticta*
142 *Macrophoma*
143 *Zythia*
144 *Phomopsis*
145 *Pyrenochaeta*
146 *Phacidiopycnis*

c conidium   py pycnidium   sm spermatia

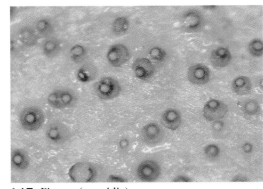

147 *Phoma* (pycnidia)

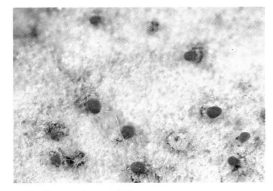

148 *Phoma* (pycnidia)

149 *Phomopsis* (pycnidia)

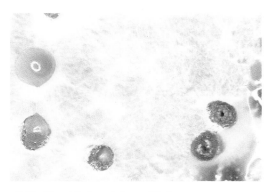

150 *Phacidiopycnis* (pycnidia)

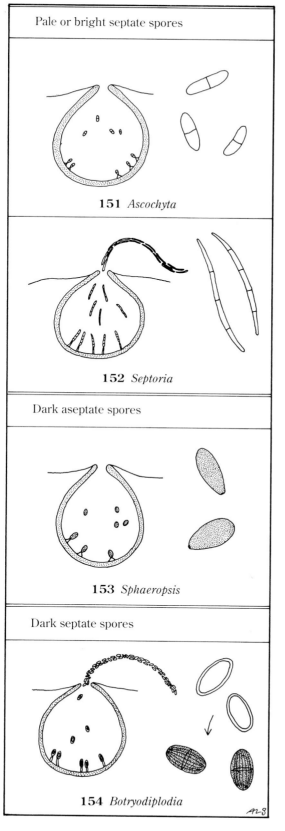

151 *Ascochyta* — Pale or bright septate spores

152 *Septoria*

153 *Sphaeropsis* — Dark aseptate spores

154 *Botryodiplodia* — Dark septate spores

155 *Ascochyta* (pycnidia)

156 *Septoria* (pycnidia)

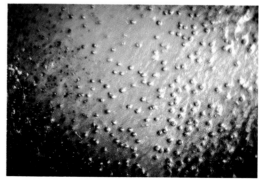

157 *Sphaeropsis* (pycnidia)

158 *Botryodiplodia* (spore tendrils)

CHAPTER 2
CITRUS FRUITS

The fruit of the genus *Citrus* is a special kind of berry (a hesperidium) composed of segments filled with spindle-shaped juice sacs. The whole is covered by white spongy tissue (albedo) and a coloured layer (flavedo) containing numerous oil glands (9). Although there is an outer waxy coating (cuticle), there are many natural openings (stomata). These various layers make up the peel or rind. The fruit having developed from a superior ovary, the calyx is to be found at the stem-end where, together with the floral disk, it forms the 'button' (6).

Citrus fruits are non-climacteric (page 14) and do not undergo a ripening process. As the fruits mature on the tree, the pulp becomes juicy and sweet (or acidic, in lemons and limes) and remains so for an extended period. In tropical growing areas the peel remains green even after the pulp has become edible, and citrus fruits are marketed green. In regions with low night temperatures (below about 13°C) the fruits undergo mild stress which is sufficient to induce internal production of ethylene. This results in destruction of chlorophyll and the development of yellow, orange and red pigments, imparting 'characteristic' colours familiar to consumers of the citrus fruits of international commerce (9).

The **sweet orange**, *Citrus sinensis* (L.) Osbeck, originated in the Far East and has been cultivated there since ancient times. It was introduced into Europe in the 15th century (long after the arrival of the **sour orange**, *Citrus aurantium* L.), and later became so popular that protected enclosures known as 'orangeries' were constructed in areas too cold for satisfactory outdoor growth. In 1493 Columbus took seeds from the Canary Islands to Hispaniola (Haiti), and a few years later citrus was planted in continental America (6). Now oranges are grown throughout the tropics and sub-tropics. Current world production exceeds 40 million tonnes, the ten leading producer countries in 1987 being Brazil, the USA, Spain, Mexico, China, Italy, India, Egypt, Pakistan and Israel. Cultivars include Washington Navel (the 'navel' being a secondary fruitlet at the distal end), Valencia, a late-maturing 'blond orange', and Moro, a 'blood orange' characterised by internal development of deep red pigment (9).

The **mandarin**, *Citrus reticulata* Blanco, is especially important in Japan, followed by Spain, Brazil, the USA, Pakistan, South Korea, Italy, Morocco, Turkey and China. World production is about 8 million tonnes. Satsumas are a type of mandarin, with pale orange skin. Tangerines as a group used to be distinguished by their deep orange skin colour (e.g. the cultivar Clementine), but the names tangerine and mandarin are often used synonymously. Many new hybrids have been produced, including crosses between tangerines and oranges, known as tangors (e.g. Temple, Ortanique) and between tangerines and grapefruits, known as tangelos (e.g. Minneola, Ugli). Modern mandarin cultivars and their hybrids have been bred so as to have few seeds and loose skins and, as a group, are often referred to as 'easy-peel citrus' (9).

The **grapefruit** or **pomelo**, *Citrus paradisi* Macf., was first recorded in the West Indies in the 18th century (6). It is now known to have been the result of a cross between the **pummelo** or **shaddock** (*Citrus maxima* (Burman) Merr.) and the sweet orange. Grapefruits are now grown extensively in the USA, where pink- and red-fleshed cultivars (e.g. Ruby Red) are especially popular. Elsewhere white-fleshed cultivars (e.g. Marsh) are most commonly grown, the other main producer countries (of grapefruits and pummelos combined) being Israel, Thailand, Cuba, China, Argentina, South Africa, Mexico, Paraguay and Cyprus. Total world production is over 4 million tonnes.

The **lemon**, *Citrus limon* (L.) Burman, probably came from India or Burma. It may have been known to the Romans, but did not become widespread in Mediterranean countries until mediaeval times. The Arabs planted lemons in North Africa and Spain, and after the Crusades cultivation became important in Italy (6). Being sensitive to extremes of temperature, the lemon grows best in mild sub-tropical regions.

The **lime**, *Citrus aurantifolia* (Christm.) Swingle, takes the place of lemon in the tropics. The ten major producers (of lemons and limes combined) are the USA, Italy, Mexico, Spain, India, Argentina, Turkey, Brazil, Greece and China, world production being about 6 million tonnes.

The characteristics of the harvested crop are determined to a very great extent by growing conditions, orchard practices and techniques of harvesting and handling. Diseases and disorders originating in the citrus grove include alternaria rot, anthracnose, black spot, brown rot, melanose, stem-end rot, frost damage, creasing and granulation (2,3,7). Control measures must therefore begin during the growing season (6,9,10). Diseases and disorders predominantly related to harvesting and handling include green and blue mould rots, sour rot, trichoderma rot and oleocellosis (2,7,10).

Although various harvesting machines have been devised (9), most citrus is picked by hand, either by twisting and pulling the fruit (so that the 'button' is left behind) or by cutting close to the stem-end by means of sharp clippers. Pulling can be advantageous as a means of avoiding subsequent stem-end rot (q.v.), but clipping is necessary for fruit prone to 'plugging', the accidental tearing of peel around the stem-end. Care during harvesting and handling is essential, since injuries predispose the fruit to infection (3).

Fruit harvested early in the season may be 'degreened' by means of manufactured ethylene released into the room (9). The appropriate regime is dictated by cultivar and growing conditions. High humidity and a temperature of 30°C permit rapid wound-healing before moulds can invade. At temperatures nearer to 20°C penicillium moulds can become established in unhealed wounds unless prevented by prior fungicide drenching (3). Ethylene treatment carries with it the disadvantage of hastening the death of green 'button' tissue, thereby predisposing the fruit to invasion by stem-end rot fungi. This undesirable effect can be counteracted by treating the fruit with a growth-regulating chemical which maintains vitality of the buttons (9).

Other packhouse procedures may include washing in a heated disinfectant, rinsing, drying, grading, dyeing, waxing and packing (9). Modifications in packhouse layout can be of great value in reducing opportunities for fungal infection, the simplest example being enclosure of the area where decaying fruits are removed (1). In view of the ease with which fungi develop strains tolerant of a given fungicide, a range of chemicals should be used, so that unrelated compounds can be applied at different stages, for example before and after degreening or before and after storage (1,3).

Optimal storage temperatures vary with cultivar, growing conditions, and the length of time for which storage is envisaged (5). Grapefruits, lemons and limes are especially susceptible to chilling injury (q.v.) and are usually stored at 10°C or above. Mandarins may be kept at 4° to 5°C, while some orange cultivars can be successfully stored at 2° to 3°C. Detailed figures for different types are given in Appendix 3. In refrigerated space, fresh air ventilation should not be excessive; one air change per hour is probably more than adequate (8). One of the purposes of ventilation is to flush out ethylene which, though produced in scant measure by sound healthy citrus, is produced in abundance by injured or mouldy citrus (8). It is for this reason that ethylene-sensitive commodities should not be stored near to citrus.

If moulds are well-controlled, then citrus fruits (except for mandarins) can be stored for several months with little change beyond dehydration. Moisture loss can be reduced by waxing or, more effectively, by heat-sealing individual fruits in high-density polyethylene film (4). Seal-packaging has the further advantage of promoting wound-healing and, at a later stage, of isolating any decaying fruits from their neighbours and thereby preventing 'soiling' by spores. The traditional way of preventing such contamination is by the use of fungicides (such as diphenyl incorporated in tissue wraps or pads) which inhibit spore production (3). The search continues for new fungicides and new techniques to extend the storage life of this important crop.

1 BANCROFT M.N., GARDNER P.D., ECKERT J.W. & BARITELLE J.L. (1984) Comparison of decay control strategies in California lemon packinghouses. *Plant Disease* **68**, 24–28.
2 CEPONIS M.J., CAPPELLINI R.A. & LIGHTNER G.W. (1986) Disorders in citrus shipments to the New York market, 1972–1984. *Plant Disease* **70**, 1162–1165.
3 ECKERT J.W. (1978) Post-harvest diseases of citrus fruits. *Outlook on Agriculture* **9**, 225–232 illus.
4 GOLOMB A., BEN-YEHOSHUA S. & SARIQ Y. (1984) High-density polyethylene wrap improves wound-healing and lengthens shelf-life of mechanically harvested grapefruit. *Journal of the American Society for Horticultural Science* **109**, 155–159.
5 MUNOZ-DELGADO J.A. (1980) Applications de la technique du froid dans la commercialisation des agrumes. *International Journal of Refrigeration* **3**, 279–287.
6 REUTHER W., CALAVAN E.C. & CARMAN G.E. (Eds) (1967–**1989**) *The citrus industry*. Revised ed. Vols 1 to 5. Riverside: University of California.
7 SMOOT J.J., HOUCK L.G. & JOHNSON H.B. (1971) Market diseases of citrus and other subtropical fruits. *Agriculture Handbook of the United States Department of Agriculture* No. 398, 101 pp illus.
8 WAKS J., CHALUTZ E., SCHIFFMANN-NADEL M. & LOMENIEC E. (1985) Relationship among ventilation of citrus storage room, internal fruit atmosphere, and fruit quality. *Journal of the American Society for Horticultural Science* **110**, 398–402.
9 WARDOWSKI W.F., NAGY S. & GRIERSON W. (1986) *Fresh citrus fruits*. Westport, Conn.: AVI Publishing Co., 571 pp illus.
10 WHITESIDE J.O., GARNSEY S.M. & TIMMER L.W. (Eds) (1988) *Compendium of citrus diseases*. St Paul, Minnesota: American Phytopathological Society, 80 pp illus.

See also:

ROUSE R.E. (Chairman) (1988) Workshop: Description and characteristics of major citrus cultivars of the world. *HortScience* **23**, 679–697.

ALTERNARIA ROT of citrus caused by
Alternaria alternata (Fr.) Keissler
Alternaria citri Ell. & Pierce

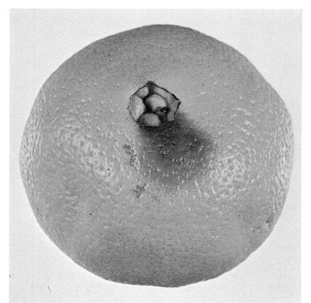

OCCURRENCE *Alternaria alternata* is ubiquitous and has many hosts; it has been recorded on citrus in India (9). *A. citri* is widely distributed in citrus-growing countries, for example the USA (1,2,10), Israel (3), South Africa (4), Australia (6) and Japan (7). Both these species, and others, have been recorded on citrus imported into France (5). Even within a single species, different strains attack citrus fruits in different ways. Alternaria rot can be a problem in consignments destined for processing, since even low levels of infection result in off-flavoured juice (8).

SYMPTOMS Detection of infected fruits is made difficult by the fact that considerable internal rotting may take place before any symptoms appear externally (2,9). Some strains of these fungi tend to grow down the axis of the fruit before invading tissue nearer the peel. In lemons the pulp becomes greyish-brown, soft and slimy (1). In oranges, grapefruits and mandarins infected internal tissue is black, giving rise to the common names of **black rot** and **black centre rot** (5,9). In mandarins (which are attacked by a distinct pathotype) lesions often develop on the side of the fruit and infected peel appears brown, hence the name **brown spot** of mandarin (6); on certain cultivars, however, symptoms consist of corky eruptions and pits (10).

BIOLOGY The spores (conidia) are disseminated throughout the grove by air currents. The fungi are unable to invade normal healthy citrus, but if the fruit has been injured (5) or weakened by unfavourable growing conditions (climatic or cultural) then it becomes susceptible to invasion. Oranges, and especially mandarins, may be infected early in the season and drop from the tree (6,10). Spores may also lodge at the stem-end or, in navel oranges, at the stylar end (9). After harvest sound healthy citrus remains resistant to invasion until the 'button' becomes moribund, whereupon the fungus progresses into the fruit (1,5). Alternaria rot is

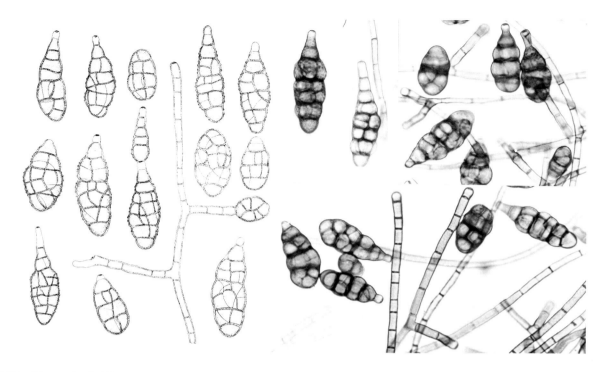

162 *Alternaria citri* (from CMI Descr. No. 242, Ellis & Holliday 1970) Conidia and conidiophores x 500

therefore most prevalent in citrus which has been degreened with ethylene (page 55) or stored for an extended period (2).

CONTROL The majority of fungicides are ineffective against *Alternaria* spp., although some successes have been reported (3,7,10). The chief means of control is prevention of injury and maintenance of fruit vitality (5). Harvesting at correct maturity is very important. Immersion in a growth-regulating chemical serves to maintain the buttons in a fresh green state, thereby delaying fungal invasion of the fruit (3,5). Such a treatment is especially necessary if the fruit is to be treated with ethylene. Eventually, however, the button will become senescent and lose its resistance, so consignments infected with this fungus should be kept under refrigeration and marketed promptly (2).

---

1 BARTHOLOMEW E.T. (1926) Alternaria rot of lemons. *Bulletin of California Agricultural Experiment Station* No. 408, 38 pp illus.
2 BROWN G.E. & McCORNACK A.A. (1972) Decay caused by *Alternaria citri* in Florida citrus fruit. *Plant Disease Reporter* **56**, 909–912 illus.
3 COHEN E. & SHUALI M. (1983) [Combined treatment with 2,4-D and thiabendazole drencher, before degreening citrus fruits, to delay drying of buttons and stem-end rot development.] *Alon Hanotea* **37**, 669–672; 750.
4 DOIDGE E.M. (1929) A study of some Alternarias affecting citrus in South Africa. *Bulletin of the Union of South Africa Department of Agricultural Science* No. 69, 29 pp.
5 JOLY P. (1967) Les pourritures noires des agrumes provoquées par les Alternaria. *Fruits* **22**, 89–95 illus.
6 KIELY T.B. (1964) Brown spot of Emperor mandarin. *Agricultural Gazette of New South Wales* **75**, 854–856 illus.
7 MASUI M. (1986) Befran (iminoctadine), a new fungicide. *Japan Pesticide Information* No. 49, 7–9.
8 PATRICK R. & HILL E.C. (1959) Microbiology of citrus fruit processing. *Bulletin of Florida Agricultural Experiment Station* No. 618, 62 pp illus.
9 SINGH R.S. & KHANNA R.N. (1966) Black core rot of mandarin oranges caused by *Alternaria tenuis* Auct. *Plant Disease Reporter* **50**, 127–131 illus.
10 WHITESIDE J.O. (1979) Alternaria brown spot of Dancy tangerines and its control. *Proceedings of Florida State Horticultural Society* **92**, 34–37.

ANTHRACNOSE of citrus caused by
Glomerella cingulata (Stonem.) Spauld. & v. Schrenk
Conidial state: *Colletotrichum gloeosporioides* (Penz.) Sacc.

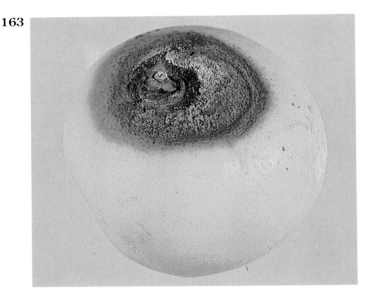

163

OCCURRENCE The common anthracnose fungus has been recorded in most citrus-growing countries, including the USA (1), Belize, Australia, Japan (5), India, South Africa, Israel (4) and Italy (3). A closely related fungus, *Gloeosporium foliicolum* Nishida attacks citrus in Japan and the USA (2). The disease is of particular importance on mandarins (1).

SYMPTOMS Fruits may be disfigured by 'tearstain' patterns on the peel (5) or they may undergo severe rotting, with the formation of dark sunken lesions (3). Humid conditions encourage the development of spore-masses which are at first salmon-pink and later brown.

BIOLOGY The fungus produces both the sexual stage (perithecia giving rise to ascospores) and the asexual stage (acervuli giving rise to conidia), and the spore-bearing bodies form in dead branches (3). Infection is chiefly by conidia disseminated by rain-water. Normal healthy citrus is resistant, but trees weakened by unfavourable growing conditions are prone to attack (3). Prolonged rain at blossom time can result in early season infection and premature fruit drop. Alternatively the infection may remain latent until after harvest, and the disease is manifest during transport and storage (2). Mature green fruit treated with ethylene (to promote colouring of the fruit) is especially vulnerable, as the ethylene stimulates fungal penetration of the intact peel.

CONTROL Anthracnose may be controlled in the grove by maintaining vigorous trees (3) and spraying with fungicide (4). Post-harvest decay is reduced if the fruit is given a pre-harvest application of ethephon; this promotes uniform ripening, allowing the crop to be picked at prime maturity (1). Following washing and fungicide treatment, degreening (if used) should be as brief as possible. Refrigerated storage serves to delay the appearance of symptoms (5).

---

1 BARMORE C.R. & BROWN G.E. (1978) Pre-harvest ethephon application reduces anthracnose on Robinson tangerine. *Plant Disease Reporter* **62**, 541–544.
2 FISHER F.E. (1970) Gloeosporium rot of citrus fruit in Florida. *Plant Disease Reporter* **54**, 173–175 illus.
3 SALERNO M. & CUTULI G. (1985) Le malattie degli agrumi: antracnosi (anthracnose). *Informatore Fitopatologico* **35**(5), 29–30 illus.
4 SOLEL Z. & OREN Y. (1978) Laboratory and field evaluation of fungicides to control anthracnose stain of citrus fruit. *Phytoparasitica* **6**, 59–64.
5 TANAKA H. (1968) [Studies on anthracnose tear stain of citrus fruits.] *Bulletin of the Horticultural Research Station of Japan B* **8**, 99–110 illus.

ASPERGILLUS BLACK MOULD ROT of citrus caused by *Aspergillus niger* v. Tieghem

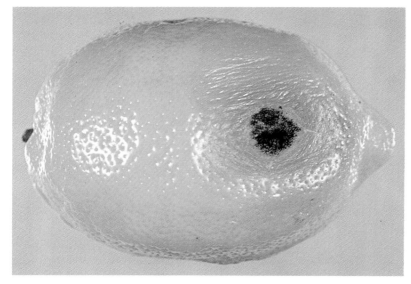

OCCURRENCE This fungus is present in soil and in the atmosphere, but it infects citrus fruits only at fairly high temperature and only if they have been weakened in some way. It has been reported on citrus in the Americas, South Africa and Mediterranean countries (4), and most particularly in India (1,2,3,5).

SYMPTOMS The rot may be predominantly internal, masses of black powdery spores being apparent only when the fruit is cut (4). Alternatively, a very soft sunken water-soaked spot may develop in the peel, later giving rise to black spores resembling soot (2,5). The decay is accompanied by an odour of fermentation.

BIOLOGY Infection has been observed whilst the fruits were still on the tree, externally sound but having suffered invasion via weak stylar ends (4). Pre-harvest infection may also occur if the fruit is injured by thorns during windy weather. Harvesting and handling wounds similarly predispose fruits to infection (5) and the fungus may invade through cuts in the skin or via a stem-end injury (2). At temperatures of 25°C and above it develops rapidly and spreads to adjacent fruits (3); in mixed infections it tends to outgrow other fungi.

CONTROL Aspergillus rot may be controlled in the grove by maintaining vigorous trees and preventing injury to the fruits. Harvested fruits should be handled and packed with care (5) and stored under refrigeration; decay is insignificant at temperatures below 15°C (3).

---

1 BABU K.J., LAXMINARAYANA P. & REDDY S.M. (1983) Evaluation of different volatile compounds in the control of fruit-rot of lemon. *Pesticides* **17**(12), 35,38.
2 BHARGAVA S.N. (1972) Aspergillus rot on *Citrus aurantifolia* fruits in the market. *Plant Disease Reporter* **56**, 64 illus.
3 KHANNA K.K. & CHANDRA S. (1975) Studies on storage diseases of fruits and vegetables. III Factors affecting storage rot of lime caused by *Aspergillus niger*. *Proceedings of the National Academy of Sciences, India* **45 B**, 109–110.
4 PASSALACQUA T. (1932) Sul 'mal della terra' dei frutti di mandarino. *Lavori del Regio Istituto Botanico di Palermo* **13**, 55–61.
5 SRIVASTAVA M.P. & TANDON R.N. (1969) Some storage diseases of orange. *Indian Phytopathology* **22**, 282–284 illus.

BACTERIAL CANKER of citrus caused by
Xanthomonas campestris pv. *citri* (Hasse) Dye

165

OCCURRENCE Citrus canker can be serious in areas where rainfall is frequent during periods of shoot emergence and early fruit development (CMI Descr. 11). It is endemic in South America (1,4), parts of Africa and in most Asian countries including Japan and India (2) (CMI Map 11). After comprehensive eradication campaigns several decades ago, involving the destruction of millions of citrus trees, the USA, South Africa, Australia and New Zealand were eventually declared free of the disease (2,4,5). The organism may inadvertently be transmitted between countries, however, and prompt eradication measures have recently been enforced in the Yemen Arab Republic, on Thursday Island (off the northern coast of Australia) and, once again, in Florida (5).

SYMPTOMS The first signs on the fruits are small, circular, slightly raised pale brown lesions with oily or water-soaked margins (2). Later these spots enlarge and become corky, with crater-like depressions traversed by fissures and cracks.

BIOLOGY Canker lesions also develop on young leaves and stems, and bacteria exude when lesions are wetted by rain or dew. The organism is dispersed in water-droplets by wind and animals. Severe infection can lead to defoliation and fruit drop (5). The disease may spread rapidly (1) during warm wet periods, especially if the weather is also windy.

CONTROL Developing fruits are vulnerable for about 90 days after petal-fall, and susceptible cultivars need protection with copper sprays during this period (1,4). Wind-breaks help to reduce the incidence of infection. If the disease is newly introduced into an area it may be possible to contain the outbreak if drastic eradication measures are taken early enough (2). Once it is established, however, there is a risk of epidemic spread (4), and it is considered preferable to make strenuous efforts to prevent its introduction. Quarantine legislation should be strictly enforced, and vigilance is required to ensure that any canker-infected fruits are intercepted at the port of entry (4,5). If it is desired to export citrus from an infested area, care should be taken to select only fruits which are free from lesions, and to wash them in chlorinated water before shipment (3). With regard to the recent outbreak of canker in Florida, there is debate as to whether it really represents a new introduction or whether it might be a remnant from the past (6). The rationale behind the costly eradication campaign has been challenged on two grounds, firstly that complete elimination of the pathogen is (and was) probably unattainable in practice and, secondly, that the Gulf Coast climate does not appear conducive to epidemic development and spread of canker (6). The issue remains controversial (5).

---

1 DANOS E., BERGER R.D. & STALL R.E. (1984) Temporal and spatial spread of citrus canker within groves. *Phytopathology* **74**, 904–908.
2 DYE D.W. (1969) Eradicating citrus canker from New Zealand. *New Zealand Journal of Agriculture* **118**(2), 20–21.
3 MONNIER E. & FERRACCI S. (1986) Contrôle du chancre citrique sur agrumes en post-récolte par trempage dans une solution d'hypochlorite de sodium. Essai eau de Javel et traitements des agrumes. *Fruits* **41**, 465–475.
4 ROSSETTI V., FEICHTENBERGER E. & SIVEIRA M.L. (1982) *Citrus canker: an analytical bibliography.* Instituto Biológico, São Paulo, Brazil, 230 pp.
5 SCHOULTIES C.L., CIVEROLO E.L., MILLER J.W., STALL R.E., KRASS C.J., POE S.R. & DuCHARME E.P. (1987) Citrus canker in Florida. *Plant Disease* **71**, 388–395 illus.
6 WHITESIDE J.O. (1988) The history and rediscovery of citrus canker in Florida. *Citrograph* **73**, 197–206.

BLACK PIT of citrus caused by
Pseudomonas syringae pv. *syringae* van Hall

166

OCCURRENCE This bacterial disease can cause a blemish of all types of citrus, but lemons are particularly susceptible (4,5). It occurs in regions possessing a cool wet season, for example parts of Australia, Japan, South Africa, certain Mediterranean countries, the Soviet Union and the United States (California but not Florida) (3).

SYMPTOMS The first symptoms are light brown spots on the skin, later becoming dark brown and eventually black and markedly sunken (2). There may be concentric rings of brown on a lighter ground, and close examination usually reveals a small wound in the centre of each pit. Under moist conditions greyish-yellow droplets exude from the lesions, but the tissue beneath remains firm (2,4). The diameter of the pits is usually not more than 1 cm, but lesions as large as 3 cm are occasionally recorded. Some enlargement may occur during storage (3,4).

BIOLOGY Black pit (known as citrus blast when it attacks the leaves) only causes problems if the growing season has been cool, wet and windy (3,4). The bacterium can attack fruits which have suffered physical injury from hail or from thorns during branch movement (2).

CONTROL Control measures which have been suggested include pre-harvest spraying, the provision of wind-breaks to reduce mechanical damage to fruit, and the planting of sturdy thornless varieties (3,4).

---

1 COHEN S., COHEN E., SCHIFFMANN-NADEL M. & VOLCANI Z. (1978) Respiration rate and ethylene evolution of lemon infected by *Pseudomonas syringae*. *Phytopathologische Zeitschrift* **91**, 355–358.
2 DOIDGE E.M. (1917) A bacterial spot of citrus. *Annals of Applied Biology* **3**, 53–80 illus.
3 KNORR L.C. (1965) Serious diseases of citrus foreign to Florida. *Bulletin of Florida Department of Agriculture* No. 5, 59 pp illus.
4 SALERNO M. & CUTULI G. (1985) Le malattie degli agrumi: batteriosi (blast and black pit). *Informatore Fitopatologico* **35**(5), 27–28 illus.
5 SMITH C.O. & FAWCETT H.S. (1930) A comparative study of the blast bacterium and some other allied organisms. *Journal of Agricultural Research* **41**, 233–246 illus.

BLACK SPOT of citrus caused by
Guignardia citricarpa Kiely
Conidial state: *Phyllosticta citricarpa* (McAlp.) van der Aa

167

OCCURRENCE Citrus black spot is an important disease in Australia (1,2,8), eastern and southern Africa (4,6,7,10) and parts of Asia (5,9) but the disease has not yet been encountered in the Americas or in Mediterranean countries (CMI Map 53).

SYMPTOMS Various symptoms may occur, the nature of the lesions being determined by fruit maturity and ambient temperature at the time of infection (3,4). Early season infection may result in an appearance resembling melanose (q.v.). Infection of immature fruit may also lead to a symptom called 'hard spot' or 'limited spot' characterised by the presence of a halo. Although chiefly observed before harvest, hard spot may also develop on early-harvested fruit after picking. 'Freckle spot' or 'speckled blotch' is seen in fruit approaching maturity, and consists of clusters of dark brown lesions which may develop further during storage. 'Virulent spot' is observed late in the season when fruit is mature and temperatures high, and is characterised by irregular reddish sunken lesions which spread rapidly after harvest. Although only the rind is affected, the disfigurement can render the fruit unmarketable (1,4).

BIOLOGY The fungus produces both the sexual stage (perithecia giving rise to ascospores) and the asexual stage (pycnidia giving rise to conidia), and both types of spore-bearing bodies develop abundantly on fallen citrus leaves (5). During moist periods in the early part of the growing season, ejected ascospores are dispersed by air currents and cause direct infection of young fruits and leaves (5,6). In addition, water-borne conidia may be splashed on to low-hanging fruit or washed down from infected twigs (10). A long period of latency follows infection (6), and the fruit may be on the tree for many months before the ambient temperature becomes high enough to induce symptom expression.

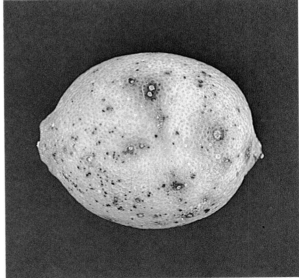

16

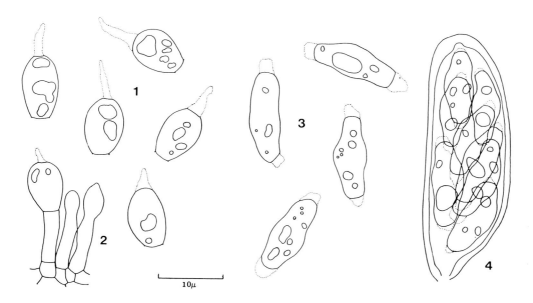

169 *Guignardia citricarpa* (from CMI Descr. No. 85, Sutton & Waterston 1966) 1 conidia 2 conidiophores 3 ascospores 4 ascus

CONTROL Control is chiefly by fungicide sprays and, since the developing fruit is susceptible for the first four to five months after petal-fall, spraying should be started early in the season if the weather is likely to be wet (1,4,7,9). Post-harvest waxing helps to reduce development of black spot during storage (8). It is inadvisable to rely on a single fungicide for pre- and post-harvest use, because the fungus can develop tolerant strains.

---

1 BERTUS A.L. (1981) Fungicidal control of black spot and melanose on coastal Valencia oranges. *Australasian Plant Pathology* **10**, 53–55.
2 KIELY T.B. (1950) Control and epiphytology of black spot of citrus on the central coast of New South Wales. *Bulletin of the New South Wales Department of Agricultural Science* No. 71, 88 pp.
3 KIELY T.B. (1960) Speckled blotch of citrus. *Agricultural Gazette of New South Wales* **71**, 474–476 illus.
4 KOTZÉ J.M. (1981) Epidemiology and control of citrus black spot in South Africa. *Plant Disease* **65**, 945–950 illus.
5 LEE Y.S. & HUANG C.S. (1973) [Effect of climatic factors on the development and discharge of ascospores of the citrus black spot fungus.] *Journal of Taiwan Agricultural Research* **22**, 135–144.
6 McONIE K.C. (1967) Germination and infection of citrus by ascospores of *Guignardia citricarpa* in relation to control of black spot. *Phytopathology* **57**, 743–746.
7 SCHÜPP H. (1961) Untersuchungen über *Guignardia citricarpa* Kiely, den Erreger der Schwartzfleckenkrankheit auf Citrus. *Phytopathologische Zeitschrift* **40**, 258–271.
8 SEBERRY J.A., LEGGO D. & KIELY T.B. (1967) Effect of skin-coatings on the development of black spot in stored Valencia oranges. *Australian Journal of Experimental Agriculture and Animal Husbandry* **7**, 593–600.
9 TSIA Y.P., SHIA O, LIAU T.H. & SUN M.H. (1977) [Field trials of citrus black spot control in Taiwan.] *Plant Protection Bulletin, Taiwan* **19**, 140–145.
10 WHITESIDE J.O. (1967) Sources of inoculum of the black spot fungus, *Guignardia citricarpa*, in infected Rhodesian citrus orchards. *Rhodesia, Zambia and Malawi Journal of Agricultural Research* **5**, 171–177.

BROWN ROT of citrus caused by *Phytophthora* spp.

170

OCCURRENCE Brown rot is a major disease of citrus and is especially prevalent when there are late season rains (3). It has been recorded in most citrus-growing countries, including the USA (3,6), Italy (7), Israel (4,9,10), South Africa (5), Australia (2) and India (8). Among the species of *Phytophthora* are *P. citricola* Sawada (3), *P. citrophthora* (Smith & Smith) Leon. (7,9), *P. hibernalis* Carne (2), *P. nicotianae* (van Breda de Haan) var. *parasitica* (Dastur) Waterh. and *P. syringae* (Kleb.) Kleb. (8).

SYMPTOMS The first symptom is a greyish-brown spot which, even after enlargement, tends to remain firm and leathery. Infected fruits give off a characteristic pungent odour (8). Under humid conditions a fine white spore-bearing mould develops on the peel and spreads by contact to neighbouring fruit (5,6). The rot develops very rapidly at 25°C but hardly at all at 4°C (9).

BIOLOGY The fungi are common in soil, and can cause a serious root rot (1,2). They survive in the sexual state (thick-walled oospores) and produce the asexual state (sporangia and/or zoospores) in moist conditions (5). During rainy weather spores are splashed on to low-hanging fruits and, if the peel remains wet for some time, the fungus is capable of direct penetration (3,8). Rotting may occur whilst the fruit is still on the tree, or fruits with incipient infections may pass unnoticed at harvest (9).

CONTROL A combination of control measures is used, starting with pre-harvest fungicides in the grove (4,5,8). It is beneficial to spray the ground beneath the trees, in order to inhibit spore-formation on the soil surface (10). Fallen fruits should be promptly removed before they can be attacked by snails, which may otherwise further distribute the fungus (7). It is advisable to disinfect the picking boxes periodically, and to ensure that they are not left outside during wet weather. After harvest, the fruits should be immersed in a hot-water bath for several minutes so that deep-seated infections may be deactivated. Heat treatment should not be delayed for more than three or four days, because brown rot can spread rapidly through the fruit (5,6). Finally the fruit may be coated with wax containing a fungicide.

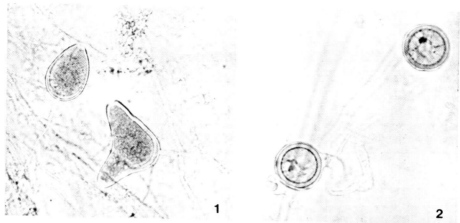

172 *Phytophthora citricola* (from CMI Descr. No. 114, Waterhouse & Waterston 1966)
1 sporangia x 300   2 oogonia (stained) x 500

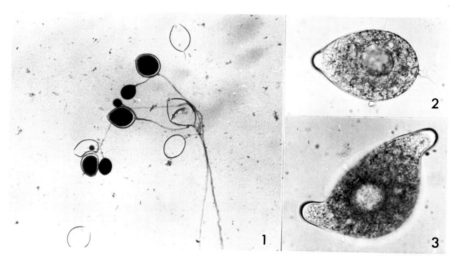

173 *Phytophthora citrophthora* (from CMI Descr. No. 33, Waterhouse & Waterston 1964)
1 sporangiophore (stained) x 200   2,3 sporangia x 750

---

1 BOCCAS B. & LAVILLE E. (1978) *Les maladies à Phytophthora des agrumes*. Paris: Institut de Recherches sur les Fruits et Légumes, 162 pp illus.
2 DOEPEL R.F. (1966) *Phytophthora* species on citrus in Western Australia. *Plant Disease Reporter* **50**, 494–496 illus.
3 FELD S.J., MENGE J.A. & PEHRSON J.E. (1979) Brown rot of citrus: a review of the disease. *Citrograph* **64**, 101–106.
4 GUTTER Y. (1983) Supplementary antimold activity of phosetyl Al, a new brown rot fungicide for citrus fruits. *Phytopathologische Zeitschrift* **107**, 301–308.
5 HOUGH J.A., KELLERMAN C.R. & FOURIE L. (1980) Pre-harvest spray for brown rot control of citrus. *Citrus and Subtropical Fruit Journal* No. 556, 10–11.
6 KLOTZ L.J. & DeWOLFE T.A. (1961) Brown rot contact infection of citrus fruits prior to hot water treatment. *Plant Disease Reporter* **45**, 268–271.
7 MAGNANO di San L., G. & PENNISI A.M. (1984) Attacchi conjiunti di limacce e di marciume bruno a frutti di arancio dolce. *Informatore Fitopatologico* **34**(11), 34–36 illus.
8 RAO N.N.R. (1985) Relative efficacy of a single pre-monsoon spray of some fungicides for the control of brown fruit-rot of citrus. *Indian Journal of Agricultural Sciences* **55**, 189–192.
9 SCHIFFMANN-NADEL M. & COHEN E. (1969) The incubation period of *Phytophthora citrophthora* in citrus fruits stored at different temperatures. *Phytopathology* **59**, 237–238.
10 SOLEL Z. (1983) Aerial versus ground application of fungicides to control brown rot of citrus fruit. *Plant Disease* **67**, 878–880.

COTTONY ROT of citrus caused by
Sclerotinia minor Jagger
Sclerotinia sclerotiorum (Lib.) de Bary

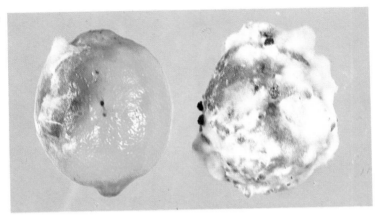

174

OCCURRENCE   Cottony rot is not common, but occasionally causes heavy losses, particularly in lemons. *Sclerotinia minor* has been recorded on lemons in Italy (4) and on grapefruit in Israel. Records of *S. sclerotiorum* include lemons in the USA (5), grapefruit in Israel, oranges in Australia, mandarins in Korea (1) and several types of citrus in Italy (3,4).

SYMPTOMS   The affected area, which may be at the stem-end or elsewhere on the fruit, is at first firm and brown, as in brown rot (q.v.) or in grey mould rot (q.v.) (3). Subsequently, however, the tissue softens and fluid is released. Under humid conditions there is copious development of a bright white cottony mould (5) which later gives rise to numerous firm white bodies, those of *S. minor* being 1–2 mm across (4) and those of *S. sclerotiorum* 6–7 mm (3). They exude drops of liquid and then become dark, dry resting bodies (sclerotia).

BIOLOGY   The sclerotia persist in the soil, and can withstand adverse conditions (5). Their subsequent 'germination' is favoured by a cool moist environment (1). *S. minor* characteristically produces strands of mould which can infect fruits that have fallen on the ground (4). In contrast, the sclerotia of *S. sclerotiorum* are more likely to produce the sexual stage (apothecia giving rise to ascospores). The spores are discharged into the air (70) and can cause infection of branches (twig blight) (3) and also of dying flowers, leading to infection of developing fruits (5). Post-harvest rotting by either fungus can be serious, since adjacent sound fruits are invaded, resulting in 'nests' of decay (4).

CONTROL   The grove should be kept free of ground cover, since weeds may harbour the fungi (5), and humid conditions near the soil surface encourage germination of sclerotia (3). These resistant bodies are difficult to kill; there are various chemical, physical and biological methods which have proved partially successful in controlling the disease in other crops (watery soft rot, q.v.) Post-harvest development of *S. sclerotiorum* in lemons has been substantially reduced by the addition of carbon monoxide to the storage atmosphere (2). Under normal circumstances, however, any consignment of citrus found to be infected should be marketed promptly, since decay can continue even under refrigeration (5).

---

1 BAI D.H. (1977) [Survey on the fruit rot occurrence and damages of shipping mandarin.] *Korean Journal of Plant Protection* **16**, 245–247.
2 EL-GOORANI M.A. & SOMMER N.F. (1979) Suppression of post-harvest pathogenic fungi by carbon monoxide. *Phytopathology* **69**, 834–838 illus.
3 SALIS E. (1967) Ulteriore contributo alla conoscenza di *Sclerotinia sclerotiorum* (Lib.) Massee come parassita degli agrumi in Sardegna. *Note Fitopatologiche per la Sardegna* No. 9, 8 pp.
4 SAVASTANO G. (1932) Richerche sperimentale sul marcio dei frutti degli agrumi. 1. Specie batteriche e fungine isolate ed alcune loro caratteristiche biologiche. *Bollettino della R. Stazione di Patologia Vegetale* NS **12**, 306–342 illus.
5 SMITH C.O. (1916) Cottony rot of lemons in California. *Phytopathology* **6**, 268–278 illus.

FUSARIUM ROT of citrus caused by
Fusarium spp.

OCCURRENCE Fusarium rot is usually considered a minor disease of citrus and may be found occasionally in any citrus-growing area, for example South Africa (1), Israel (4,5), India (2) and Australia (3). In Israel it appears to have increased in importance in recent years, being often associated with alternaria rot (q.v.) of the stem-end. Species recorded in various countries include *F. lunulosporum* Gerlach, *F. moniliforme* Sheld.(**248**), *F. oxysporum* Schlecht.: Fr. and *F. solani* (Mart.) Sacc.

SYMPTOMS Lesions are generally dark brown, leathery and sunken, sometimes becoming soft and water-soaked in the centre. Under humid conditions a white or pinkish mould forms on the surface. Internally there may be a purplish or reddish-brown discoloration of the flesh (4).

BIOLOGY These fungi live in the soil, and spores (conidia) are blown about in the atmosphere. Fruit is attacked only if weakened in some way, either physically or physiologically. The author has observed fusarium rot of the stylar end in consignments of navel oranges, the fungus probably having penetrated via cracks in the navel. In all types of citrus fruit the rot may also begin at the stem-end, after the fungus has colonised the button (3,4,5). Decay is slow, even at temperatures favourable to the fungi (4).

CONTROL The crop should be harvested at optimal maturity, and care should be taken to discard fruits with navel-end splits. A growth regulator is of value in maintaining freshness of the stem-buttons, especially in fruit which is to be degreened. Fruits with moribund buttons, and fruits which have been waxed, show increased susceptibility to fusarium rot (5). Prolonged storage should be avoided.

1 GERLACH W. (1977) *Fusarium lunulosporum* spec. nov. von Grapefruit aus Südafrika, ein Fruchtfäuleerreger. *Phytopathologische Zeitschrift* **88**, 280–284 illus.
2 GHATAK P.N. (1938) Investigations on orange rot in storage. 1. Orange rot due to two strains of *Fusarium moniliforme* Sheldon. *Journal of the Indian Botanical Society* **17**, 141–148.
3 HUELIN F.E. (1942) The handling and storage of Australian oranges, mandarins and grapefruit. *CSIRO Bulletin* No. 154, 60 pp illus.
4 JOFFE A.Z. & SCHIFFMANN-NADEL M. (1972) Les espèces de *Fusarium* isolées des fruits d'agrumes d'Israël. *Fruits* **27**, 117–119 illus.
5 WAKS J., SCHIFFMANN-NADEL M., LOMANIEC E. & CHALUTZ E. (1985) Relation between fruit waxing and development of rots in citrus fruit during storage. *Plant Disease* **69**, 869–870.

GREASY SPOT RIND BLOTCH of citrus caused by
Mycosphaerella citri Whiteside
Conidial state: *Stenella citri-grisea* (Fisher) Sivanesan

OCCURRENCE Greasy spot is important only in regions where there are prolonged periods of hot humid weather. The disease occurs in south-east USA (Florida and Texas) (4,5) and in parts of South America (1) and the Caribbean (3) (CMI Map 524). In Japan, only leaf symptoms have been described; *M. citri* occurs on the island of Okinawa (2), while on Honshu and Kyushu the causal fungus is said to be distinct and has long been known as *M. horii* Hara (10).

SYMPTOMS In 'greasy spot rind blotch' small specks develop on the fruit, associated with the stomatal openings which lie between the oil glands (6,7). The specks, which are usually more prevalent on the stylar-end half of the fruit, are at first pink in colour but later become brown and then black. Larger lesions may be intermingled with the specks. Surrounding cells retain their green colour (even after ethylene treatment) and this symptom is characteristic of the disease. Lesions may become slightly sunken with age (hence the original name of 'pink pitting') and the blemish is especially disfiguring on grapefruit (6).

BIOLOGY In the grove there is sparse development of the asexual stage (conidia). The sexual stage (perithecia giving rise to ascospores) is of more significance; the spore-bearing bodies develop in abundance on citrus leaves, leading to defoliation (5). Rain, dew or irrigation water on the fallen leaves encourages the discharge of ascospores, which are carried by air currents and impinge on the developing fruits (4,5). Infection occurs via the stomata (7), and symptoms are manifest before the fruit is harvested. The citrus rust mite has been implicated in greasy spot disease, originally as the primary cause and subsequently as a predisposing factor; however, it now appears that there is no direct relationship between infestation and disease (4,5,9).

CONTROL The disease can be controlled in the grove by an application of fungicide alone (1) or mixed with oil (3,8); it may be advisable to spray the ground as well as the trees. Ploughing-in of infected leaf litter has also been recommended (5). Some acaricides (used for control of citrus rust mite) have a slight fungicidal action, though not to the extent of being able to replace traditional fungicides (9).

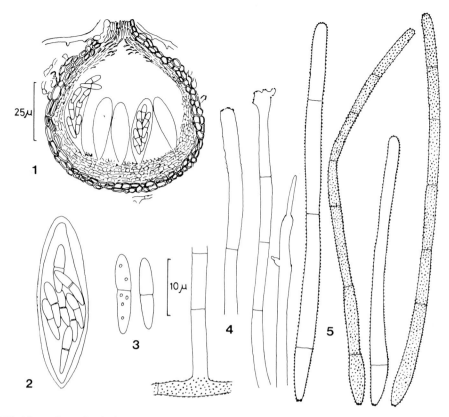

177 *Mycosphaerella citri* (from CMI Descr. No. 510, Sivanesan & Holliday 1976)
1 pseudothecium 2 ascus 3 ascospores 4 conidiophores 5 conidia

1 FOGUET J.L., GONZÁLEZ J.L. & CAMPO S.A. (1979) Relación entre la época de aplicación de tratamientos fungicidas y la eficacia del control de manchas de los pomelos. *Revista Industrial y Agrícola de Tucumán* **56**(1), 11–15.
2 IEKI H. (1986) The causal fungus of citrus greasy spot in Okinawa district of Japan. *Annals of the Phytopathological Society of Japan* **52**, 484–487 illus.
3 SAENZ M. & LEWIS F. (1985) Resultados preliminares sobre la posibilidad del uso del aceite y sus mezclas con varios fungicidas para el control de la mancha grasienta *Mycosphaerella citri*. *Ciencia y Técnica en la Agricultura, Protección de Plantas* **8**(1), 7–15.
4 TIMMER L.W., REEVE R.J. & DAVIS R.M. (1980) Epidemiology and control of citrus greasy spot on grapefruit in Texas. *Phytopathology* **70**, 863–867.
5 WHITESIDE J.O. (1970) Etiology and epidemiology of citrus greasy spot. *Phytopathology* **60**, 1409–1414 illus.
6 WHITESIDE J.O. (1972) Blemishes on citrus rind caused by *Mycosphaerella citri*. *Plant Disease Reporter* **56**, 671–675 illus.
7 WHITESIDE J.O. (1972) Histopathology of citrus greasy spot and identification of the causal fungus. *Phytopathology* **62**, 260–263 illus.
8 WHITESIDE J.O. (1982) Timing of single-spray treatments for optimal control of greasy spot on grapefruit leaves and fruit. *Plant Disease* **66**, 687–690.
9 WHITESIDE J.O. (1983) Fungicidal effects of some acaricides on *Mycosphaerella citri*. *Plant Disease* **67**, 864–866.
10 YAMADA S. (1956) [Studies on the greasy spot (black melanose) of citrus. 2. Morphological characters of the causal fungus (*Mycosphaerella horii* Hara).] *Bulletin of Tokai-Kinki Agricultural Experiment Station (Horticultural Division)* No. 3, 49–62 illus.

GREEN MOULD ROT of citrus caused by
Penicillium digitatum Sacc.
BLUE MOULD ROT of citrus caused by
Penicillium italicum Wehmer

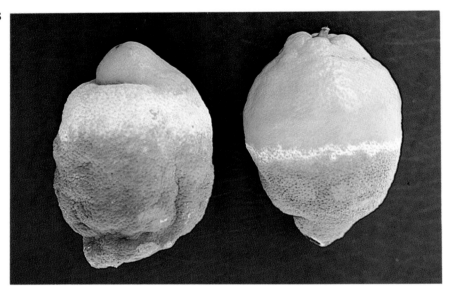

178

OCCURRENCE Green and blue mould rots occur in all citrus-growing areas and often constitute the predominant type of decay (5). Investigations have been carried out in many countries, including the USA (1,3,5), Cuba (4), South Africa (9), Israel (6), Egypt (2) and Australia (8,10).

SYMPTOMS In both diseases the earliest symptom is a soft water-soaked area on the peel of the fruit, which soon becomes covered with white mould. Coloured spores form at the centre of the lesion; in green mould rot there is usually a broad band of white beyond the sporing area, whereas in blue mould rot the white margin is generally not more than about 2 mm wide. The two fungi frequently appear together; during short-term transport and storage green mould usually predominates because at moderate temperatures it grows the more rapidly (3).

BIOLOGY Spores (conidia) of these fungi are invariably present in the atmosphere of citrus-growing areas, and initial infection of fruit is characteristically via wounds in the peel (7). Fruit-piercing insects such as the Mediterranean fruit fly can cause injuries whilst the fruits are still on the tree, permitting pre-harvest infection which may be undetectable at the time of picking (9). Infection also occurs via wounds sustained during harvesting and handling, the fruit being especially susceptible during periods of wet weather. With regard to the nature of the wound, cuts and cracks are more likely to become infected than are abrasions (4). Post-harvest development of decay is chiefly determined by temperature, and at 20° to 25°C a fruit can be completely rotten within a few days (3). Blue mould is capable of spreading directly into uninjured healthy fruits in contact, causing 'nesting', and this can have

179 'Nesting' by blue mould

Conidial heads of *Penicillium italicum* (stained with cotton blue in lactophenol) x 400

serious consequences in long-term storage. In contrast, green mould is generally unable to infect adjacent fruits unless they are injured (1). Nevertheless, even straightforward 'soiling' by clouds of spores can render sound fruit unmarketable (8). Furthermore, the ethylene evolved by an infected fruit (page 55) has the effect of hastening 'ageing' in the rest of the fruit.

182 *Penicillium italicum* (from CMI Descr. No. 99, Onions 1966) Conidial heads x 500

181 'Soiling' by green mould

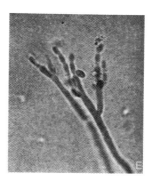

183 *Penicillium digitatum* (from CMI Descr. No. 96, Onions 1966) Conidial head x 500

184 Inhibition of spore production

CONTROL   Control begins in the grove, with sprays to control fruit-piercing insects (9). Since correlations have been demonstrated between the concentration of *Penicillium* spores in the atmosphere and the incidence of post-harvest decay, it is of great importance to minimise spore build-up in the vicinity. Fallen fruits should be removed from the grove every few days and destroyed before they can support colonies of spores. Likewise, fruits culled in the packing house should be promptly segregated; changes in packhouse layout can be very effective in reducing the spore population (5). It is unwise to harvest shortly after rain, since the peel will be turgid and thus more prone to injury and consequent infection. If the fruit is to be degreened it can be advantageous to carry out the process at high temperature and humidity; at 30°C and 95% RH wounds tend to heal before the fungi can become established.

The use of suitable post-harvest fungicides is essential, and treatment should be carried out within 24 hours of harvest, before the fungi penetrate too deeply (2). Because of the evolution of tolerant fungal strains, new chemicals are constantly being evaluated (6,10). In view of the ease with which *Penicillium* can evolve, it is important that the post-harvest fungicide be chemically unrelated to the pre-harvest spray (10). Likewise, if citrus is stored before shipment, different fungicides should be used before and after storage. A combination of fungicides is usually necessary (5,6). Diphenyl has remained useful over many years, its most important function being to suppress spore production on infected fruits, thus reducing the losses due to 'soiling' (8). Refrigeration serves to curtail the development of green and blue moulds, but decay continues slowly and rapid deterioration resumes at ordinary temperatures (3).

---

1 BARMORE C.R. & BROWN G.E. (1982) Spread of *Penicillium digitatum* and *P. italicum* during contact between citrus fruits. *Phytopathology* **72**, 116–120 illus.
2 EL-TOBSHY Z.M., ABDEL-LATIF M.F. & MAHMOUD M.B. (1982) Influence of chemical application time on green mould control of navel orange. *Agricultural Research Review, Cairo* **60**(2), 233–243.
3 FAWCETT H.S. & BARGER W.R. (1927) Relation of temperature to growth of *Penicillium italicum* and *P. digitatum* and to citrus fruit decay produced by these fungi. *Journal of Agricultural Research* **35**, 925–931.
4 FRESNEDA J.A. & CAMACHO J.L. (1984) Influencia del tipo de herida en la infección de naranja y toronja por *Penicillium digitatum*. *Ciencias de la Agricultura* No. 18, 33–38
5 GARDNER P.D., ECKERT J.W., BARITELLE J.L. & BANCROFT M.N. (1986) Management strategies for control of *Penicillium* decay in lemon packing-houses: economic benefits. *Crop Protection* **5**, 26–32.
6 GUTTER Y. (1985) Combined treatment with thiabendazole and 2-aminobutane for control of citrus fruit decay. *Crop Protection* **4**, 346–350.
7 KAVANAGH J.A. & WOOD R.K.S. (1967) The role of wounds in the infection of oranges by *Penicillium digitatum* Sacc. *Annals of Applied Biology* **60**, 375–383.
8 LEGGO D. & SEBERRY J.A. (1965) The effect of diphenyl wraps and case liners on green mould wastage in stored oranges. *Australian Journal of Experimental Agriculture and Animal Husbandry* **5**, 91–96.
9 ROTH G. (1967) Citrus fruit decay in South Africa caused by *Penicillium digitatum* Sacc. *Phytopathologische Zeitschrift* **58**, 383–396 illus.
10 WILD B.L. (1983) Double resistance by citrus green mould *Penicillium digitatum* to the fungicides guazatine and benomyl. *Annals of Applied Biology* **103**, 237–241.

GREY MOULD ROT of citrus caused by
Botryotinia fuckeliana (de Bary) Whetzel
Conidial state: *Botrytis cinerea* Pers.

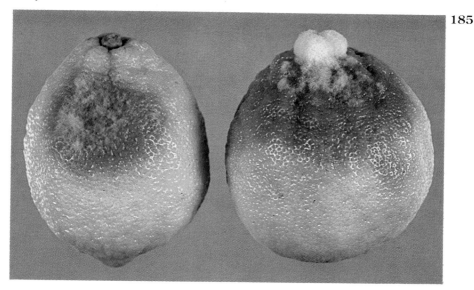

OCCURRENCE Grey mould rot is more usually found on lemons and mandarins than on other types of citrus. It has been reported from several countries, including the USA (3,4), the USSR (2), Spain, Kenya, Korea and Australia (1). It has also been observed in consignments of Algerian oranges imported into France (5).

SYMPTOMS Rotting may begin at the stem-end or on the side of the fruit (1,5). The lesion is at first mid-brown and firm, but later darkens and becomes soft. 'Nests' of decay develop and, under humid conditions, a mass of grey-brown spores is formed (1,4,5).

BIOLOGY The fungus survives on plant debris in the grove, and spores are dispersed by rain, wind and insects (4). Infection of trunk and branches may occur after frost (5). In moist conditions the fungus colonises dying flower parts, after which it remains quiescent in the developing fruit (4,5). However, infected petals falling on to fruits can initiate decay immediately (4,5). Further infections may occur via wounds sustained during harvesting and handling (5). Substantial rotting occurs even in refrigerated storage (2), and decay spreads from fruit to fruit (1,5). The fungus is also capable of spreading from decaying leaves and damp wooden boxes, and in this way can cause direct infection of healthy fruit (4). Decay is favoured by high humidity following a period of cold storage (5).

CONTROL Blossom infection can be prevented by the use of fungicide sprays (4). To minimise infection at harvest-time, picking boxes should not be left in the grove during moist weather (4). Gentle handling, strict hygiene, and hot water treatment can also be beneficial (4). Application of a growth-regulating chemical (to prolong button vitality) delays the onset of stem-end rotting (3). Good air circulation is necessary during storage and transport, so that the atmosphere does not become saturated (5).

1 ADAM D.B. (1923) Experiments in citrus fruit storage. *Journal of the Department of Agriculture, Victoria* **21**, 307–317 illus.
2 BOCHAROVA Z.Z. (1941) [Diseases of citrus fruit in storage. 1. *Botrytis cinerea* on stored citrus fruit.] *Mikrobiologiya* **8**, 1187–1194.
3 ERICKSON L.C., DeWOLFE T.A. & BRANNAMAN B.L. (1958) Growth of some citrus-fruit pathogens as affected by 2,4-D and 2,4,5-T. *Botanical Gazette* **120**, 31–36.
4 KLOTZ L.J., CALAVAN E.C. & ZENTMYER G.A. (1946) The effect of *Botrytis* rot on lemons. *California Citrograph* **31**, 247, 262 illus.
5 MOREAU C. (1960) 'La moisissure grise' des oranges. *Fruits* **15**, 69–71 illus.

MELANOSE of citrus caused by
Diaporthe citri Wolf
Conidial state: *Phomopsis citri* Fawcett

OCCURRENCE The fungus has been recorded in most citrus-growing countries but the disease is serious only in regions having abundant rainfall during the early stages of fruit development. Thus melanose, and also stem-end rot (page 80) caused by the same fungus, are particularly prevalent in Florida and the West Indies, and are relatively unimportant in drier areas such as California, Arizona and Mediterranean countries. Melanose is also reported from Japan (4,8), Taiwan (7) and coastal regions of Australia (1,3). All types of citrus are susceptible but the disease is especially disfiguring to grapefruits because of their smooth skin and pale colour (6).

SYMPTOMS Small, raised, reddish-brown to black pustules are present on the peel, making it rough to the touch. The individual spots are approximately 1 mm across and may be randomly distributed or exhibit a 'tearstain' pattern. When lesions are very numerous they coalesce, forming brown crusted areas on the fruit surface (5).

BIOLOGY In the grove the fungus produces both the sexual stage (perithecia giving rise to ascospores), to be found beneath dead bark, and the asexual stage (pycnidia giving rise to conidia) which occurs on living twigs and leaves as well as dead branches and bark (10). Airborne ascospores serve to disseminate the fungus widely (8), but infection is chiefly by conidia washed down on to the young fruit (5). If it remains wet for some hours, the peel is subject to direct penetration by the fungus (2,4). It is spore-bearing raindrops rolling over the shoulder of the fruit which results in the 'tearstain' pattern of infection. Melanose is essentially a non-progressive skin blemish since, as the fruit develops on the tree, the fungus dies back and causes no further damage; this is in marked contrast to its behaviour when infecting the stem-end (page 80).

CONTROL The incidence of melanose tends to increase with the age of the orchard, no doubt a reflection of the increasing amount of dead wood available for colonisation by the fungus (5). Early season pruning is beneficial, provided it is followed by a fungicide spray programme whilst the fruit is young and susceptible. During wet seasons frequent spraying may be necessary (7) but, with careful timing, a single application may suffice (9). Once the fruit has attained a certain size (e.g. approximately 7 cm in diameter for grapefruit), the peel is immune to attack (6).

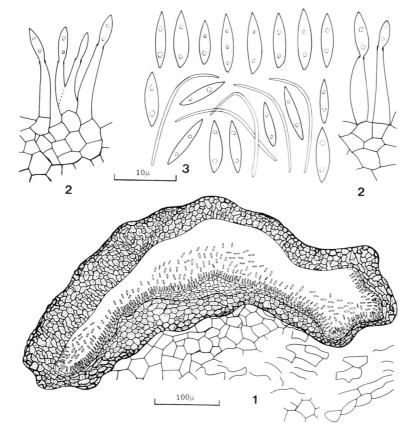

188 *Diaporthe citri* (conidial state: *Phomopsis citri*) (from CMI Descr. No. 396, Punithalingam & Holliday 1973) 1 pycnidium 2 part of pycnidial wall, conidiogenous cells and a-conidia 3 a and b-conidia

1 ANON. (1972) Melanose of citrus. *New South Wales Department of Agriculture Plant Disease Bulletin* No 22.
2 BACH W.J. & WOLF F.A. (1928) The isolation of the fungus that causes citrus melanose and the pathological anatomy of the host. *Journal of Agricultural Research* **37**, 243–252 illus.
3 BERTUS A.L. (1981) Fungicidal control of black spot and melanose on coastal Valencia oranges in New South Wales. *Australasian Plant Pathology* **10**, 53–55.
4 HOMMA Y. & YAMADA S.I. (1969) [Factors influencing infection and development of citrus melanose caused by *Diaporthe citri* (Faw.) Wolf.] *Bulletin of the Horticultural Research Station, Japan B* **9**, 85–96.
5 RUEHLE G.D. & KUNTZ W.A. (1940) Melanose of citrus and its commercial control. *Bulletin of Florida Agricultural Experiment Station* No. 349, 54 pp illus.
6 TIMMER L.W. & FUCIK J.E. (1976). The relationship of rainfall distribution, fruit growth and fungicide application to the incidence of melanose on grapefruit in Texas. *Plant Disease Reporter* **60**, 565–568.
7 TSAI Y.P. (1983). [Citrus melanose and its control.] *Plant Protection Bulletin, Taiwan* **25**, 63–67.
8 USHIYAMA K. (1973) [Studies on citrus melanose disease of satsuma orange. 2. Perfect stage of the causal fungus, *Diaporthe citri* (Fawc.) Wolf and inoculations with ascospores.] *Annals of the Phytopathological Society of Japan* **39**, 120–125.
9 WHITESIDE J.O. (1980) Timing of fungicide spray treatments for citrus melanose control. *Proceedings of Florida State Horticultural Society* **93**, 21–24.
10 WOLF F.A. (1926) The perfect stage of the fungus which causes melanose of citrus. *Journal of Agricultural Research* **33**, 621–625 illus.

SCAB of citrus caused by *Elsinoë* and *Sphaceloma* spp.

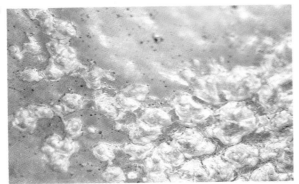

189

OCCURRENCE There are three types of scab, and they are important only in areas having a substantial summer rainfall. Thus, in the USA, **common scab** (or **sour orange scab**), caused by *Elsinoë fawcettii* Bitanc. & Jenkins, is common in Florida and Texas but does not occur in the more arid regions such as California and Arizona (3). For the same reason the disease is more prevalent in Japan (5) than in the Mediterranean region (CMI Map 125). It also occurs in West Africa. **Sweet orange scab**, caused by *Elsinoë australis* Bitanc. & Jenkins, is important in South American countries (1) and Sicily (CMI Map 55), while **Tryon's scab**, caused by *Sphaceloma fawcettii* var. *scabiosa* (McAlp. & Tryon) Jenkins, is found in Australia and parts of Asia and Africa (CMI Map 161).

SYMPTOMS Common or sour orange scab is characterised by irregular protuberant corky outgrowths up to 3 mm across and deeply cracked or fissured (4). The disease is more likely to occur on sour oranges, lemons, mandarins or grapefruits than on sweet oranges and limes. In sweet orange scab the lesions are flatter and smoother and tend to be circular, up to 4 mm in diameter. In Tryon's scab (affecting lemons and mandarins) the lesions are typically crater-like, and usually only about 1 mm across.

BIOLOGY These fungi produce the asexual state (acervuli giving rise to conidia) and *E. australis* and *E. fawcettii* also produce the sexual stage (pseudothecia giving rise to ascospores). The spore-bearing bodies persist in twig and leaf lesions, and spore dispersal is by wind, water, insects and mites (2). The optimal temperature for germination of asexual spores is about 25°C, but more important than temperature is the presence of moisture, whether rain, irrigation water or dew (2,3). In wet conditions infection of young leaves occurs (mature tissue being immune) (2,5). Similarly, developing fruits are susceptible while mature fruits are resistant to attack. The scabbed area does not increase after harvest, but remains as a superficial disfigurement.

CONTROL It is beneficial to remove and destroy old infected leaves and branches before the start of new growth (2). Fungicide sprays should be applied before flowering, at late petal fall, and a few weeks after fruit set (5). Screening of Indian cultivars has shown that some 'rough lemons' (*Citrus jambhiri*) are immune, whilst all limes are susceptible.

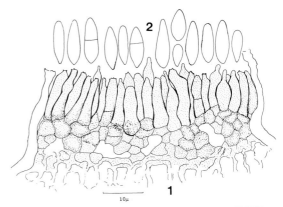

190 *Sphaceloma fawcettii* var. *scabiosa* (from CMI Descr. No. 437, Sivanesan & Critchett 1974) 1 acervulus 2 conidia (see also CMI Descr. Nos. 438, 440)

---

1 BITANCOURT A.A. & JENKINS A.E. (1937) Sweet orange fruit scab caused by *Elsinoe australis. Journal of Agricultural Research* **54**, 1–18 illus.
2 BRUN J. (1971) Les scab des agrumes. *Fruits* **26**, 759–767.
3 WHITESIDE J.O. (1975) Biological characteristics of *Elsinoë fawcettii* pertaining to the epidemiology of sour orange scab. *Phytopathology* **65**, 1170–1175 illus.
4 WHITESIDE J.O. (1986) Semiselective media for the isolation of *Elsinoë fawcettii* from citrus scab pustules. *Plant Disease* **70**, 204–206.
5 YAMADA S. (1961) [Epidemiological studies on the scab disease of satsuma orange caused by *Elsinoë fawcettii* Bitancourt & Jenkins and its control.] *Special Bulletin of Tokai-Kinki Agricultural Experiment Station (Horticultural Division)* No. 2, 56 pp illus.

SEPTORIA SPOT of citrus caused by
Septoria spp.

OCCURRENCE This disease has occasionally caused serious problems in Australia (inland New South Wales) (1), Argentina (2), Italy (4), Greece (5) and the USA (California) (3). Species involved include *Septoria citri* Pass., *S. depressa* McAlp. and *S. limonum* Pass.

SYMPTOMS The spots are at first circular, 1–2 mm in diameter, dark and sunken, and may be surrounded by a reddish brown halo (3). Lesions may also exhibit a 'tearstain' pattern or may coalesce to form large blemished areas (4). These have a characteristic purplish hue and eventually give rise to minute black bodies (pycnidia) (3).

BIOLOGY The fungus exists in the asexual state (pycnidia giving rise to conidia), and the spore-bearing structures form in dead branches in the grove (3). During cool wet weather or sprinkler irrigation (4) spores are washed or blown on to the fruit and are capable of causing direct infection of the rind. The infection remains quiescent, however, until resistance of the fruit is lowered, for example by frost, when the fungus is able to invade the weakened tissue (1,3). During storage the lesions increase in size (3).

CONTROL The disease is controlled in the grove by protective fungicide sprays (2,3). Post-harvest treatments are ineffective.

---

1 ANON. (1949) Septoria spot of citrus fruit. *Agricultural Gazette of New South Wales* **60**, 85–87 illus.
2 GONZÁLEZ J.L., STEIN B.E. & VINCIGUERRA H. (1983) Ensayo comparativo de diferentes fungicidas cúpricos para el control de septoriosis en limones. *Revista Industrial y Agrícola de Tucumán* **60**(2), 49–52.
3 KLOTZ L.J. & DeWOLFE T.A. (1969) Septoria spot of citrus. *California Citrograph* **54**, 530–531 illus.
4 SALERNO M. & CUTULI G. (1985) Le malattie degli agrumi: septoriosi (septoria spot). *Informatore Fitopatologico* **35**(6), 25–26 illus.
5 ZACHOS D.G. (1957) Septoria spots of citrus in Greece. *FAO Plant Protection Bulletin* **6**, 41.

SOUR ROT of citrus caused by
Geotrichum candidum Link

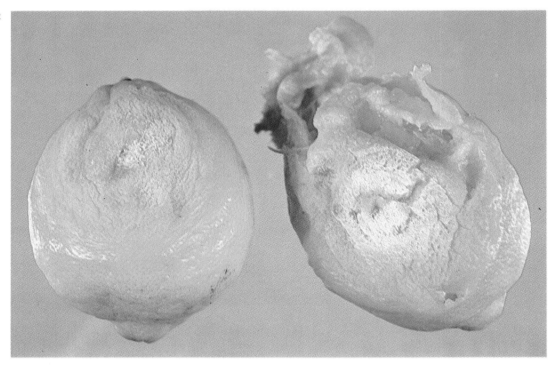

OCCURRENCE Sour rot has been reported from most citrus-growing areas, including the USA (1,2,4), Cuba (3), Israel (10), Turkey (8), India (5) and Australia (9). The disease is especially prevalent after warm wet seasons, and can cause serious losses. Particularly vulnerable are lemons and grapefruits stored for extended periods at moderate temperatures.

SYMPTOMS The first symptom is a watersoaked spot in which the affected tissue is extremely soft but not discoloured. The surface of the lesion becomes covered with slimy off-white spores (2), and the fruit tissue beneath is broken down into a sour-smelling watery mass (3) which is very attractive to fruit flies (7).

BIOLOGY The fungus survives in the soil and, during wet or windy weather, spores (conidia) are splashed or blown on to the surface of developing fruits (7,8). Pre-harvest infection occurs via injuries caused by insects (especially if the fruit surface remains wet), and may be undetected at picking time. Post-harvest infection occurs via wounds sustained during harvesting and handling (3,7), mature turgid fruit being especially susceptible (1). Spore-laden juices leaking from a rotten fruit can contaminate and infect healthy uninjured fruit, and also cause wetting and collapse of cardboard cartons. The fungus grows best at 25° to 30°C, and develops only slowly at temperatures below 10°C (7).

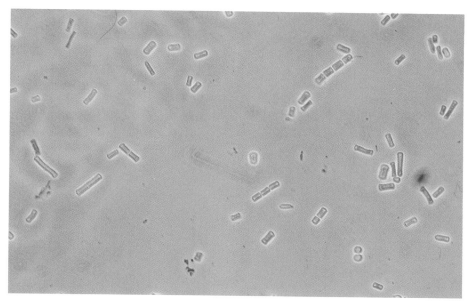

194 Conidia of *Geotrichum candidum*

CONTROL The first requirement for reducing infection levels is to control insects in the grove (7), and to pick the fruit at the correct stage of maturity, avoiding wet periods. Gentle handling at harvesting and in the packing house helps to prevent injury to the rind (3). Much effort is being put into the search for post-harvest fungicides effective against *Geotrichum* (5,6,10). In this regard it is useful to note that fungicidal suppression of *Penicillium* spore-formation is also likely to be beneficial in reducing the rate of development of *Geotrichum* (7,9). The use of vented cartons, coupled with prompt refrigeration, helps to hold the disease in check (4).

---

1 BAUDOIN A.B.A.M. & ECKERT J.W. (1985) Development of resistance against *Geotrichum candidum* in lemon peel injuries. *Phytopathology* **75**, 174–179 illus.
2 BUTLER E.E., WEBSTER R.K. & ECKERT J.W. (1965) Taxonomy, pathogenicity, and physiological properties of the fungus causing sour rot of citrus. *Phytopathology* **55**, 1262–1268 illus.
3 CASAMAYOR R., GARCÍA R. & TOLEDO J.L. (1983) Observaciones sobre el control del hongo *Geotrichum candidum* causante de pudriciones pos cosecha en frutos cítricos. *Centro Agrícola* **10**(3), 55–64 illus.
4 HARDING P.R. & SAVAGE D.C. (1962) Investigation of the ability of *Geotrichum candidum* and *Trichoderma viride* to spread by contact infection in non-vented versus vented cartons of lemons. *Plant Disease Reporter* **46**, 804–807.
5 KAUL J.L., SHARMA R.L. & SHANDILYA T.R. (1977) Efficacy of different chemicals against *Geotrichum candidum* causing sour rot of citrus fruits. *Indian Journal of Mycology and Plant Pathology* **7**, 78–79.
6 KITAGAWA H. & KAWADA K. (1984) Effect of sorbic acid and potassium sorbate on the control of sour rot of citrus fruits. *Proceedings of Florida State Horticultural Society* **97**, 133–135.
7 LAVILLE E. (1974) La pourriture des agrumes due à *Geotrichum candidum* Link var. *citri-aurantii* (Ferr.) R. Cif. & F. Cif. Etude bibliographique. *Fruits* **29**, 35–38 illus.
8 MAHMOOD T. (1970) *Geotrichum candidum*, causing sour rot of lemon in Turkey. *Plant Disease Reporter* **54**, 881–882 illus.
9 MORRIS S.C. (1982) Synergism of *Geotrichum candidum* and *Penicillium digitatum* in infected citrus fruit. *Phytopathology* **72**, 1336–1339.
10 SCHACHNAI A. & BARASH I. (1982) Evaluation of the fungicides CGA 64251, guazatine, sodium *o*-phenylphenate, and imazalil for control of sour rot on lemon fruits. *Plant Disease* **66**, 733–735.

STEM-END ROTS of citrus caused by
Botryosphaeria ribis **Grossenb. & Duggar**
Conidial state: *Dothiorella gregaria* Sacc.
Diaporthe citri **Wolf**
Conidial state: *Phomopsis citri* Fawcett
Physalospora rhodina **(Berk. & Curt.) Cooke**
Conidial state: *Botryodiplodia theobromae* Pat.

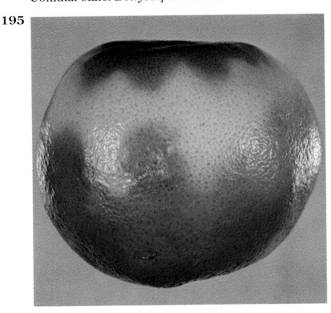

OCCURRENCE All three fungi are widely distributed, but stem-end rot is important only in areas having abundant rainfall during the growing season. Thus in the USA stem-end rot is much more common in Florida and Texas (3,6) than in semi-arid regions such as Arizona and California. Botryodiplodia and phomopsis rots frequently occur in the same lots of fruit, the former being more prevalent in consignments which have been subjected to ethylene degreening treatments (2). In Japan and Australia phomopsis rot appears to be the more important (5,9), whereas in South Africa and Israel botryodiplodia rot predominates (10). Dothiorella rot has been reported from the USA and also from Israel, Italy (8) and Trinidad.

SYMPTOMS The first sign is a slight softening of the tissue around the button. The affected rind turns brown, and the flesh beneath may darken. In botryodiplodia rot the decay spreads rapidly down the axis of the fruit, soon becoming visible at the stylar end. The fungus also tends to grow more quickly along tissue dividing the fruit segments, and this is reflected in the pattern of decay in the rind. Dothiorella rot may exhibit similar symptoms. In phomopsis rot the decayed tissue tends to shrink, resulting in a ridge of healthy tissue bordering the affected area; under humid conditions there may be a superficial growth of fine white mould. Infection may also occur via injuries on any part of the peel. It is often impossible to distinguish the rots caused by these fungi (which may occur separately or in combination), hence the necessity to isolate and culture the causal organism(s) so that the spores can be examined microscopically (for drawings see also pages 75, 97).

BIOLOGY Although these fungi may produce the sexual state (perithecia giving rise to ascospores) this form is not commonly found. It is the asexual state (pycnidia giving rise to conidia) which is important in infection, and the spore-bearing structures are produced in dead wood (8). The phomopsis stage of *Diaporthe citri* is also found in living stem and leaf tissue (5), and is the cause of melanose (q.v.). Spores are washed down by rain and tend to lodge underneath the button at the point of attachment of the fruit. Immature fruit is resistant to invasion but eventually, usually after harvest, the fungus progresses from the button into the fruit and initiates rotting (2). Particularly susceptible are fruits of low vitality resulting from adverse growing conditions or overmaturity, and fruits with moribund buttons following ethylene treatment (1,9). During

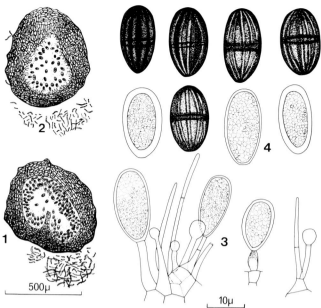

197 *Botryodiplodia theobromae* (from CMI Descr. No. 519, Punithalingam 1976) 1,2 pycnidia 3 conidiogenous cells and conidia 4 conidia

prolonged storage, *Botryodiplodia theobromae* can spread into surrounding healthy fruits (7).

CONTROL Where stem-end rotting is likely to occur before harvest, control may be effected by pruning out dead wood and spraying the trees with fungicide (4). More usually, however, stem-end infections are undetectable at the time of picking and packing, and are manifest only during transport and storage (7). A simple control measure is to harvest the fruit by pulling instead of clipping, the aim being to leave behind the stem-buttons (which may be infected), thereby pre-empting fungal invasion of the fruit itself (6). If, owing to the risk of tearing the peel, the buttons are left in place, they can be kept in a fresh condition by application of a growth-regulating chemical either before or after the degreening process (9). Prolonging button vitality postpones fungal advance into the fruit (1). Post-harvest treatments with hot water and fungicides also serve to control the development of decay (3).

---

1 BARMORE C.R. & BROWN G.E. (1985) Influence of ethylene on increased susceptibility of oranges to *Diplodia natalensis*. *Plant Disease* **69**, 228–230.

2 BROOKS C. (1944) Stem-end rot of oranges and factors affecting its control. *Journal of Agricultural Research* **68**, 363–381 illus.

3 BURGER D.W. & DAVIS R.M. (1982) New fungicides for the postharvest control of stem-end rot in Texas grapefruit. *HortScience* **17**, 976–977.

4 GRASSO S. (1984) Saggi *in vitro* di alcuni fungicidi contro *Phomopsis citri*. *Tecnica Agricola* **36**, 33–38.

5 HOMMA Y. & YAMADA S.I. (1969) [Mechanisms of infection and development of citrus stem-end rot caused by *Diaporthe citri* (Faw.) Wolf. 1. The process of entry of the causal fungus to host tissues.] *Bulletin of the Horticultural Research Station, Japan B* **9**, 99–115.

6 MECKSTROTH G.A. (1944) Pulling versus clipping of Florida oranges. *Citrus Industry* **25**, 9, 12, 18.

7 NADEL-SCHIFFMANN M. (1953) Contact transmission of *Diplodia* rot in Shamouti orange fruits. *Palestine Journal of Botany, Rehovot Series* **8**, 193–196.

8 SAVASTANO G. (1932) Una gommosi del limone causata da *Dothiorella*. *Bolletino della R. Stazione di Patologia Vegetale NS* **12**, 245–274 illus.

9 SINGH G., RIPPON L.E. & GILBERT W.S. (1977) 2,4-D residues in stored lemons from post-harvest treatments. *Australian Journal of Experimental Agriculture and Animal Husbandry* **17**, 167–170.

10 ZAUBERMANN G. & BARKAI-GOLAN R. (1975) Changes in respiration and ethylene evolution induced by *Diplodia natalensis* in orange fruit. *Phytopathology* **65**, 216–217.

TRICHODERMA ROT of citrus caused by
Trichoderma viride Pers. ex S.F. Gray

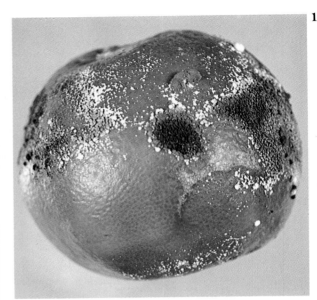

OCCURRENCE Trichoderma rot can be an important cause of loss in some citrus-growing regions, for example South Africa (9) and Israel (5). It is also reported from the USSR (10), China and India (7). In the USA the disease is prevalent in California (chiefly on lemons) but rare in Florida and Texas (6,8). The author has observed it on Uruguayan mandarins and Honduran grapefruits imported into the UK.

SYMPTOMS Infected peel remains pliable (5), eventually becoming dark brown. The veins on the interior surface of the peel also turn brown, but the flesh beneath is unchanged in colour. Usually associated with trichoderma rot is a characteristic odour of coconuts (3). Under humid conditions a white mould spreads over the surface, and gives rise to quantities of yellow-green to dark green spores. Frequently, however, this fungus is overgrown by other organisms, notably the common green mould *Penicillium digitatum*, with which *T. viride* is probably often confused (1,4,9).

BIOLOGY *T. viride* is common in soil and its spores are found in the atmosphere (5). Infection of citrus fruits often follows severe mechanical injury to the peel, but may also occur at the stem- or stylar-end (1,2,4). Incidence of trichoderma rot is often enhanced in consignments affected by penicillium moulds (9). This is probably because the latter evolve ethylene which hastens senescence in the fruit (especially the button), thereby predisposing it to attack by the weakly parasitic *T. viride*. This fungus may (6,8) or may not (1,4) spread into healthy uninjured fruits, causing 'nesting'. It is capable of colonising damp wooden storage boxes (2,8) and invading subsequent batches of fruit via injuries. The optimal growth temperature of the fungus lies between 25° and 30°C, the minimum being about 5°C (5).

CONTROL Storage boxes should be steam-sterilised, disinfected with a fungicide (8), or lined with film. It is important to minimise injury to the fruit during picking and packing (4). The use of diphenyl-impregnated pads or wrappers gives partial control of *T. viride*; in the absence of diphenyl treatment it is preferable to pack the fruit in vented rather than non-vented cartons (6). Refrigeration serves to arrest decay and, at 10°C and below, no new infections take place (6).

200 Conidia of *Trichoderma viride*

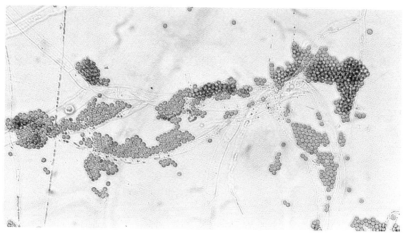

201 Conidia of *Trichoderma viride*

---

1 COLE A.L.J. & WOOD R.K.S. (1970) The infection of oranges by *Trichoderma viride* and mixed infection by *Trichoderma viride* and *Penicillium digitatum*. *Annals of Applied Biology* **66**, 75–82.
2 DOIDGE E.M. & PLANK J.E. van der (1936) The fungi which cause rots of stored citrus fruits in South Africa. *Science Bulletin of the South African Department of Agriculture and Forestry* No. 162, 23 pp.
3 FAWCETT H.S. & WEINDLING H. (1934) Types of *Trichoderma* rot of lemons and oranges. *Phytopathology* **24**, 1144 (Abstr.).
4 GUTTER Y. (1961) On the pathogenicity of *Trichoderma viride* to citrus fruits. *Bulletin of the Research Council of Israel, D* **10**, 157–164 illus.
5 GUTTER Y. (1963) Contributions to the biology of *Trichoderma viride* Pers. ex Fries, pathogenic to citrus fruits. *Israel Journal of Botany* **12**, 27–40.
6 HARDING P.R. & SAVAGE D.C. (1962) Investigation of the ability of *Geotrichum candidum* and *Trichoderma viride* to spread by contact infection in nonvented versus vented cartons of lemons. *Plant Disease Reporter* **46**, 804–807.
7 KANAUJIA R.S. (1979) Fruit rot of mandarin orange. *Indian Phytopathology* **32**, 450.
8 KLOTZ L.J. & DeWOLFE T.A. (1952) Steam sterilization of citrus boxes. *California Citrograph* **38**, 68.
9 PIENAAR R.S. (1969) Types of decay on European markets during 1968. *South African Citrus Journal* No. 422, 25–27.
10 TZERETELI L.Y. & TCHANTURIA N.N. (1939) [Diseases of citrus fruits in storage.] *Sovetskaya Botanika* **3**, 111–115.

OTHER DISEASES of citrus

ASPERGILLUS ROTS may be caused by species other than the black mould *Aspergillus niger* (page 160). In India it is reported that oranges may be rotted by *A. flavus* Link ex Fr., *A. niveus* and *A. variecolor* (8).

CERATOCYSTIS ROT has been observed in Indian oranges, the causal fungus being *Ceratocystis fimbriata* Ell. & Halsted. Infection occurs only in wounded fruit. Lesions are initially pale, soft and watery, but later darken with the formation of asexual spores (conidia) and subsequently black spore-bearing bodies (perithecia giving rise to ascospores). Rotting is accompanied by a rancid odour (9).

CERCOSPORA SPOT was first described in Angola, and the causal fungus was named *Cercospora angolensis*. It is now called *Phaeoramularia angolensis* (de Carvalho & O. Mendes) P.M. Kirk. In Nigeria it causes a serious problem known locally as 'brown spot disease' (but see page 56). Infection can occur at any stage of fruit development; in mature fruit the lesions are flat and brown, each surrounded by a yellow halo (3). Under moist conditions the lesions darken as spores (conidia) are formed.

CHARCOAL ROT is caused by *Macrophomina phaseolina* (page 273). It has been observed in Israel (5), and is described as a stem-end rot which initially resembles botryodiplodia rot (q.v.) in the brown colour of the affected peel and the pattern of decay. Later, however, when the peel has become black, there is a copious production of minute (0.1 mm) resting bodies (sclerotia).

FRECKLE, recorded on mandarins, is a superficial disfigurement which has caused economic losses in Japan (10). Fruit is attacked in the grove and, on mature fruit, lesions are brownish grey and irregular in shape, sometimes resembling wind-injury. The causal organism was found to be *Ascochyta pisi* and other species of *Ascochyta* from peas. The disease can be controlled with fungicide sprays, but it is preferable to refrain from growing peas near the grove and from using pea haulm as a mulch (10).

PINK MOULD ROT has been reported on oranges in India (1) and the USSR, the causal fungus being *Trichothecium roseum* (**341**). Lesions are initially brown but may later be covered with a pale pink mould bearing characteristic spores.

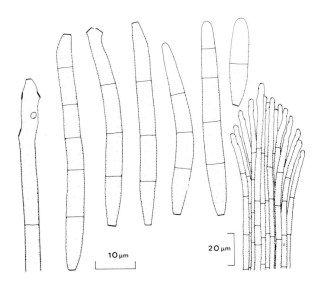

202 *Phaeoramularia angolensis* (from CMI Descr. No. 843, Kirk 1986) Conidia and conidiophores

203 Pink mould rot

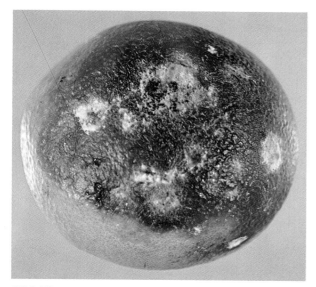

204 Pleospora rot

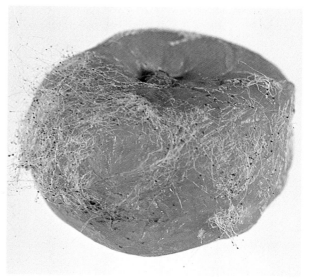

205 Rhizopus rot

PLEOSPORA ROT is caused by *Pleospora herbarum* (Pers.) Rabenh. and has been described on Italian lemons (2). The firm lesions each possess a dark brown centre supporting greyish mould growth which later gives rise to dark green asexual spores (conidia) (**116**). The author has observed pleospora rot on mandarin (origin unknown) and also on grapefruit from Cyprus. In each fruit the firm, dark, slow-growing lesion was partially covered by a buff-coloured mould; after darkening and producing conidia, the fungus also formed the sexual stage (perithecia giving rise to ascospores).

RHIZOPUS ROT is caused by *Rhizopus stolonifer* (page 231), which occasionally invades wounded fruits (especially mandarins), producing a soft water-soaked sunken lesion which is not discoloured. On split fruits in a humid atmosphere there is copious formation of coarse white mould bearing minute spherical black spore heads.

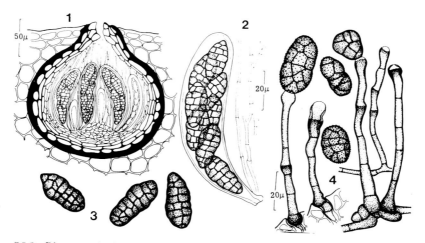

206 *Pleospora herbarum* (from CMI. Descr. No. 150, Booth & Pirozynski 1967)
1 pseudothecium 2 ascus 3 ascospores 4 conidia and conidiophores

207 Sooty mould

SOOTY BLOTCH is a superficial disfigurement which has been ascribed to *Gloeodes pomigena* (Schw.) Colby in Australia and to *Stomiopeltis citri* in Brazil and South Africa (7). These fungi are not capable of invading citrus rind, but under moist conditions in the grove they produce dark smudges on the surface of mature fruit. The blemish does not spread during storage, though it may become darker. Sooty blotch can be removed by brushing; alternatively, the fruit can be dipped in a bleaching solution and then rinsed (7).

SOOTY MOULD may be one of several species of *Capnodium*, *Cladosporium* or *Meliola*, which partially envelop the fruit in a dark fungal mat, necessitating diligent brushing in the packhouse (6). Sooty moulds, which are very common, live on the honeydew produced by insects such as aphids and mealybugs, and the main control measure is therefore to use insecticide sprays or to encourage natural predators of the insects. In some situations oil sprays are appropriate and serve to loosen existing mould deposits as well as killing the insects.

YEASTY ROT is caused by *Candida krusei* (Castell.) Berkhout. The disease is described on Californian lemons, as a soft watery rot similar in general appearance to sour rot (q.v.) but having a yeasty odour and lacking superficial mould growth (4). The disease has also been recorded on mandarins in India.

---

1 CHEEMA S.S., MUNSHI G.D. & SHARMA B.D. (1981) Laboratory evaluation of fungicides for the control of *Trichothecium roseum* Link, a new fruit rot pathogen of sweet orange. *Hindustan Antibiotics Bulletin* **23**, 27–29.
2 COCCHI F. (1931) Un marciume dei limoni dovuto a *Pleospora herbarum* (Pers.) Rabenh. *Bolletino della R. Stazione di Patologia Vegetale NS* **11**, 179–213 illus.
3 EMECHEBE A.M. (1981) Brown spot disease of citrus caused by *Phaeoisariopsis* sp. *Annals of Applied Biology* **97**, 257–262 illus.
4 HARDING P.R. (1968) *Candida krusei* as a primary pathogen in lemons. *Plant Disease Reporter* **52**, 433–435.
5 MINZ G. & GUTTER J. (1949) *Sclerotium bataticola* as a potential pathogen of various citrus fruits. *Palestine Journal of Botany, Rehovot Series* **7**, 182–184 illus.
6 NATH D.K. (1973) Insect transmission of sooty mould (*Capnodium* sp.) to orange orchards at Darjeeling district, West Bengal. *Science and Culture* **39**, 262–263.
7 PLANK J.E. van der (1945) The use of hypochlorous acid and its salts in citrus packhouses for bleaching sooty blotch and as disinfectants against mould. *Science Bulletin of the South African Department of Agriculture* No. 241, 60 pp illus.
8 SHARMA R.B., MONGA A., ROY A.N. & GUPTA M.N. (1981) Pathophysiological studies on three new postharvest fruit rot diseases of *Citrus sinensis*. *Indian Journal of Mycology and Plant Pathology* **11**, 309–310.
9 SINGH A.K. & BASU CHAUDHARY K.C. (1974) Ceratocystic soft rot of sweet orange. *Current Science* **43**, 726–727.
10 TANAKA S. (1977) A new fungal disease of satsuma mandarin. *Proceedings of the International Society for Citriculture* Vol. 3, 1006–1007 illus.

CHILLING INJURY of citrus

OCCURRENCE In general, grapefruits, lemons and limes are most susceptible, oranges least, while mandarins are intermediate. The critical temperature, below which there is a risk of injury, lies between 10° and 15°C for grapefruits, lemons and limes, and between 1° and 5°C for oranges and mandarins. However, a cultivar grown in one climatic zone may differ in susceptibility from the same cultivar grown elsewhere. Chilling injury is quite distinct from freezing injury (q.v.), since it occurs at temperatures well above freezing point.

SYMPTOMS Considerable variability is shown, possibly reflecting different temperature ranges. Severely affected fruit possesses an off-taste, and is predisposed to decay (2). The most common symptom is a **brown pitting** of the rind, the discrete depressed lesions being several mm across and sometimes surrounded by a diffuse brown halo. The discoloured areas may be confined to one side of the fruit or may extend over much of the surface. **Watery breakdown** is another manifestation of chilling injury, in which internal and external tissues are soft and water-soaked, with an odour of fermentation. **Brown staining** or **scald** is a superficial discoloration of the peel, seen especially in grapefruits and mandarins, while **oil-gland darkening** is characteristic of grapefruits and tangelos. Symptoms peculiar to lemons include **albedo browning**, a discoloration of the white spongy rind tissue which may show externally as a slight darkening of the peel, and **membranous stain**, in which the tissue between the segments is discoloured brown.

BIOLOGY The nature of chilling injury is complex and is still being elucidated; basically the affected tissue is still alive but the biochemical changes are proceeding along abnormal pathways. The period of exposure to chilling temperatures determines whether or not the disruption shows as visible symptoms (4). Susceptibility to post-harvest chilling injury is influenced by many factors, including the weather during the growing season (3), the position of the fruit in the tree canopy (2,5), the date of harvest (3), the degree of moisture loss from the peel (1,5) and the use of certain growth regulators and fungicides (2).

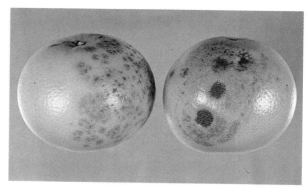

208

CONTROL Chilling injury can be avoided by knowing the characteristics of fruit from specific harvests (3) and storing it above the critical temperature (see Appendix 3). If storage at a lower temperature is required (for example, to comply with plant quarantine regulations for fruit fly control) (2), certain measures can be taken to lower the chill threshold. These include reduction of moisture loss by applying oil, wax or film (1,2,5), and 'conditioning' of the fruit to chilling temperatures by prior storage for a few days at a higher temperature (2,5). Some fungicides (e.g. diphenyl) may increase susceptibility, while others (e.g. thiabendazole) decrease it (2).

---

1 ALJUBURI H.J. & HUFF A. (1984) Reduction in chilling injury to stored grapefruit (*Citrus paradisi* Macf.) by vegetable oils. *Scientia Horticulturae* **24**, 53–58 illus.
2 CHALUTZ E., WAKS J. & SCHIFFMANN-NADEL M. (1985) Reducing susceptibility of grapefruit to chilling injury during cold treatment. *HortScience* **20**, 226–228.
3 KAWADA K., GRIERSON W. & SOULE J. (1978) Seasonal resistance to chilling injury of 'Marsh' grapefruit as related to winter field temperature. *Proceedings of Florida State Horticultural Society* **91**, 128–130.
4 McDONALD R.E., HATTON T.T. & CUBBEDGE R.H. (1985) Chilling injury and decay of lemons as affected by ethylene, low temperature, and optimal storage. *HortScience* **20**, 92–93.
5 PURVIS A.C. (1985) Relationship between chilling injury of grapefruit and moisture loss during storage: amelioration by polyethylene shrink film. *Journal of the American Society for Horticultural Science* **110**, 385–388 illus.

FREEZING INJURY of citrus

209

OCCURRENCE This disorder is caused by temperatures below the freezing point which, for most citrus fruits, is in the region of $-1.5°C$. Freezing may occur before harvest during frosty weather in the grove, or after harvest as a result of improper conditions of carriage or storage.

SYMPTOMS Fruits which have been severely frozen may exhibit browning of the peel or a mushy breakdown, but in most instances there are no external symptoms (1). However, if a thawed fruit is pulled apart, the membranes between the segments appear water-soaked and may show numerous white crystals (of hesperidin or narangin) resulting from the freezing. In mandarins crystals also form in the pulp, and in grapefruits the pulp takes on a milky appearance. Tissue subjected to pre-harvest freezing subsequently dries out, and cavities may develop in the flesh (5).

BIOLOGY In some parts of the world (for example, Florida) citrus trees are occasionally subject to various types of frosty weather. During a radiative frost, when the air is fairly still, the fruit loses most heat from its upper surface (the stem-end) by radiation to the sky; subsequent injury tends to be concentrated in this region of the fruit (2,5). During a windy freeze, individual fruits may be damaged on one side only (2); fruits on the leeward side of the tree may escape injury entirely (5). If freezing occurs during carriage or storage, the worst damage is likely to be in fruits nearest the air delivery vent or those exposed to freezing ambient air. In contrast to chilling injury (q.v.), freezing has a worse effect on the pulp than on the peel. It is the membranes and juice sacs which are particularly vulnerable, and there may be considerable internal distortion (2,5). There is enhanced evaporation of moisture through the peel (4) and, even if this is prevented, the damaged pulp suffers dehydration and shrivelling (2). Freeze-damaged fruits show an increased tendency to decay in store (3).

CONTROL There are various means of protecting trees from frost; the beneficial application of irrigation water is an alternative to the use of orchard heaters (5). Protective measures are not always effective, however, and it may be necessary to cull freeze-damaged fruits from the harvested crop (5). There being no external signs, the process relies on the fact that damaged fruits tend to be less dense than sound fruits, on account of the loss of water that has taken place. They can thus be separated by flotation in a specially prepared solution (5). Post-harvest freezing is avoided by careful attention to the conditions of storage and carriage.

---

1. HATTON T.T., SMOOT J.J & HALE P.W. (1977) Internal freeze damage in Florida grapefruit held in Florida and similar fruit shipped and held in Japan during late spring. *Proceedings of Florida State Horticultural Society* **90**, 154–156.
2. PURVIS A.C., BROWN G.E. & CARTER R.D. (1985) Postharvest water loss from freeze-damaged citrus fruit. *HortScience* **20**, 756–757 illus.
3. SCHIFFMANN-NADEL M., WAKS J. & CHALUTZ E. (1975) Frost injury predisposes grapefruit to storage rots. *Phytopathology* **65**, 630.
4. SYVERTSEN J.P. (1982) Dehydration of freeze-damaged oranges. *HortScience* **17**, 803–804.
5. WARDOWSKI W.F., GRIERSON W. & JOHNSON M. (1986) Separation and grading of freeze-damaged citrus fruit. In *Fresh citrus fruits* (Ed. by W.F. Wardowski, S. Nagy & W. Grierson), pp 275–286 illus. Westport, Conn.: AVI Publishing Co.

OLEOCELLOSIS or RIND-OIL SPOTTING of citrus

210

OCCURRENCE This blemish is especially common on lemons and limes (1) but may also occur on oranges (2,5) and mandarins (4). Investigations have been carried out in several countries, including the USA (1), Italy (3), Israel (2), Egypt (5) and Japan (4).

SYMPTOMS Injured areas are firm, irregular in shape, and discoloured yellow, green or brown, the oil glands appearing prominent owing to a slight sinking of the tissue between them (1). Lesions may vary in size from a few mm to several cm, and may cover much of the fruit surface.

BIOLOGY If citrus fruits are subjected to certain stresses, oil may be expressed from the glands in the flavedo (page 54), with toxic effects on the surrounding tissue. Oil spotting sometimes occurs before harvest (3,4), either as a result of compression of fruits in contact with one another in the same bunch, or following a sudden spell of cold weather. Most commonly, however, it is induced by harvesting and handling which cause physical injury to the peel (1,5). Immature, turgid fruit is especially susceptible (1), and the problem tends to be most severe in fruits harvested during cool, wet periods early in the season (2). In green fruit the injured areas fail to colour properly during the degreening process (page 55), especially if humidity in the room is below optimal. Fruit allowed to colour on the tree becomes less susceptible with increasing sponginess of the peel (2), but injury may still occur, the affected areas becoming brown.

CONTROL Pre-harvest sprays of an ethylene-releasing compound render the fruit less vulnerable to oleocellosis (2). It may be advisable to refrain from irrigating during the month before harvest. A knowledge of soil moisture and atmospheric humidity provides a basis for predicting the water status of the fruit (1). Instruments are available for measuring the 'rind oil rupture pressure' (2), and fruit should not be harvested if too turgid. This generally means avoiding picking too early in the season, or during the early hours of the morning, or when the fruit is wet (5). Care should be taken at all stages to protect the fruit from rough treatment (1,5). Special indicator papers (or an aluminium foil fruit wrap) can be used to identify points in the handling chain at which physical injury is being sustained.

---

1 CAHOON G.A., GROVER B.L. & EAKS I.L. (1963) Cause and control of oleocellosis on lemons. *Proceedings of the American Society for Horticultural Science* **84**, 188–198 illus.
2 ERNER Y. (1982) Reduction of oleocellosis damage in Shamouti orange peel with ethephon pre-harvest spray. *Journal of Horticultural Science* **57**, 129–133.
3 LO GIUDICE V. & CATARA A. (1972) Oleocellosi da contatto in frutti di agrumi. *Tecnica Agricola* **24**, 253–258 illus.
4 SAWAMURA M., MANABE T., OONISHI S., YASUOKA K. & KUSUNOSE H. (1984) Effects of rind oils and their components on the induction of rind spot in citrus species. *Journal of Horticultural Science* **59**, 575–579.
5 SHAHEIN A.H., EISSA A.M. & ETMAN A.A. (1981) Some factors affecting oil spotting on fruits during storage of two orange varieties. *Egyptian Journal of Horticulture* **8**, 123–130.

OTHER DISORDERS of citrus

BRONZING and **RUSSETING** may be caused by rust mites, minute spider-like creatures whose feeding habits result in a brownish or purplish stain, apparent before harvest. 'Tearstains' caused by the rust mite are distinguished from melanose (q.v.) by their smoothness. Mites are controlled by spraying the trees with an acaricide.

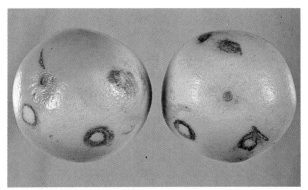

211 Chemical injury

CHEMICAL INJURY may be caused by improper use of post-harvest fungicides (4). It can take the form of circular discoloured areas resulting from retention of liquid between adjacent fruits, or alternatively there may be generalised browning or reddening of the peel (7). Control involves careful adjustment of such factors as pH of the solution, temperature of the bath, and time of immersion.

CREASING is an important disorder of mature oranges and mandarins, in which cracks develop in the albedo (page 54), leading to 'puffiness' (q.v.) and rupture of the peel (9). The first signs are present at harvest, and affected fruits should not be packed. The disorder is controlled by pre-harvest applications of a growth-regulating chemical (9).

DEFORMATION is especially important in large grapefruit and can lead to substantial reduction in value. Fruit which is handled roughly, or which becomes dehydrated after harvest, loses elasticity and may suffer permanent deformation when compressed in a pack (10). Control involves careful harvesting and handling, protection from sun and wind after picking, maintenance of high humidity during degreening, and appropriate packaging techniques.

GRANULATION is a condition in which the contents of some juice vesicles turn from liquid into a gel, causing part of the fruit to appear deficient in juice (4). Valencia oranges, mandarins and mandarin hybrids tend to be affected only at the stem-end, navel oranges along the central axis. Granulation has pre-harvest origins, being related to advancing maturity, but it may not develop until after harvest. Studies in South Africa, where the disorder causes significant losses, suggest a connection with late-season rains (3).

PETECA is a type of pitting of the rind, in which the edges of the depressions slope gently rather than abruptly, as the underlying albedo (page 54) collapses and becomes discoloured. The occurrence of peteca seems to be largely determined by orchard and climatic conditions, although the symptoms usually develop only after harvest. Predisposing factors include cold wet weather, heavy oil spraying, and post-harvest waxing (4).

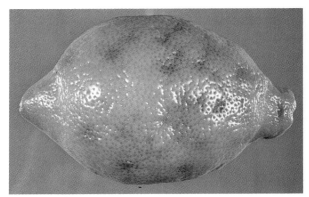

212 Peteca

PUFFINESS, most prevalent in mandarins, is a condition in which an air space develops between peel and pulp, making the fruit liable to split when handled. It may follow 'creasing' (q.v.). The disorder is controlled by pre-harvest sprays of a growth-regulating chemical (6) and avoidance of excessive humidity in store.

RIND-STAINING has been used to describe several disorders characterised by irregularly shaped brown areas on the peel. There are many possible causes, including unusual weather before harvest, chemical spray injury, chilling injury (q.v.), heat damage, brush injury from packhouse operations, and 'ethylene burn' which sometimes follows degreening. Diagnosis is difficult unless details of pre-shipment history are known.

SCALE INSECTS tend to adhere tightly to the peel, surviving the cleaning processes in the packhouse. If numerous they cause gross disfigurement, being dark and shiny, about 1 to 2 mm in length. Furthermore, the presence of scale insects inhibits uniform degreening. Clearly, it is essential to control these pests in the grove.

STEM-END RIND BREAKDOWN or AGEING is especially important in oranges, and involves the development of brown areas on the peel at the stem-end. Characteristic of this disorder is a narrow (2 to 5 mm) ring of undamaged tissue immediately surrounding the 'button'. Symptoms generally develop only after packing, and affected fruit is predisposed to decay. Control involves minimising moisture loss after harvest, by protecting the fruit from sun and wind and keeping it in a humid atmosphere until it can be waxed (8).

213 Stem-end rind breakdown

STYLAR-END BREAKDOWN of limes first appears as a water-soaked discoloured area which spreads from the stylar-end. It results from the rupture of juice vesicles in the pulp, leading to contact between juice and rind. Only the larger sizes of lime are vulnerable to the disorder, which is aggravated by rough handling and exposure to high temperatures at harvest (1). The weakened tissue is usually colonised by penicillium moulds (q.v.).

214 Thrips damage

THRIPS DAMAGE is caused by small insects which feed on developing fruits, resulting in the formation of a characteristic roughened brown band around the stem-end or stylar-end. Occasionally other areas of the fruit may be affected. Thrips can be controlled by insect sprays in the grove.

WIND-SCARRING is the result of abrasion of developing fruits by leaves or twigs (2). Corky tissue, which may have a silvery appearance, develops over these superficial wounds.

---

1 DAVENPORT T.L. & CAMPBELL C.W. (1977) Stylar-end breakdown in 'Tahiti' lime: aggravating effects of field heat and fruit maturity. *Journal of the American Society for Horticultural Science* **102**, 484–486.
2 FREEMAN B. (1976) Rind blemish of citrus. 1. Initiation and development. *Scientia Horticulturae* **4**, 317–327 illus.
3 GILFILLAN I.M. & STEVENSON J.A. (1977) Postharvest development of granulation in South African export oranges. *Proceedings of the International Society for Citriculture* Vol. 1, pp 299–303.
4 KLOTZ L.J. (1973) *Color handbook of citrus diseases*. Riverside: University of California, 122 pp illus.
5 KNORR L.C. (1973) *Citrus diseases and disorders*. Gainesville: University of Florida, 163 pp illus.
6 KURAOKA T., IWASAKI K. & ISHII T. (1977) Effect of GA_3 on puffing and levels of GA-like substances and ABA in the peel of satsuma mandarin (*Citrus unshu* Marc.). *Journal of the American Society for Horticultural Science* **102**, 651–654.
7 LONG J.K. & ROBERTS E.A. (1958) The phytotoxic and fungicidal effects of sodium o-phenylphenate in controlling green mould wastage in oranges. *Australian Journal of Agricultural Research* **9**, 609–628 illus.
8 McCORNACK A.A. & GRIERSON W. (1965) Practical measures for control of stem-end rind breakdown of oranges. *Florida Agricultural Extension Circular* No. 286.
9 MONSELISE S.P., WEISER M., SHAFIR N., GOREN R. & GOLDSCHMIDT E.E. (1976) Creasing of orange peel – physiology and control. *Journal of Horticultural Science* **51**, 341–351.
10 RIVERO L.G., GRIERSON W. & SOULE J. (1979) Resistance of 'Marsh' grapefruit to deformation as affected by picking and handling methods. *Journal of the American Society for Horticultural Science* **104**, 551–554.

CHAPTER 3
MISCELLANEOUS TROPICAL AND SUBTROPICAL FRUITS

The fruits considered in this chapter include a diverse range of unrelated species, from the mango which is borne on immense trees (3,6,7) to the passion fruit which grows on a vine (8,9,10). Most of these fruits are climacteric (page 14), the pineapple being a notable exception (1). What they all have in common is their perishability; in contrast to some of the temperate fruits (such as apples, pears and grapes) which may be stored for several months, tropical fruits have a storage life of a few weeks only, even under optimal conditions (4,5). Prolonging storage life by refrigeration is limited by the fact that tropical produce is especially susceptible to chilling injury (q.v.) (2,8).

These constraints have not prevented the development of important trades between distant countries (5,10). The international banana trade is long-established, this commodity accounting for the largest tonnage of any perishable foodstuff carried by sea. Marine carriage is entirely appropriate for bananas, in view of their year-round availability in each growing area, together with the existence of purpose-built refrigerated ships. Most banana cargoes are carried in conventional vessels, but container carriage is used on certain routes. The other tropical and subtropical fruits are moved in very much smaller quantities and, because of seasonality and fluctuations in supply and demand, can often justify the cost of air freight (5,8). If, by this means, fruits can be got to distant markets within a

215 Senescent rambutans

few days, they may be harvested at an advanced stage of maturity, permitting full development of aroma and flavour, as is possible when consumed locally. For surface shipment, fruits must be harvested at an earlier stage of maturity, which may or may not permit attainment of optimal eating quality (5).

The main problems of tropical fruits are the establishment of practical methods of assessing harvest maturity (8), the instigation of gentle handling practices (4), and the logistics of co-ordinating picking (10), treatment (4), packing and despatch to the appointed carrier (5). For bananas and pineapples, grown in carefully regulated plantations and shipped out on a weekly basis, such problems have been minimised, but for the other fruits they are sometimes a limiting factor. With increasing demand for exotic fruits in the markets of temperate countries, more attention is being paid to their special requirements (2,5,6,7).

---

1 BONDAD N.D. (1976) Response of some tropical and subtropical fruits to pre- and post-harvest applications of ethylene. *Economic Botany* **30**, 67–80.
2 BROWN B.I. (1986) Temperature management and chilling injury of tropical and subtropical fruit. *Acta Horticulturae* No. 175, 339–342.
3 COOK A.A. (1975) *Diseases of tropical and subtropical fruits and nuts.* New York: Hafner Press (Macmillan), 317 pp illus.
4 ECKERT J.W. & OGAWA J.M. (1985) The chemical control of post-harvest diseases: subtropical and tropical fruits. *Annual Review of Phytopathology* **23**, 421–454.
5 MARRIOTT J. & PROCTOR F.J. (1978) Transportation and conservation of tropical fruits. *Outlook on Agriculture* **9**, 233–239 illus.
6 MARTIN F.W., CAMPBELL C.W. & RUBERTE R.M. (1987) Perennial edible fruits of the tropics. *Agriculture Handbook of the United States Department of Agriculture* No. 642, 247 pp illus.
7 MORTON J.F. (1987) *Fruits of warm climates.* Winterville, North Carolina: Creative Resource Systems, 505 pp illus.
8 PANTASTICO E.B. (Ed.) (1975) *Post-harvest physiology, handling and utilization of tropical and subtropical fruits and vegetables.* Westport, Conn.: AVI Publishing Co., 560 pp illus.
9 POPENOE W. (1924) *Manual of tropical and subtropical fruits, excluding the banana, coconut, pineapple, citrus fruits, olive, and fig.* New York: Macmillan, 474 pp illus.
10 SAMSON J.A. (1986) *Tropical fruits.* 2nd ed. Harlow, UK: Longman Group, 336 pp illus.

Avocados

The **avocado**, *Persea americana* Miller, originated in South America, where it has been an important part of the diet for thousands of years. There are three main types, namely Mexican, Guatemalan and 'West Indian', regarded as subtropical, semi-tropical and tropical, respectively. The crop is now grown in many parts of the world, including the Mediterranean basin. Western hemisphere countries predominate, however, the ten leading producers being Mexico, the USA, Brazil, the Dominican Republic, Indonesia, Haiti, Peru, Israel, Venezuela and Colombia. World production is nearly 2 million tonnes, of which a small proportion is grown for export, mainly in California (6), Florida (9), Israel (5) and South Africa (4,7).

Cultivars vary greatly in the size, shape and colour of their fruits. The avocados of international trade tend to be pear-shaped and little over 10 to 15 cm in length. Fuerte (a Guatemalan–Mexican hybrid) possesses a thin, smooth, green skin while that of Hass (a Guatemalan type) is thick, woody and knobbled, turning purplish black as the fruit ripens. Common to all avocados is a rich, smooth, oily pulp which is eaten as a nutritious savoury. In the centre is a large stone, the seed.

Avocados can be left hanging on the tree for several weeks after reaching maturity, and yet they will not ripen. Ripening is possible only after detachment from the plant (8). On the other hand, if picked immature, avocados do not have the capacity to ripen normally after harvest. Maturity indices include dry weight, oil content and picking date, the optimal date of harvest in some localities varying only slightly from year to year (8).

At harvest the fruits are quite hard. They can be kept in the preclimacteric state (page 14) for up to several weeks, depending on cultivar (1). 'Cold-tolerant' cultivars may be stored at 4° to 5°C, while others, having a greater susceptibility to chilling injury (q.v.), must be held at 12° to 13°C. Since these moderate temperatures favour development of any latent anthracnose (q.v.) which may be present (2,4,10), it can be advantageous to use controlled atmosphere storage (page 23). This has the added benefit of reducing susceptibility of avocados to chilling injury, enabling even the most 'cold-sensitive' cultivars to tolerate temperatures of 7° to 8°C (9).

Waxing (which also modifies gaseous exchange) has variable effects, perhaps according to formulation and method of application. In one study, on Fuerte stored at 5°C and then ripened at 20°C, waxing was found to be of little benefit in delaying softening, though it was of value in reducing moisture loss at ambient temperature (5). In another study on the same cultivar, held under similar conditions (7), waxing extended the preclimacteric period significantly but induced unacceptable pulp discoloration. Such 'internal darkening' (q.v.) may have other causes; pre-shipment, shipboard and post-discharge factors have all been implicated.

If refrigeration is unavailable, sealed packs may be the only means of delaying ripening. The modified atmosphere created by polyethylene bags is effective provided that the storage time does not exceed a week or so and the ambient temperature is not too high (3). At 30°C and above, abnormal ripening is likely whether the avocados are wrapped or not; the skin may become pitted and off-flavours develop (6). Avocados ripen best at temperatures between 15° and 25°C, and ethylene may be used to ensure that all the fruits are 'triggered' simultaneously (6). Once ripe, they are able to withstand a lower temperature than can be tolerated by pre-climacteric fruits.

---

1 ARRIOLA M.C. de, MENCHU J.F. & ROLZ C. (1976) *Caracterización, manejo y almacenamiento de aguacate*. Guatemala: ICAITI, 67 pp.
2 CAPPELLINI R.A., CEPONIS M.J. & LIGHTNER G.W. (1988) Disorders in avocado, mango, and pineapple shipments to the New York market, 1972–1985. *Plant Disease* **72**, 270–273.
3 CHAPLIN G.R. & HAWSON M.G. (1981) Extending the life of unrefrigerated avocados (*Persea americana* Mill.) fruit by storage in polyethylene bags. *Scientia Horticulturae* **14**, 219–226.
4 DARVAS J.M., KOTZE J.M. & WEHNER F.C. (1987) Pathogenicity of fungi causing pre- and post-harvest diseases of avocado fruit. *Phytophylactica* **19**, 489–493 illus.
5 DURAND B.J., ORCAN L., YANKO U., ZAUBERMAN G. & FUCHS Y. (1984) Effects of waxing on moisture loss and ripening of 'Fuerte' avocado fruit. *HortScience* **19**, 421–422.
6 LEE S.K. & YOUNG R.E. (1984) Temperature sensitivity of avocado fruit in relation to C_2H_4. *Journal of the American Society for Horticultural Science* **109**, 689–692.
7 LELYVELD L.J. van & BOWER J.P. (1984) Enzyme reactions leading to avocado fruit mesocarp discoloration. *Journal of Horticultural Science* **59**, 257–263 illus.
8 LEWIS C.E. (1978) The maturity of avocados – a general review. *Journal of the Science of Food and Agriculture* **29**, 857–866.
9 SPALDING D.H. & REEDER W.F. (1975) Low-oxygen high-carbon dioxide controlled atmosphere storage for control of anthracnose and chilling injury of avocados. *Phytopathology* **65**, 458–460.
10 ZENTMYER G.A. (1984) Avocado diseases. *Tropical Pest Management* **30**, 388–400.

ANTHRACNOSE of avocados caused by
Glomerella cingulata (Stonem.) Spauld. & v. Schrenk
Conidial state: *Colletotrichum gloeosporioides* (Penz.) Sacc.

216

217

OCCURRENCE This is an important disease in most avocado-growing countries, for example the USA (4), Puerto Rico (1), Argentina (2), Israel (4), India, South Africa (5) and Australia (3).

SYMPTOMS Signs of anthracnose typically appear when the fruit begins to soften, and infected ripening fruits exhibit circular brown to black spots on the skin. The lesions are later sunken and, in humid conditions, support the growth of masses of pink spores, while the flesh beneath undergoes rapid decay (2).

BIOLOGY The fungus may produce both the sexual state (perithecia giving rise to ascospores) and the asexual state (acervuli giving rise to conidia); the latter plays the more important role in infection (1). Spores are washed down from dead branches by rain, and the disease is associated with persistent wet conditions in the orchard (3). Initial infection may occur at any time during the growing season, and the fungus can attack sound developing fruits. In some localities symptoms may be produced whilst the fruit is still on the tree, particularly in the wake of cercospora spot (q.v.). Anthracnose more usually remains quiescent, however, and the fungus effects complete penetration of the skin only after ripening has commenced (4). Thus infected avocados may be harvested in apparently healthy condition, and the disease is manifest during storage, transport and marketing.

CONTROL It may be necessary to apply orchard fungicide sprays from the time of fruit-set up till harvest (2,3). Post-harvest fungicide treatments are also beneficial (5), and storage in controlled atmosphere (reduced oxygen, increased carbon dioxide) can delay the development of symptoms.

---

1 NOLLA J.A.B. (1926) The anthracnoses of citrus fruits, mango and avocado. *Journal of the Department of Agriculture of Porto Rico* **10**(2), 25–50 illus.
2 OSTE C.A. & RAMALLO N.E.V. de (1974) Podredumbre apical de la palta producida por *Colletotrichum gloeosporioides* Penz. *Revista Industrial y Agrícola de Tucumán* **51**(1), 37–40.
3 PETERSON R.A. & INCH A.J. (1980) Control of anthracnose on avocados in Queensland. *Queensland Journal of Agricultural and Animal Sciences* **37**, 79–83.
4 PRUSKY D., KEEN N.T. & EAKS I. (1983) Further evidence for the involvement of a preformed antifungal compound in the latency of *Colletotrichum gloeosporioides* on unripe avocado fruits. *Physiological Plant Pathology* **22**, 189–198.
5 ROWELL A.W.G. (1983) Post-harvest disease control in avocados using prochloraz. *Yearbook of the South African Avocado Growers' Association* **6**, 19.

CERCOSPORA SPOT or BLOTCH of avocados caused by
Pseudocercospora purpurea (Cooke) Deighton

218

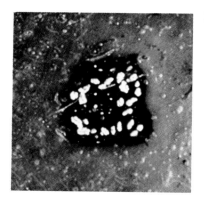

219

OCCURRENCE   This blemish is common on avocados from wet growing areas in the USA (Florida but not California), the Caribbean (5), Brazil (1), Cameroon (3) and South Africa (2).

SYMPTOMS   Spots on the fruit are at first small greenish-white dots, later developing into slightly sunken irregular brown blotches which, under humid conditions, support tufts of greyish mould. Mature spots are rarely larger than 0.5 cm, but the cracks and fissures which develop in the lesions permit the entry of other organisms, in particular the anthracnose fungus (q.v.).

BIOLOGY   The fungus responsible for cercospora spot exists in two forms, a rarely recorded sexual state (perithecia and ascospores of the *Mycosphaerella* type) and the more common asexual state (conidia) which is of importance in the infection process. The disease is perpetuated in old leaf infections, and spore production and dissemination follow periods of wet weather (3,4,5). Fruits which are either very young or almost mature are immune to attack, but at intermediate stages of fruit development this fungus is capable of direct penetration of the intact skin.

CONTROL   Since infection occurs in the developing fruits, cercospora spot is controlled by field applications of a suitable systemic fungicide (2). Different avocado cultivars show great variation in susceptibility to this disease.

1 ALBUQUERQUE F.C. (1962) Mancha parda do abacate. *Revista de la Sociedad de Agronomía e Veterinaria de Pará (Brazil)* **8**, 35–41 illus.
2 DARVAS J.M. (1982) Chemical control of *Cercospora* spot disease of avocados. *Yearbook of the South African Avocado Growers' Association* **5**, 58–59.
3 GAILLARD J.P. (1971) Lutte contre le Cercospora de l'avocatier au Cameroun. *Fruits* **26**, 225–230.
4 HINO T. & TOKESHI H. (1976) Varietal resistance of avocado to cercosporiosis and some observations on the disease cycle. *Summa Phytopathologica* **2**, 127–132.
5 MOREZ H. (1962) Notes sur l'avocatier en Guadeloupe. *Fruits* **17**, 179–184.

DOTHIORELLA ROT of avocados caused by
Botryosphaeria ribis Grossenb. & Duggar
Conidial state: *Dothiorella gregaria* Sacc.

OCCURRENCE This fungus is reported to cause occasional damage to avocados in Israel, South Africa (4), the USA (8) and parts of South America (6,10). In New Zealand the causal organism is *Botryosphaeria parva* Pennycook & Samuels (2) and in Australia *Dothiorella aromatica* (Sacc.) Petrak & Sydow (5).

SYMPTOMS Various symptoms have been described. In some localities the disease takes the form of numerous purplish-brown spots on the skin of ripening fruits, with a shallow decay of the flesh beneath (10). Sometimes there is internal browning and abnormal softening of the flesh (2). Alternatively the fungus may cause severe stem-end rotting, often in association with other fungi (5). Infected skin appears pimpled owing to the formation of minute spherical bodies (pycnidia), and in humid conditions a greyish fungal 'mat' develops (7). A rank odour accompanies the decay.

BIOLOGY The fungus may produce both the sexual state (perithecia giving rise to ascospores) and the asexual state (pycnidia giving rise to conidia) and the spore-bearing bodies are formed in bark, dead twigs and senescent leaves (1,6). During wet weather spores are washed down by rain and cause a latent infection via stomata or the stem-end tissue. It is reported that symptoms usually appear only when the fruit begins to soften, some time after harvest (4). However, the author has isolated this fungus from substantial lesions on hard unripe avocados (imported from California). Furthermore, in inoculation experiments, it was reported to cause rotting of young fruits on the tree (7). The explanation is probably that when infection occurs via wounds there is no latent period.

CONTROL In wet growing areas the disease is controlled by orchard fungicide sprays (4,6); post-harvest fungicide treatment is also beneficial (5). However, failure of both sprays and dips has been reported (2).

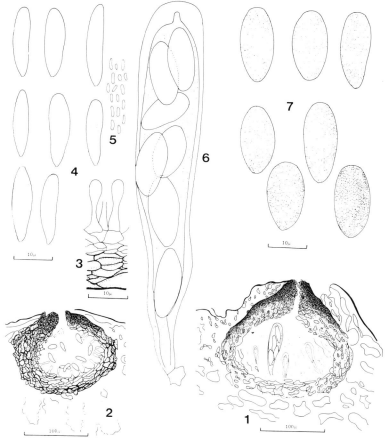

222 *Botryosphaeria ribis* (from CMI Descr. No. 395, Punithalingam & Holliday 1973)
1 pseudothecium 2 pycnidium 3 conidiogenous cells 4 macroconidia
5 microconidia (spermatia) 6 ascus 7 ascospores

1 FREZZI M.J. (1952) Presencia de *Botryosphaeria ribis* en la República Argentina y su importancia económica. *Revista de Investigaciones Agrícolas, Buenos Aires* **6**, 247–262 illus.
2 HARTILL W.F.T., MANNING M.A. & ALLEN D.J. (1986) Control of postharvest diseases of avocado. In *Proceedings of the 39th New Zealand Weed and Pest Control Conference,* pp 158–161.
3 HORNE W.T. & PALMER D.F. (1935) The control of *Dothiorella* rot on avocado fruits. *Bulletin of California Agricultural Experiment Station* No. 594, 16 pp.
4 LABUSCHAGNE N. & ROWELL A.W.G. (1983) Chemical control of post-harvest diseases of avocados by pre-harvest fungicide application. *Yearbook of the South African Avocado Growers' Association* **6**, 46–47.
5 MUIRHEAD I.F., FITZELL R.D., DAVIS R.D. & PETERSON R.A. (1982) Post-harvest control of anthracnose and stem-end rots of Fuerte avocados with prochloraz and other fungicides. *Australian Journal of Experimental Agriculture and Animal Husbandry* **22**, 441–446.
6 PINTO de T., A., ALVAREZ A., M. & TOBAR C., G. (1986) Pudrición de frutos y cancros en ramas de palto, causados por *Dothiorella* sp., en la V Región de Chile. *Agricultura Técnica* **46**, 499–501.
7 SMITH C.O. (1934) Inoculations showing the wide host range of *Botryosphaeria ribis*. *Journal of Agricultural Research* **49**, 467–476 illus.
8 STEVENS H.E & PIPER R.B. (1941) Avocado diseases in Florida. *Circular of the United States Department of Agriculture* No. 582, 46 pp illus.
9 ZAUBERMAN G., SCHIFFMANN-NADEL M., FUCHS Y. & YANKO U. (1975) La biologie et la pathogenicité des champignons causant les pourritures de l'avocat après la récolte. *Fruits* **30**, 499–502.
10 ZENTMYER G.A. (1961) Avocado diseases in the Americas. *Ceiba* **9**, 61–79.

SCAB of avocados caused by
Sphaceloma perseae Jenkins

223

OCCURRENCE This disease occurs in North, Central and South America, and has also been recorded in the Caribbean islands of Guadeloupe, Cuba, Haiti and Puerto Rico (1). It also occurs in west and southern Africa and the Philippines (CMI Map 232).

SYMPTOMS The fungus can only attack young developing fruits; they become immune on reaching a certain size. Raised corky brown spots are produced on the skin, sometimes leading to a russeted appearance which mars the appearance of the fruit even before it is harvested (3). Subsequent cracking may allow secondary organisms to penetrate the fruit.

BIOLOGY Only the asexual state of this fungus has been observed (acervuli giving rise to conidia) (1). Leaves and twigs are attacked, and spores are washed down by rain (4). The fungus is capable of direct penetration of susceptible young fruits.

CONTROL Since the damage is caused during the growing period it is essential in wet areas to apply fungicide sprays early in the season (2,3,4,5). Some cultivars of avocado are highly resistant to scab, however, making control measures unnecessary.

---

1 JENKINS A.E. (1934) *Sphaceloma perseae* the cause of avocado scab. *Journal of Agricultural Research* **49**, 859–869 illus.
2 McMILLAN R.T. (1970) Effectiveness of copper when combined with Nu-Film 17 for control of avocado scab. *Proceedings of Florida State Horticultural Society* **83**, 386–388.
3 RAMALLO N.E.V. de (1969) Sarna o verrugosis del palto en Tucumán. *Revista Industrial y Agrícola de Tucumán* **46**(2), 27–30 illus.
4 RONDON G.A., FIGUEROA M. & GUILLEN P. (1975) La sarna o verrugosis del aguacate (*Persea americana* Mill) en Venezuela y su control. *Agronomía Tropical* **25**, 381–384 illus.
5 RUEHLE G.D. & CONOVER R.A. (1962) Ferbam as a control for avocado scab. *Proceedings of Florida State Horticultural Society* **75**, 363–364.

STEM-END ROTS of avocados

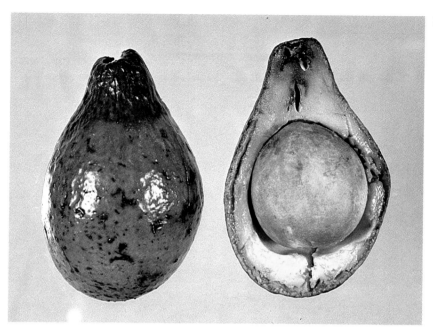

OCCURRENCE *Botryodiplodia theobromae* (**197**) has caused losses in Australia (3), South Africa (1), the Ivory Coast (2), Israel (4) and the USA (5). Other fungi may be associated with stem-end rots, either alone or in combination, for example *Dothiorella* spp. (page 96), *Phomopsis perseae* Zerova (3) and *Thyronectria pseudotrichia* (Schw.) Seeler (1).

SYMPTOMS The rot begins at the stem-end of the fruit as a small ring of firm brown tissue, and the decay eventually spreads throughout the fruit (5).

BIOLOGY The stem-end rot fungi inhabit dead branches and bark of avocado trees, and their spores are dispersed by rain (3). *Botryosphaeria ribis* and *Diaporthe citri* are capable of causing latent infection in developing fruits (3). However, *Botryodiplodia theobromae* is characteristically a wound parasite, and it is likely that most infections occur at the time of harvest (2). Eventually these fungi cause a serious decay of the flesh.

CONTROL Orchard spraying with a range of fungicides can give partial control (1). It is also wise to destroy leaf litter (1). There are several strategies for minimising post-harvest decay. Infected fruits harvested with short stalks (2–3 mm) may suffer stem-end rotting before ripening is completed; in contrast, fruits harvested with longer stalks (6–10 mm) are able to ripen before decay progresses into the fruit. Other control methods include field-dipping of the cut stem in a fungicide wax, and treatment of the entire fruit with a post-harvest fungicide dip (2). Alternatively, the use of an antioxidant chemical helps to maintain fruit resistance to fungal invasion, thereby prolonging storage- and shelf-life for a few days (4). Careful handling and rapid cooling are also recommended (2).

---

1 DARVAS J.M., KOTZÉ J.M. & WEHNER F.C. (1987) Field occurrence and control of fungi causing post-harvest decay of avocados. *Phytophylactica* **19**, 453–455.
2 FROSSARD P. (1964) La pourriture pédonculaire des avocats en Côte d'Ivoire. Influence de la température sur le développement de *Diplodia natalensis*. *Fruits* **19**, 401–403.
3 PETERSON R.A. (1978) Susceptibility of Fuerte avocado fruit at various stages of growth to infection by anthracnose and stem-end rot fungi. *Australian Journal of Experimental and Animal Husbandry* **18**, 158–160.
4 PRUSKY D. (1988) The use of antioxidants to delay the onset of anthracnose and stem end decay in avocado fruits after harvest. *Plant Disease* **72**, 381–384.
5 STEVENS H.E. & PIPER R.B. (1941) Avocado diseases in Florida. *Circular of the United States Department of Agriculture* No. 582, 46 pp illus.

OTHER DISEASES of avocados

ALTERNARIA ROT sometimes becomes a problem when other disease organisms have been suppressed by the use of a post-harvest fungicide. In Israel it was observed that *Alternaria* sp. caused rotting of the stem-end and other parts of the fruit, and its development was not checked by post-harvest chemical treatment (9).

BACTERIAL SOFT ROT, caused by *Erwinia carotovora*, has been recorded on unripe Israeli avocados (6). Soft, dark, flat spots appear, eventually coalescing to cover the entire fruit. Infected flesh is soft and discoloured, and gives off a putrid odour. Another bacterium, *Pseudomonas syringae* pv. *syringae*, causes a dry rot of avocado fruits (6).

BLUE MOULD ROT is caused by *Penicillium expansum* (page 178) and has been recorded on avocados in the USA (3) and the West Indies (7). It is usually associated with wounds.

FUSARIUM ROT may be caused by several species of *Fusarium*, for example *F. pallidoroseum* (page 110). The disease is important in Israel (8) and has also been recorded in South Africa (1), the USA (3) and the West Indies (7). Infection of avocados usually takes place via the stem-end or through injuries, and the effect is an acceleration in the rate of softening and ripening (8).

225 Various moulds on senescent avocados

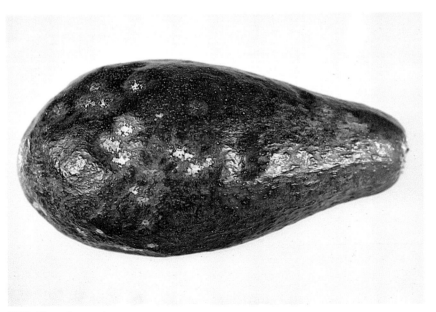

226 Fusarium rot

PESTALOTIOPSIS ROT is a minor post-harvest disease, occasionally encountered in South Africa (1). It is caused by *Pestalotiopsis versicolor* (Speg.) Steyart.

PHYTOPHTHORA ROT has recently been reported from Mexico (2) and the USA (4). The causal fungus, *Phytophthora citricola* (page 64) is capable of attacking fruits hanging low on the tree, by means of spores splashed from the soil. A rapid black rot ensues.

PINK MOULD ROT, caused by *Trichothecium roseum* (page 191), is a minor disease, usually found only on avocados which have been stored for some time (3,10).

RHIZOPUS ROT, usually caused by *Rhizopus stolonifer* (**401**), is characterised by rapid decay of ripe avocados, and the formation of a coarse white mould with black spore-heads. Liquid may leak from the fruit; the odour is peculiar and unpleasant, although unlike the rancid smell of dothiorella rot (q.v.) or the putrid odour of bacterial spoilage (3). Infection is associated with wounds and, in one study, appeared only on fruit which had undergone cold storage (1). Rhizopus rot has been recorded in South African, American and Israeli avocados (1,10).

SOOTY BLOTCH is caused by a species of the fungus *Akaropeltopsis*. The disease occurs in South Africa and is characterised by smoky blotches on the skin (5). Control is by pre-harvest fungicide sprays (1).

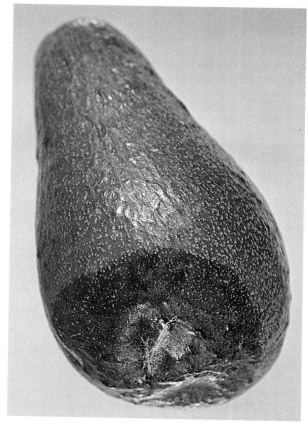

227 Rhizopus rot

---

1 DARVAS J.M. & KOTZÉ J.M. (1987) Fungi associated with pre- and post-harvest diseases of avocado fruit at Westfalia Estate, South Africa. *Phytophylactica* **19**, 83–85 illus.
2 FUCIKOVSKY L. & LUNA I. (1987) Avocado fruit diseases and their control in Mexico. *Yearbook of the South African Avocado Growers' Association* **10**, 119–121 illus.
3 HORNE W.T. (1934) Avocado diseases in California. *Bulletin of California Agricultural Experiment Station* No. 585, 72 pp illus.
4 KOIKE S.T., OUIMETTE D.G. & COFFEY M.D. (1987) First report of avocado fruit rot caused by *Phytophthora citricola*. *Plant Disease* **71**, 1045 illus (cover).
5 SMITH E.M., KOTZÉ J.M. & WEHNER F.C. (1985) Sooty blotch of avocado caused by *Akaropeltopsis* sp. *Phytophylactica* **17**, 101–102 illus.
6 VOLCANI Z. (1959) Bacterial soft rot of avocado fruit. 2. *Bulletin of the Research Council of Israel D* **7**, 39–42 illus.
7 WARDLAW C.W. (1934) Preliminary observations on the storage of avocado pears. *Tropical Agriculture* **11**, 27–35.
8 ZAUBERMAN G. & SCHIFFMANN-NADEL M. (1979) Physiological response of avocado fruit to infection by different *Fusarium* species. *Phytopathologische Zeitschrift* **89**, 359–365.
9 ZAUBERMAN G., SCHIFFMANN-NADEL M., FUCHS Y. & YANKO U. (1975) La lutte contre les pourritures de l'avocat et son effet sur le changement de la flore des champignons pathogènes des fruits. *Fruits* **30**, 503–504.
10 ZENTMYER G.A., PAULUS A.O., GUSTAFSON C.D., WALLACE J.M. & BURNS R.M. (1965) Avocado diseases. *Circular of California Agricultural Experiment Station* No. 534, 11 pp illus.

CHILLING INJURY of avocados

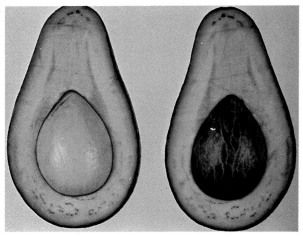

228

OCCURRENCE Low temperature injury has been studied in several cultivars grown in various parts of the world, for example Fuerte and Hass in California (1,2), Israel (5), Cyprus, South Africa and Australia (3). In Florida a range of cultivars has been studied, including Taylor (4).

SYMPTOMS The first external signs are dark patches on the skin (5). If the fruit is cut in half, a general greyish discoloration of the pulp may be apparent (3). Cutting the fruit transversely permits examination of the vascular strands which, in chilled avocados, appear as dark brown 'dots' in the pulp. The distal end of the fruit is usually affected before the stem-end. Texture and taste may be abnormal (2). It should be noted that other factors besides chilling injury may be responsible for internal darkening (q.v.) and failure to ripen properly.

BIOLOGY Susceptibility to chilling injury varies with cultivar, season and maturity of the fruit. Especially sensitive are avocados having a low calcium content. Temperatures below the critical value can result in abnormal metabolism which, if prolonged, leads to the development of visible signs of injury (1). Avocados are very susceptible during the climacteric rise (page 14) and especially at the peak of respiratory activity (2). The presence of ethylene increases their sensitivity. The ripening process may be impaired, and injured tissue is predisposed to decay. On the other hand, fruits which have been allowed to ripen normally are able to withstand temperatures which are harmful to preclimacteric avocados.

CONTROL The appropriate storage temperature is determined by such factors as cultivar and intended duration of storage (1). However, since growing conditions also exert an influence, local trials should be carried out if possible. Approximate values for unripe avocados are 4° to 5°C for Lula and Taylor, 5° to 8°C for Fuerte and Hass, and 10° to 13°C for Fuchs, Pollock and Waldin. Controlled atmosphere storage permits the use of slightly lower temperatures or longer storage periods (4). Similar benefits are obtained with the use of sealed polyethylene bags (3). Removal of ethylene from the storage atmosphere may help to reduce chilling injury. On the other hand, ethylene at ambient temperature can be used to ensure uniform ripening in a batch of avocados, which may then be held for a limited time at below chilling temperatures without injury (2).

---

1 EAKS I.L. (1983) Effects of chilling on respiration and ethylene production of 'Hass' avocado fruit at 20°C. *HortScience* **18**, 235–237.
2 KOSIYACHINDA S. & YOUNG R.E. (1976) Chilling sensitivity of avocado fruit at different stages of the respiratory climacteric. *Journal of the American Society for Horticultural Science* **101**, 665–667.
3 SCOTT K.J. & CHAPLIN G.R. (1978) Reduction of chilling injury in avocados stored in sealed polyethylene bags. *Tropical Agriculture* **55**, 87–90.
4 VAKIS N.W., GRIERSON W. & SOULE J. (1970) Chilling injury in tropical and subtropical fruits. 3. The role of CO_2 in suppressing chilling injury of grapefruit and avocados. *Proceedings of the American Society for Horticultural Science (Tropical Region)* **14**, 89–100.
5 ZAUBERMAN G., FUCHS Y. & AKERMAN M. (1985) Peroxidase activity in avocado fruit stored at chilling temperatures. *Scientia Horticulturae* **26**, 261–265.

OTHER DISORDERS of avocados

'**CUKES**' are small, malformed, seedless avocados, so called because of their resemblance to miniature cucumbers. During some seasons and in certain cultivars (e.g. Fuerte) these abnormal fruits comprise a substantial proportion of the crop (2). This is one disorder which occasions no losses, however, since in the market 'cukes' fetch a premium price.

INTERNAL DARKENING is a greyish or even a black discoloration of the pulp, starting at the distal end and around the stone (1,5). The vascular strands may or may not be conspicuous. Symptoms can be induced by chilling (q.v.), waxing, or storage in a modified atmosphere. Investigations have been carried out in South Africa, following serious problems in the export trade, affecting consignments of avocados shipped in refrigerated containers (1). An unfavourable atmosphere can build up if the container is not properly ventilated with fresh air (page 25). It was found that avocados from water-stressed trees are predisposed to internal darkening, thus providing a plausible explanation for the observed variability in incidence and severity. Control involves attention to irrigation during the early stages of fruit development, coupled with an awareness of the need for container ventilation to continue after unloading from the vessel for onward transport (1). Since various factors can induce similar internal symptoms, it can be difficult to distinguish between disorders (4). Correct diagnosis and control may involve detailed monitoring of pre- and post-harvest conditions.

RIND DISCOLORATION may occur with or without pulp darkening. Lesions are brown with a definite margin and may have a random distribution. In contrast to those of anthracnose (q.v.) and in common with those of chilling injury (q.v.) they remain superficial. Sometimes the affected area encircles the stem-end. The pulp often appears normal but, on softening, off-flavours develop and

229 'Cukes'

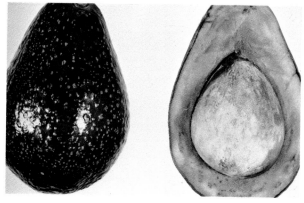

230 Internal darkening

there is an odour of fermentation. The disorder can be induced by an atmosphere deficient in oxygen and/or rich in carbon dioxide (3). Such an environment may develop in refrigerated containers which have been charged with a modified atmosphere before shipment (page 25). Since the initial atmosphere is subsequently altered by respiration of the fruit, care is needed to ensure proper pre-cooling of avocados so as to maintain a low rate of respiration during transit. Moderate ventilation may also be appropriate.

---

1 BOWER J.P. & LELYVELD L.J. van (1985) The effect of stress history and container ventilation on avocado fruit polyphenol oxidase activity. *Journal of Horticultural Science* **60**, 545–547.
2 SAROOSHI R.A., BLUNDELL D.R. & PEASLEY D. (1979) Blemishes and abnormalities of avocado fruit. *Agricultural Gazette of New South Wales* **90**(2), 18–20 illus.
3 SPALDING D.H. & MAROUSKY F.J. (1981) Injury to avocados by insufficient oxygen and excessive carbon dioxide during transit. *Proceedings of Florida State Horticultural Society* **94**, 299–301 illus.
4 SWARTS D.H. (1984) Post-harvest problems of avocados – let's talk the same language. *Yearbook of the South African Avocado Growers' Association* **7**, 15–19 illus.
5 VAKIS N.J. (1982) Storage behaviour of Ettinger, Fuerte and Hass avocados grown on Mexican rootstock in Cyprus. *Journal of Horticultural Science* **57**, 221–226.

Bananas

Bananas are the fruits of certain cultivars of the genus *Musa*. Having originated in South East Asia, the plant was taken in early times to the Pacific Islands and to Africa. In the 15th century the banana was taken from West Africa to the New World, and the crop is now of great importance throughout the tropics (9). World production is about 40 million tonnes, the leading producer countries being Brazil, India, the Philippines, Ecuador, Indonesia, Thailand, Mexico, Vietnam, Colombia and Honduras. Of the bananas which enter international trade (more than one sixth of total production), most come from Ecuador, Colombia, Honduras, Costa Rica and Panama, but there are also substantial exports from the Philippines to Japan, from the West Indies to the UK, from the Canary Islands to the Spanish mainland, and from West Africa and the French Antilles to France.

Plantains, and also 'cooking bananas', are closely related crops which are important staples, especially in East and West Africa. World production is about 27 million tonnes, and Uganda is the major producer. An insignificant proportion enters world trade. In contrast to dessert bananas, plantains and cooking bananas have a substantial starch content even when fully ripe. In most other respects (except where stated), the characteristics of bananas and plantains are similar.

The plant is a tree-like herb with large leaves which emerge in a rolled position from the centre of the pseudostem (9). It produces a single flowering shoot consisting of several flower clusters, each cluster having 12 to 20 flowers in two rows, covered by a large reddish bract. In cultivated species the fruits arise without fertilisation and are seedless; their weight makes the shoot hang down, so that the developing 'hands' of bananas curve upwards. The hands at the proximal end mature slightly earlier than those towards the distal end of the bunch.

Banana and plantain fruits show a climacteric pattern of respiration (page 14), and plantains in particular produce substantial quantities of ethylene during ripening (1). Fruit intended for local consumption can be left on the plant until fully developed. Harvested bananas (but not plantains) may fail to turn yellow at tropical temperatures, even though they are ripe inside (7). Fruit for distant markets must be cut immature and shipped green. Maturity can be measured by counting the number of days from 'shooting', and in commercial banana plantations a system of colour-coded tags may be used. Fruit can be assessed visually by the shape of the 'fingers'; the immature fruit has an angular cross-section which gradually becomes more rounded as the fruit develops. The diameter (expressed in various units) of the middle outside finger of a particular hand gives the 'grade' (9). More sophisticated methods are also available (5). The appropriate maturity stage for cutting depends partly on the clone or cultivar of banana and partly on the duration of the proposed journey. Thus, for example, Ecuadorean Cavendish bananas destined for the Eastern US are picked at grade 44–46 (the diameter in $1/32''$) while the same type of banana intended for Mediterranean ports must be picked 'thinner' (grade 41–43) because it must withstand a longer voyage. Whether the higher or lower end of the range is used is determined by the season of the year and the climatic conditions during the growing period (3). The physiological state of bananas is influenced not only by the weather but also by the incidence of sigatoka leaf spot disease (q.v.). Fruit from a diseased plant tends to be physiologically more mature than is indicated by its physical size, and hence liable to abnormal and premature ripening (4,10). Preventive measures against sigatoka, and also against latent anthracnose infection, must be taken during the growing period.

At cutting time, and subsequently, careful handling is essential, firstly to prevent the fruit from being stained by latex (which exudes copiously from cut surfaces), and secondly to minimise the incidence of cuts and bruises, the effects of which will show when the fruit ripens (8). Injury leads to an increased respiration rate, and also predisposes fruit to direct fungal attack, for example by the anthracnose fungus (4,10). Now that bananas are shipped in cartons rather than (as previously) on the stem, special precautions must be taken to disinfect the vulnerable fresh wound on the 'crown', which results from severance of each hand from the stem (8). If dehanding is done in the field, each cut crown is immediately covered with a 'crown pad', a small strip of absorbent material impregnated with a fungicide. If, as is more usual, dehanding is done in the packing station, then, after latex has been washed off, hands are treated with fungicide either by dipping or by spraying the crowns (8). Good control of crown rot is essential; it is currently the most important post-harvest disease (8).

Since the early 1960s all export fruit has been packed in vented cartons, mostly as hands or clusters (part hands) but occasionally, for certain markets, as single fingers. A polyethylene bag or wrap encloses the fruit within, serving to protect it from abrasion and moisture loss. Evaporation of moisture leads to weight loss, and furthermore water-stressed fruit is more likely to ripen prematurely (1). Sealed film packaging of suitable formula and gauge (and preferably including an ethylene absorbent) can be used to modify the atmosphere surrounding the fruit. An increased concentration of carbon dioxide

231 Premature ripening on board ship

has the effect of reducing respiration rate and so prolongs storage life (2). A similar effect can be achieved with various types of coating applied directly to the fruit (8).

During long distance transport bananas must be kept in the pre-climacteric state, so that ripening can later be induced artificially, under carefully controlled conditions, after arrival in the ripening room. (In this way, fruit in prime condition can be released on to the market in an orderly fashion, according to demand.) If premature ripening is to be avoided, then, besides cutting the fruit at the appropriate maturity, three other requirements must be fulfilled. First, there must be no delay between cutting and loading the fruit into refrigerated space; the aim should be to start cooling the fruit within 24 hours of harvest. Secondly, the cooling process itself must be as rapid as possible. In contrast to many other commodities, bananas are not usually pre-cooled before being offered for shipment, and it is the task of the ship to remove the 'field heat' and maintain the appropriate carriage temperature. The third requirement is the prevention of ethylene accumulation in the atmosphere surrounding the bananas. A concentration as low as 1 part per million may be sufficient to trigger the ripening process (2).

Thus, care must be taken to segregate bananas from other commodities with the capacity to produce ethylene, such as mangoes, melons, citrus, pineapples etc. In addition, the small quantities of ethylene produced by the bananas themselves must be flushed out by operation of the vessel's fresh air ventilation system. During the first critical 24 hours of cooling, fresh air ventilation should be restricted, so as to permit maximal refrigeration efficiency. Subsequent policy is dictated by circumstances, but it would appear that fresh air intake is often excessive (2 or even 3 air changes per hour), leading to copious condensation of moisture from the warm, moist, introduced air. The ideal method is to estimate ethylene levels by monitoring carbon dixode concentration and ventilating so as to keep the latter below 0.3%. In the absence of a carbon dioxide meter, a rate of 1 air change per hour is probably sufficient, a continuous slight intake being preferable to periodic opening and closing of vents.

The appropriate carriage temperature is determined by such factors as cultivar and growing conditions, and is limited by the susceptibility of all types of banana to chilling injury (q.v.). In marketing terms the delivery of chilled fruit presents an even greater problem than the arrival of 'ship-ripes' (hands which have ripened during the voyage). Ripening and chilling often go together, however, since fruit entering the climacteric is especially susceptible and may be injured by temperatures that pre-climacteric fruit can withstand. Ripening during the voyage may also be caused by a failure to cool the cargo sufficiently quickly, perhaps because of poor stowage, faulty refrigeration equipment or incorrect technique.

An extensive review of the international banana trade, 'Le marché international de la banane, étude géographique d'un système commercial' by J.-C. Maillard (1984, 1985, 1986), has been published in twenty parts in the journal *Fruits* **39**(9) to **41**(5).

---

1 GEORGE J.B. & MARRIOTT J. (1985) The effect of some storage conditions on the storage life of plantains. *Acta Horticulturae* No. 158, 439–447.
2 LIU F.W. (1976) Banana responses to low concentrations of ethylene. *Journal of the American Society for Horticultural Science* **101**, 222–224.
3 MARRIOTT J. (1980) Bananas – physiology and biochemistry of storage and ripening for optimum quality. CRC *Critical Reviews in Food Science and Nutrition* **13**, 41–88.
4 MEREDITH D.S. (1970) Major banana diseases: past and present status. *Review of Plant Pathology* **49**, 539–554.
5 NOLIN J. (1985) État de maturité des bananes (cv. Giant Cavendish) à la récolte: une nouvelle méthode de mesure. *Fruits* **40**, 623–631.
6 SCHIFFMANN-NADEL M. & MICHAELI H. (1984) Effect of fungal infection and temperature on physiological changes in banana fruit. *Phytopathologische Zeitschrift* **111**, 179–183.
7 SEYMOUR G.B., THOMPSON A.K. & JOHN P. (1987) Inhibition of degreening in the peel of bananas ripened at tropical temperatures. 2. Role of ethylene, oxygen and carbon dioxide. *Annals of Applied Biology* **110**, 153–161.
8 SLABAUGH W.R. & GROVE M.D. (1982) Post-harvest diseases of bananas and their control. *Plant Disease* **66**, 746–750 illus.
9 STOVER R.H. & SIMMONDS N.W. (1987) *Bananas*. London: Longman, 468 pp illus.
10 WARDLAW C.W. (1972) *Banana diseases including plantain and abaca*. London: Longman, 878 pp illus.

ANTHRACNOSE of bananas caused by
Colletotrichum musae (Berk. & Curt.) v. Arx

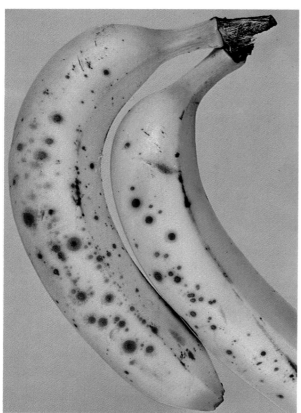

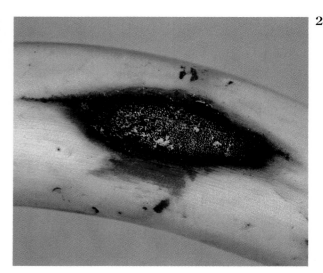

OCCURRENCE This is one of the most important diseases of bananas and is known in all producing countries (4).

SYMPTOMS There are two types of symptom, resulting from different modes of infection. Anthracnose lesions on green fruit are generally dark brown to black with a pale margin, lenticular in shape, slightly sunken and with dimensions of several cm. On ripening fruits the typical symptoms are numerous small dark circular spots which enlarge, coalesce and become sunken. The lenticular lesions may also be present in ripening fruit. On both types of lesion salmon-pink spore masses are eventually produced.

BIOLOGY The large lesions are the result of infection following physical injury, the fungus gaining entry via wounds sustained during harvesting and handling (8). The circular spots are the result of pre-harvest infections initiated in uninjured immature fruit (9). The fungus exists in the asexual state (acervuli giving rise to conidia) and the spore-bearing structures develop on leaf debris in the plantation. Spores are liberated by rainsplash or irrigation water, and dispersed by air currents and by insects visiting the developing fruits. If the fruit surface remains moist the spores germinate, each producing a swelling (appressorium) which adheres to the skin (9). Some appressoria remain viable, though quiescent, until the fruit ripens and becomes susceptible to invasion (page 17) (1,5,10). The fungus gives off ethylene, and hence can induce premature ripening in banana fruit (6). The same organism is one of the constituents of the crown rot complex (q.v.).

CONTROL Strict sanitation is necessary in plantation and packhouse, in order to minimise the number of spores available for infection. The fruit must be harvested at the correct stage of maturity and handled carefully to prevent injury. In situations where chemical control is necessary, post-harvest treatment with a systemic fungicide is more effective than pre-harvest spraying (2,3,7). The fruit should be cooled as soon as possible after harvest. Ripening times should be as brief as possible, in hygienic ripening rooms. Some cultivars (tetraploid clones) show resistance to wound anthracnose (8).

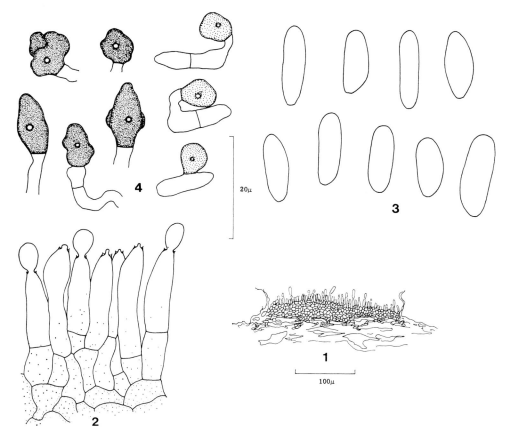

234 *Colletotrichum musae* (from CMI Descr. No. 222, Sutton & Waterston 1970)
1 acervulus   2 conidiophores   3 conidia   4 appressoria

1 BROWN A.E. & SWINBURNE T.R. (1980) The resistance of immature banana fruits to anthracnose (*Colletotrichum musae* (Berk. & Curt.) Arx). *Phytopathologische Zeitschrift* **99**, 70–80.

2 FROSSARD P. (1969) Action du thiabendazole et du benlate sur l'anthracnose des bananes et son champignon pathogène: *Colletotrichum musae*. *Fruits* **24**, 365–379.

3 GRIFFEE P.J. & BURDEN O.J. (1974) Incidence and control of *Colletotrichum musae* on bananas in the Windward Islands. *Annals of Applied Biology* **77**, 11–16.

4 MEREDITH D.S. (1960) Studies on *Gloeosporium musarum* Cke & Massee causing storage rots of Jamaican bananas. 1. Anthracnose and its chemical control. *Annals of Applied Biology* **48**, 279–290.

5 MUIRHEAD I.F. & DEVERALL B.J. (1981) Role of appressoria in latent infection of banana fruits by *Colletotrichum musae*. *Physiological Plant Pathology* **19**, 77–84.

6 PEACOCK B.C. & MUIRHEAD I.F. (1974) Ethylene production by *Colletotrichum musae*. *Queensland Journal of Agricultural and Animal Sciences* **31**, 249–252.

7 RAM V. & VIR D. (1983) Evaluation of benzimidazole and other fungicides against post-harvest spoilage of banana fruits caused by *Colletotrichum musae* (Berkeley & Curtis) von Arx. *Pesticides* **17**(5), 28–29.

8 SHILLINGFORD C.A. & SINCLAIR J.B. (1977) Susceptibility of five banana cultivars to anthracnose and crown rotting fungi. *Plant Disease Reporter* **61**, 797–801.

9 SIMMONDS J.H. (1963) Studies in the latent phase of *Colletotrichum* species causing ripe rots of tropical fruits. *Queensland Journal of Agricultural Science* **20**, 373–424 illus.

10 SWINBURNE T.R. & BROWN A.E. (1983) Appressoria development and quiescent infections of banana fruit by *Colletotrichum musae*. *Transactions of the British Mycological Society* **80**, 176–178.

CIGAR-END ROT of bananas caused by
Trachysphaera fructigena Tabor & Bunting
Verticillium theobromae (Turc.) Mason & Hughes

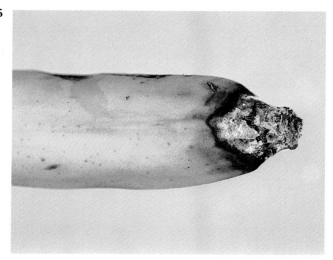

235

OCCURRENCE *Trachysphaera fructigena* can cause a destructive rot in the plantations of West and Central Africa (1,3,6). *Verticillium theobromae* is more widespread, occurring in most banana-growing regions, including the West Indies (7,10), India (2), Iran (5), Egypt (4), South Africa and the Canary Islands (CMI Map 146). In West Africa the two fungi (and others) may be associated in the same rot (7).

SYMPTOMS One or all fingers may be affected by a tip-end rot which starts with localised darkening and wrinkling of the skin (7). In trachysphaera rot the surface of the lesion becomes covered with white spores which later turn pink or brown; internally the pulp may undergo a wet rot or, in the absence of secondary organisms, may become dry and mummified (3). In verticillium rot the tissue is characteristically dry and fibrous, and the spores are grey and powdery. In both diseases the symptoms bear a resemblance to the ashy end of a burnt cigar.

BIOLOGY These fungi tend to be prevalent during wet or humid growing seasons, and airborne spores (conidia) can cause infection of the dying flower parts (7). *T. fructigena* is favoured by moderate temperatures (around 20°C) following a hot period (9). The optimal growth temperature for *V. theobromae* is approximately 25°C (4); rotting does not progress further after harvest and premature ripening does not occur. In contrast, *T. fructigena* continues to develop. Furthermore, new infections can occur in the washing tanks (3) via the cut crown, and later, in contaminated ripening rooms, via injuries.

CONTROL The chief method is frequent manual removal of dead flower parts, followed by 'bagging' of the developing fruits (1). Fungicide sprays may be necessary in some seasons (8). In the packing station care should be taken to cull infected hands in order to avoid contaminating the washing water with spores (3).

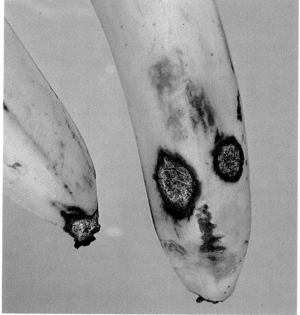

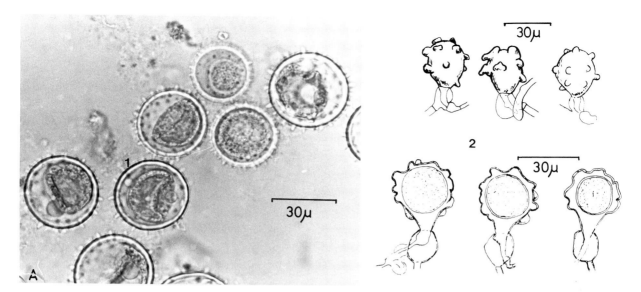

237 *Trachysphaera fructigena* (from CMI Descr. No. 229, Holliday 1970) 1 conidia 2 oogonia and antheridia (surface view and section)

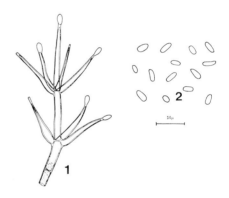

238 *Verticillium theobromae* (from CMI Descr. No. 259, Hawksworth & Holliday 1970) 1 conidiophore 2 conidia

---

1 BEUGNON M., BRUN J. & MELIN P. (1970) La lutte contre *Trachysphaera fructigena*, parasite des bananes au Cameroun. *Fruits* **25**, 187–197.

2 BHANGALE G.T. & PATIL L.K. (1983) Studies on cigar end rot of banana. 1. Occurrence, pathogenicity and etiology. *National Academy of Science Letters* **6**, 125–126.

3 BRUN J. (1970) Un nouvel aspect des pourritures de bananes 'Poyo' encours de transport. *Fruits* **25**, 781–784 illus.

4 EL-HELALY A.F., IBRAHIM I.A. & EL-AROSI H.M. (1954) Studies on some factors affecting the prevalence and distribution of cigar-end disease of banana in Egypt. *Alexandria Journal of Agricultural Research* **11**(2), 9–28 illus.

5 ERSHAD D. (1972) The occurrence of cigar-end disease of banana in Iran. *Iranian Journal of Plant Pathology* **8**, 7–16.

6 MARAMBA P. & CLERK G.C. (1974) Survival of conidia of *Trachysphaera fructigena*. *Transactions of the British Mycological Society* **63**, 391–393.

7 MEREDITH D.S. (1965) Tip rot of banana fruits in Jamaica. 2. *Verticillium theobromae* and *Fusarium* spp. *Transactions of the British Mycological Society* **48**, 327–336 illus.

8 MONTCEL H.T. du (1981) Perspectives nouvelles dans la lutte chimique contre *Trachysphaera fructigena* du bananier au Cameroun. *Fruits* **36**, 3–8.

9 MONTCEL H.T. du & LAVILLE E. (1977) Influence des conditions climatiques sur le développement du *Trachysphaera fructigena* sur bananier dans le sud-ouest du Cameroun. *Fruits* **32**, 77–85.

10 WARDLAW C.W. (1931) Banana diseases. 2. Notes on 'cigar-end' (*Stachylidium theobromae* Turc.). *Tropical Agriculture* **8**, 293–298.

CROWN MOULD, CROWN ROT AND FINGER-STALK ROT of bananas caused by
Acremonium spp.
Botryodiplodia theobromae Pat.
Ceratocystis paradoxa (Dade) Moreau
Colletotrichum musae (Berk. & Curt.) v. Arx
Fusarium pallidoroseum (Cooke) Sacc.
Verticillium theobromae (Turc.) Mason & Hughes

OCCURRENCE Crown rot is characteristically a disease complex caused by several fungi, sometimes in association with bacteria (5); different organisms predominate according to locality, time of year and other factors (6). In Central America and the Caribbean the most prevalent fungi are *Colletotrichum musae, Fusarium pallidoroseum* and *Verticillium theobromae* (2,3,4,10). *Acremonium* sp. is becoming more prominent. In South America *Ceratocystis paradoxa* is a major cause of crown rot (2), while in Africa and Asia *Botryodiplodia theobromae* and *C. paradoxa* are frequently found instead of or in addition to some of the above-mentioned fungi (1,7,8). *Glomerella cingulata* has been recorded on banana fruit in Australia, Taiwan (7) and Central America (5). Many other organisms have been described (2,8,10).

SYMPTOMS White, grey or pink mould may form on the surface of the cut crown. Infected tissue turns black and the rot may advance into the finger stalks, causing the fingers to drop off when handled (1). Finger-stalk rot may occur directly, in the absence of crown rot, if the stalks are injured through flexing of the fingers (6). Severe infection induces premature ripening.

BIOLOGY Many of the causal fungi survive on leaf debris in the plantation (7). Their spores are dispersed either by wind or rainsplash, and impinge on all parts of the developing bunch (6). When the harvested fruit is cut into hands and washed, the newly exposed tissue is vulnerable to infection; spores which have accumulated in the washing water can be drawn several mm into the wound (2). If the site of infection remains beyond the reach of fungicides, decay ensues and the combined effects of the various organisms can lead to rapid rotting (5).

CONTROL Preventive measures begin in the plantation with regular removal of leaf trash. Hygiene is also important in the packing station, and washing water should be changed frequently before it becomes heavily contaminated with spores. Dehanding should be done carefully with a sharp

241 *Fusarium semitectum* (from CMI Descr. No. 573, Booth 1978)   Conidia and conidiophores x 750

knife so as to avoid leaving a ragged cut. In some contexts 'crown bevelling' has proved beneficial, entailing a further trimming of the crown after the washing process (6). Post-harvest fungicide application is essential, and systemic fungicides are invaluable because of their ability to be carried into the internal peel tissues (9). As an alternative to packhouse treatment, field-packing of bananas can be more efficient. The fruit is dehanded in the plantation and a pad impregnated with fungicide is immediately adpressed to the cut crown, in order to stop the flow of latex and disinfect the wound. The hands or clusters can then be packed directly into cartons, without the need for washing and further handling. Regardless of the method employed to protect the crown, it is of great importance to minimise the time between cutting and cooling the fruit, especially if a long voyage is scheduled. Rapid reduction of temperature serves to slow down development of crown decay (2).

---

1 ABDEL-SATTAR M.A., SATOUR M.M. & EL-SHEHEDI A.A. (1977) *Ceratocystis* fruit rot disease of banana in Egypt. *Agricultural Research Review, Cairo* **55**(2), 79–85 illus.
2 GREENE G.L. & GOOS R.D. (1963) Fungi associated with crown rot of boxed bananas. *Phytopathology* **53**, 271–275 illus.
3 GRIFFEE P.J. (1976) Pathogenicity of some fungi isolated from diseased crowns of banana hands. *Phytopathologische Zeitschrift* **85**, 206–216.
4 KNIGHT C., CUTTS D.F. & COLHOUN J. (1977) The role of *Fusarium semitectum* in causing crown rot of bananas. *Phytopathologische Zeitschrift* **89**, 170–176.
5 LUKEZIC F.L., KAISER W.J. & MARTINEZ M.M. (1967) The incidence of crown rot of boxed bananas in relation to microbial populations of the crown tissue. *Canadian Journal of Botany* **45**, 413–421.
6 MEREDITH D.S. (1971) Transport and storage diseases of bananas: biology and control. *Tropical Agriculture* **48**, 35–50.
7 OGAWA J.M. (1970) Post-harvest diseases of bananas in China (Taiwan). *FAO Plant Protection Bulletin* **18**, 31–42.
8 ROTH G. & LOEST F.C. (1965) Collar rot of banana hands and its associated micro-organisms. *Technical Communication of the Department of Agricultural Technical Services of South Africa* No. 44, 14 pp illus.
9 SHILLINGFORD C.A. & SINCLAIR J.B. (1978) Uptake and translocation of systemic fungicides by banana fruits as determined by assay. *Plant Disease Reporter* **62**, 1107–1111.
10 WALLBRIDGE A. (1981) Fungi associated with crown-rot disease of boxed bananas from the Windward Islands during a two-year survey. *Transactions of the British Mycological Society* **77**, 567–577.

FINGER ROT of bananas caused by
Botryodiplodia theobromae Pat.

OCCURRENCE   This disease has been reported from most growing areas, for example Central America (2), the Caribbean, India (5), the Philippines (3) and Taiwan (4). It also occurs in Egypt (1) and has recently been recorded on banana fruits in the United Arab Emirates.

SYMPTOMS   Rotting usually begins at the tip of one of the fingers or at a wound site, and within a few days the entire finger may be soft, dark brown, and pimpled with minute black bodies (pycnidia) (5). Microscopic examination of the spores may be necessary to distinguish botryodiplodia rot from dothiorella rot (q.v.). The skin becomes wrinkled and the pulp is reduced to a semi-liquid. Under high humidity there is copious growth of dark grey mould.

BIOLOGY   The fungus is a common inhabitant of decaying vegetation in banana plantations, where it generally exists in the asexual state (pycnidia giving rise to conidia). Spores are disseminated by air currents and rain and, if they lodge in the dying flower parts, are later able to invade the fruit as it ripens. Infection can also occur via injuries. The fungus grows very slowly at temperatures below 20°C, its optimal growth temperature being approximately 30°C.

CONTROL   The disease can be held in check by ensuring that fruit is not over-mature when cut, and also by prevention of injury, treatment with a systemic fungicide and rapid reduction of temperature after harvest. This disease is chiefly important in local markets where refrigeration is not available.

---

1   EL-HELALY A.F., IBRAHIM I.A. & ELAROSI H.M. (1955) Studies on *Botryodiplodia* fruit-rot on banana in Egypt. *Alexandria Journal of Agricultural Research* **3**(2), 109–121.
2   GOOS R.D., COX E.A. & STOTZKY G. (1961) *Botryodiplodia theobromae* and its association with *Musa* species. *Mycologia* **53**, 262–277 illus.
3   LANTICAN M.T. & QUIMIO T.H. (1976) Pathogenicity and cultural characteristics of *Botryodiplodia* spp. causing fruit rots. *Philippine Phytopathology* **12**, 66–74.
4   TSAI Y.P. (1978) Influence of micro-organisms on development of banana black rot. *Plant Protection Bulletin, Taiwan* **20**, 39–45.
5   WILLIAMSON D. & TANDON R.N. (1966) Some pathological studies on *Botryodiplodia theobromae* Pat. causing banana rot. *Mycopathologia et Mycologia Applicata* **29**, 245–253 illus.

JOHNSTON SPOT or PITTING DISEASE of bananas caused by *Magnaporthe grisea* (Hebert) Barr

Conidial state: *Pyricularia grisea* (Cooke) Sacc.

243

OCCURRENCE This disease occurs in South America, Central America (1,2,3,5) and the Caribbean, the Canary Islands, Australia and parts of Asia. In Korea the causal fungus is *Pyricularia angulata* Hashioka (4).

SYMPTOMS Small reddish sunken spots may appear on the skin of maturing green fruit and, during transport and ripening, these develop into characteristic shallow black pits about 0.5 cm in diameter. Smaller pits form on the finger-stalk and crown, and may lead to finger-dropping. Serious post-harvest pitting can develop on fruit harvested in apparently sound condition.

BIOLOGY The fungus rarely produces the sexual state (perithecia giving rise to ascospores) and generally exists in the asexual state (conidia). During wet periods spores are produced in abundance on dying leaves and are disseminated by wind and water. Fruit is infected when still immature but symptoms do not develop until much later (5).

CONTROL Leaf and bract trash should be removed regularly, especially during rainy periods. It is essential to apply fungicide sprays at an early stage (1) and then to protect the emerging bunch with a polyethylene sleeve.

---

1 FRAIRE V.G. (1974) Evaluación de fungicidas en la prevención de las manchas de origen fungoso en el fruto del plátano. *Agricultura Técnica en México* **3**, 292–294.
2 GUYON M. (1970) Essais de lutte chimique contre la 'Johnson fruit spot' au Nicaragua. *Fruits* **25**, 685–691 illus.
3 HALMO S S. (1970) Inoculum sources of *Pyricularia grisea*, the cause of pitting disease of bananas. *Phytopathology* **60**, 183–184.
4 KIM W.G., KIM C.K. & LEE E.J. (1987) Banana blast caused by *Pyricularia angulata* Hashioka. *Korean Journal of Plant Pathology* **3**, 114–119.
5 MEREDITH D.S. (1963) *Pyricularia grisea* (Cooke) Sacc. causing pitting disease of bananas in Central America. 1. Preliminary studies on pathogenicity. *Annals of Applied Biology* **52**, 453–463.

SIGATOKA DISEASE of bananas caused by *Mycosphaerella* spp.
Conidial state: *Pseudocercospora* spp.

OCCURRENCE Sigatoka disease is named after a valley in Fiji where the problem first attracted attention in the early part of the century (3). Subsequently, epidemics were reported in Australia, Asia, Africa, Central America and the Caribbean. The causal organism is *M. musicola* Mulder. It has not yet been reported from the producing areas of the Canary Islands, Egypt and Israel. The potential of this fungal disease is such that a flourishing banana industry can be destroyed within a few years. In recent years there has appeared a similar but more virulent fungus, *Mycosphaerella fijiensis* Morelet, which is gradually spreading and becoming the dominant species (5).

SYMPTOMS Fruit symptoms include buff-coloured pulp, premature ripening, increased susceptibility to chilling injury (q.v.), and development of abnormal flavour and aroma.

BIOLOGY The disease is different from others in that the causal organism is not itself present in the fruit but yet has profound effects on fruit development (5). Sigatoka is essentially a leaf-spotting disease which can cause premature death of large areas of the plant's leaf surface. Photosynthesis is thereby drastically reduced, sometimes to the extent that fruit does not mature at all. In less severe outbreaks the size of bunches and individual fingers is reduced, with the result that fruit may be considerably more mature than is indicated by its size and appearance. This has important repercussions in the selection of bananas for distant markets. The fungi produce both the sexual stage (perithecia giving rise to ascospores) and the asexual stage (conidia), and during warm wet weather spores are formed in abundance on the leaves (5). Conidia are dispersed in water droplets to other leaves, and infection results in a linear pattern of lesions called 'line-spotting'. Ascospores are released by rainsplash (4) but can then be disseminated by air currents even in dry weather, producing lesions on the leaf margins known as 'tip-spotting'. *M. fijiensis* causes **black leaf streak**, in which especially severe symptoms result from ascospore infection, while *M. fijiensis* var. *difformis* causes **black sigatoka**, with similar symptoms (5).

CONTROL Strict quarantine laws exist in an attempt to curb the spread of these fungi. In countries where the disease is established, control is by cultural and chemical methods. Correct spacing of plants, efficient drainage and weed control all help to reduce humidity in the plantation. Badly spotted leaf trash should be removed and destroyed. Sigatoka can be checked by spraying the plantation with oil or an oil-fungicide mixture; aircraft are commonly used (5). Frequent applications may be needed and, in order to delay the appearance of tolerant fungal strains, different fungicides are used in rotation (2). A forecasting method has been devised, based on the effect of temperature on ascospores, and this can assist in the optimal timing of sprays. In view of the seriousness of this disease, work is in progress to breed resistant banana cultivars (5).

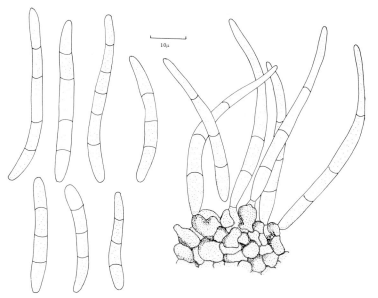

246 *Mycosphaerella musicola* (from CMI Descr. No. 414, Mulder & Holliday 1974)   Conidia and conidiophores

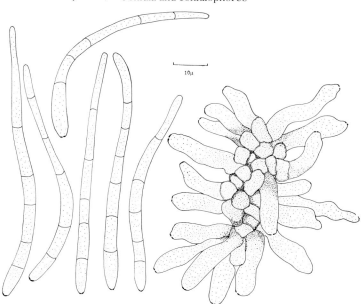

247 *Mycosphaerella fijiensis* (from CMI Descr. No. 413, Mulder & Holliday 1974)   Conidia and conidiophores

1. BUREAU E. (1985) Efficacité comparée de deux fongicides: le propiconazole et l'imazalil sur la cercosporiose à *Mycosphaerella musicola* du bananier en Guadeloupe. *Fruits* **40**, 315–319.
2. CRONSHAW D.K. (1984) Monitoring of sigatoka disease (*Mycosphaerella musicola*) in the Windward Islands for tolerance to fungicides. *Tropical Pest Management* **30**, 225–229.
3. MEREDITH D.S. (1970) Banana leaf spot disease (sigatoka) caused by *Mycosphaerella musicola* Leach. *Phytopathological Papers* No. 11, 147 pp.
4. PÉREZ VICENTE L., MOREIRA M. & MAURY F. (1983) Biología de la fase ascospórica de *Mycosphaerella musicola*, agente causal de la sigatoka en el plátano. *Agrotecnia de Cuba* **15**, 95–106.
5. STOVER R.H. (1980) Sigatoka leaf spots of bananas and plantains. *Plant Disease* **64**, 750–755 illus.

OTHER DISEASES of bananas

BLACK HEART, reported from Israel, is caused by *Fusarium moniliforme* Sheldon, the conidial state of *Gibberella fujikuroi* (Saw.) Ito. Infected fingers appear normal, but when cut longitudinally show a dark brown discoloration of the centre from tip-end to stem-end. The disease may induce premature ripening. Since infection occurs at flowering time, control is achieved by removal of the flower parts a few days after the bunch has 'shot' (page 104), followed immediately by application of a fungicide spray (2).

BLACK TIP, essentially a plantation disease, is incited by *Deightoniella torulosa* (Sydow) M.B. Ellis which also causes 'speckle' (q.v.). The fungus exists in many banana-growing areas but black tip is reported infrequently. The disease was first observed in Bermuda and later in Trinidad; it occurs in Mozambique during the rainy season (8). A slowly advancing black lesion appears at the flower-end of one or more fingers, characteristically affecting one side of the fruit in particular. The diseased area is bounded by a narrow grey or pale yellow margin; in old lesions the surface tends to rupture and under

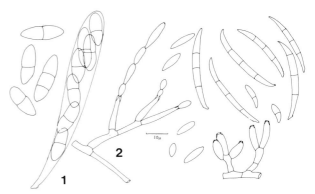

248 *Gibberella fujikuroi* (from CMI Descr. No. 22, Booth & Waterston 1964) 1 ascus and ascospores 2 micro- and macroconidia and conidiophores

moist conditions a pale brown mould develops. Infection by *Verticillium theobromae* may follow, converting the symptoms to those of 'cigar-end rot' (q.v.). Black tip can be controlled by improving hygiene and drainage in the plantation.

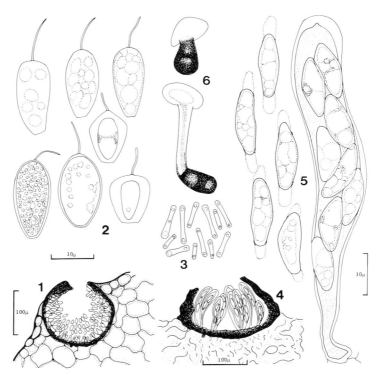

249 *Guignardia musae* (from CMI Descr. No. 467, Punithalingam & Holliday 1975) 1 pycnidium 2 conidia 3 spermatial cells 4 pseudothecium 5 ascus and ascospores 6 appressoria

BROWN SPOT or **DIAMOND SPOT** is caused by *Cercospora hayi* Calpouzos, sometimes accompanied by *Fusarium* species, and the disease has been reported from South and Central America and the Philippines. Raised yellow spots (3 to 5 mm across) appear on the peel of developing fruits. Infected cells are unable to expand as the fruit grows, and consequently in each lesion a longitudinal crack forms, surrounded by a yellow halo. As the lesion enlarges it takes on a diamond shape, and sometimes the pulp is exposed. More spots may appear during shipping and ripening. Disease incidence is reduced by prompt removal of hanging leaf trash from the plantation, especially during the rainy season. Full control is achieved with an early fungicide spray followed immediately by 'bagging' of the developing bunch.

DOTHIORELLA ROT has been reported from Israel (2), the causal fungus being *Botryosphaeria ribis* (**222**). Decay characteristically begins at the tip-end (from infected flower parts) and develops as in finger rot (q.v.). Minute black bodies (pycnidia) form in the skin, and whitish spores are exuded. A white to grey mould growth may also develop.

FRECKLE is a superficial but disfiguring blemish and is caused by *Guignardia musae* Racib. (conidial state: *Phyllosticta musarum* (Cooke) van der Aa). The disease can be severe in Hawaii, India, the Philippines and Taiwan, but is not common in the Americas. Initial symptoms are reddish-brown spots, usually circular and up to 2 mm in diameter, each lesion surrounded by a halo of dark green watersoaked tissue. Later, minute black spore-bearing bodies (pycnidia) project through the skin, making it rough to the touch. Since infection and symptom expression occur before the fruit is harvested, control depends on fungicide sprays in the plantation (7).

PHYTOPHTHORA FINGER ROT has been reported from South Africa, where the causal fungus was identified as *Phytophthora nicotianae* var. *parasitica* (page 64). The rot is firm and black, advancing rapidly from either the stalk-end or tip-end of the fingers. In advanced stages a white spore-bearing mould may form on the surface of the lesion. Although mild infection occurs occasionally in the plantation, this disease is only a problem in unhygienic packing stations, and can be avoided by frequent changes of washing water.

PINK MOULD ROT, caused by *Trichothecium roseum* (**341**), tends to be a superficial decay, following physical injury of the peel (9).

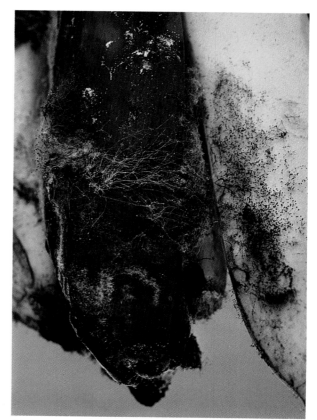

250 Rhizopus rot

RHIZOPUS ROT can cause losses in bananas kept at high temperatures, for example in local markets in India and Nigeria (1). It is caused by *Rhizopus stolonifer* (**401**), which is able to colonise bruised areas resulting from rough handling.

SOOTY MOULD is a superficial blemish noted on bananas from South and Central America and the

251 Sooty mould

117

252 Close-up of sooty mould

Philippines (10). The commonest sooty mould is *Cladosporium cladosporioides* (Fres.) de Vries. Symptoms consist of blackish-brown blotches, occurring characteristically on the concave side of the 'hand' and at the stalk-end of the fingers. Sooty moulds live on honeydew secreted by aphids and mealybugs, and control therefore depends on the use of insecticide sprays.

SPECKLE or SWAMP SPOT has been investigated in India (5) and Jamaica, where there have been sporadic severe outbreaks of the disease, caused by *Deightoniella torulosa* (which also causes 'black tip', q.v.). Symptoms develop before harvest and affected fruit should normally be rejected at the packing station. The spots are approximately 1 mm in diameter, reddish-brown to black with a dark green watersoaked halo, and they tend to be most profuse near the tips and on the inner sides of individual fingers. This pattern of distribution suggests that the infective spores are disseminated by rain, and indeed the disease is important only during wet seasons. Control methods include fungicide sprays and 'bagging' of developing bunches, though in some localities removal of leaf trash from the plantation may afford sufficient control.

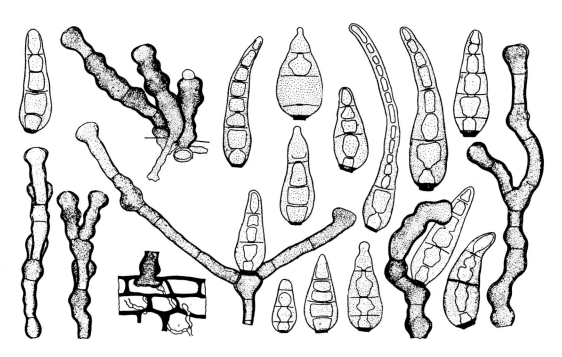

253 *Deightoniella torulosa* (from CMI Descr. No. 165, Subramanian 1968) Conidia and conidiophores x 500

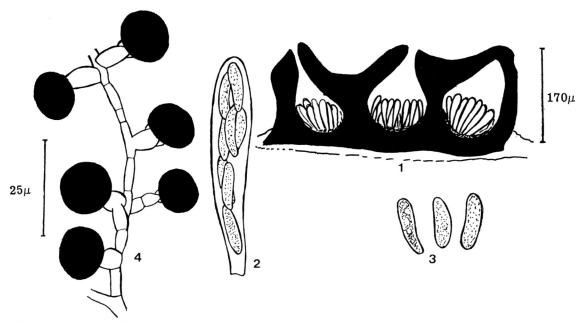

254 *Khuskia oryzae* (from CMI Descr. No. 311, Sivanesan & Holliday 1971) 1 perithecia 2 ascus 3 ascospores 4 conidiophore and conidia

SQUIRTER DISEASE is caused by *Nigrospora sphaerica* (Sacc.) Mason, sometimes considered synonymous with *Nigrospora oryzae* (Berk. & Broome) Petch, the conidial state of *Khuskia oryzae* Hudson. During the cool season the disease can be a problem in parts of Australasia where bananas are packed as single fingers. The fungus enters the cut stem-end and, although no symptoms are apparent in green fruit, rotting is evident soon after removal from the ripening room (3). A dark band forms a few cm from the stem-end, and if the finger is cut longitudinally the central portion is seen to have rotted from the stem-end. Eventually all the pulp is reduced to liquid, which is released if pressure is applied. The disease can be severe in chilled fruit requiring extended ripening periods. Packing the fruit as 'hands' or 'clusters' reduces the incidence of squirter disease. The fungus has developed a tolerance to certain systemic fungicides which used to provide control.

WATERY SOFT ROT, caused by *Sclerotinia sclerotiorum* (page 66), has been recorded on banana fruit in Costa Rica (6).

---

1 ADISA V.A. (1983) Storage rots of banana fruits in some Nigerian markets. *Fitopatologia Brasileira* **8**, 29–36.
2 CHORIN M. & ROTEM J. (1961) Experiments on the control of tip rot in banana fruits. *Israel Journal of Agricultural Research* **11**, 185–188.
3 FITZELL R.D. & ALLEN R.N. (1976) A postharvest decay of banana fruit not controlled by benzimidazole fungicides. *Australasian Phytopathological Society Newsletter* **5**(1), 7–8.
4 KAISER W.J. & LUKEZIC F.L. (1966) Influence of certain environmental conditions on spore dispersal and survival of *Cercospora hayi* from banana. *Phytopathology* **56**, 1290–1293.
5 KAPOOR I.J. & TANDON R.N. (1968) Pathological studies on *Deightoniella torulosa* (Syd.) Ell. causing leaf spot and fruit rot of banana (*Musa paradisiaca* L.). *Phytopathologische Zeitschrift* **63**, 337–340.
6 LAGUNA I.G. & SALAZAR L.G. (1984) Pudrición del fruto del banano causada por *Sclerotinia sclerotiorum* (Lib.) de Bary en Costa Rica. *Turrialba* **34**, 105–106 illus.
7 MEREDITH D.S. (1968) Freckle disease of banana in Hawaii caused by *Phyllostictina musarum* (Cke.) Petr. *Annals of Applied Biology* **62**, 329–340 illus.
8 NORONHA A. do R. (1970) Doenças da banana causadas por fungos em Moçambique. *Agronomía Moçambicana* **4**, 161–170 illus.
9 SRIVASTAVA M.P. & TANDON R.N. (1971) Postharvest diseases of banana in India. *Indian Phytopathology* **24**, 115–118 illus.
10 STOVER R.H. (1975) Sooty moulds of bananas. *Transactions of the British Mycological Society* **65**, 328–330 illus.

CHILLING INJURY of bananas

255

OCCURRENCE This is an important disorder of bananas, and can cause a lowering in market value or a total loss. It is induced by temperatures below the critical temperature, which is generally in the region of 12° to 14°C, depending on cultivar and other factors. Green fruit may suffer chilling injury in plantation, ship or port of discharge, while ripe fruit is liable to be damaged by open-air marketing during winter weather (1). Even a few hours at chilling temperatures can be sufficient to induce irreversible changes (2).

SYMPTOMS If green bananas are severely chilled there may be dark water-soaked areas on the peel. More commonly, however, the only indication of injury is a brown, under-peel discoloration, visible as streaks in a longitudinal cut or as a ring of brown 'dots' in a transverse section through a finger (5). In 'turning' or ripe fruit the general appearance is one of dullness, the peel having an almost greyish cast; however, the pulp may be unaffected. Severely chilled bananas eventually turn black and the pulp has an off-taste.

BIOLOGY Exposure to chilling temperatures results in abnormal respiration (3). Brown substances (polyphenols) accumulate in the vascular tissue within the peel, and ethylene production may increase as a result of stress (4). Susceptibility varies with cultivar, climate in the growing area, and the state of maturity of the banana, full fruit being more sensitive than threequarters-full fruit (2). Chilling temperatures are particularly damaging at the time when bananas are beginning to ripen, because of the complex biochemical changes which accompany the climacteric rise in respiration (page 14). Conversion of starch to sugar is inhibited, and normal ripening is unable to proceed (4).

CONTROL Since the banana is so sensitive to chilling temperatures, great care is necessary to protect the fruit at all stages of production and handling. In the cooler growing areas 'bagging' the developing bunches can provide insulation against low temperatures in the plantation. With regard to marine transport of bananas, it was established many years ago (5) that chilling injury is not caused by rapid cooling; the important proviso is that the air delivery temperature should not be allowed to fall below the critical value. Chilling injury at temperatures near 12°C can be prevented by the use of an oil-based dip or a coating which reduces moisture loss; a storage atmosphere close to saturation has the same effect. Cartons are designed so that, when the lid is placed on the base in one direction, the ventilation holes in the sides are occluded; this position is used when bananas are distributed in winter, the reverse position being used in warm weather when aeration is desirable.

1 JONES R.L., FREEBAIRN H.T. & McDONNELL J.F. (1978) The prevention of chilling injury, weight loss reduction and ripening retardation in banana. *Journal of the American Society for Horticultural Science* **103**, 219–221.
2 MATTEI A. (1978) La frisure (chilling) de la banane. *Fruits* **33**, 51–56.
3 MURATA T. (1969) Physiological and biochemical studies of chilling injury in bananas. *Physiologia Plantarum* **22**, 401–411 illus.
4 OLORUNDA A.O., MEHERIUK M. & LOONEY N.E. (1978) Some factors associated with the occurrence of chilling injury in banana. *Journal of the Science of Food and Agriculture* **29**, 213–218.
5 WARDLAW C.W. & McGUIRE L.P. (1930) The behaviour and diseases of the banana in storage and transport, with special reference to chilling. *Tropical Agriculture* **7**, 183–189.

OTHER DISORDERS of bananas

CORKY SCAB results from the activities of small insects (thrips) which, during hot dry weather, tend to migrate to the outer part of the developing hands. As they feed and lay eggs, the peel is scarred and, by the time the fruit has matured, affected areas have become greyish-brown, rough and corky. In seasons when thrips are prevalent, the use of an insecticide is recommended (4).

FINGER DROP, in which individual fingers become detached, is especially undesirable in markets where hands of bananas are displayed on hooks. The tendency is not confined to bananas with fingerstalk rot or crown rot (q.v.), but is also associated with rapid ripening precipitated by too high a temperature in the ripening room (3). Some of the newer banana clones (tetraploids), while possessing many favourable attributes, are susceptible to finger drop and should be ripened at slightly lower temperatures than are used for Cavendish and Valery (3).

HIGH TEMPERATURE INJURY At temperatures slightly above 30°C the peel may fail to turn yellow but the pulp nevertheless is capable of normal ripening. Between 35° and 40°C ethylene production is inhibited and ripening cannot be induced by ethylene from an external source (5). If bananas are held at 40°C for several days, they may be irreversibly damaged and fail to ripen on return to moderate temperature (1,5).

'SINKERS' are hands which, instead of floating, sink to the bottom of the de-latexing tanks. Even though they may later be retrieved, the inevitable abrasion results in heavy scarring of the peel. The

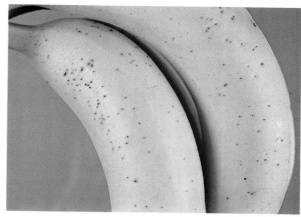

256 Corky scab

problem is reported to be seasonal, occurring for example in Honduras only during March, April and May. The buoyancy of the fruit is related to nutritional factors during its development, and the occurrence of 'sinkers' can be prevented by the application of potash to the soil (2).

SPLITTING of the peel is the result of ripening at high temperature in a saturated atmosphere, such as may develop within polyethylene packaging. It is sometimes observed in bananas which have ripened prematurely on board ship.

YELLOW PULP is a disorder in which the flesh has a honey-coloured tinge, and the fruit tends to ripen prematurely. The most probable cause is pre-harvest stress resulting from such factors as drought, nutritional deficiencies and sigatoka disease (page 114).

---

1 DICK E. & MARCELLIN P. (1985) Effets des températures élevées sur l'évolution des bananes après récolte. Tests prophylactiques. *Fruits* **40**, 781–784.
2 JOHNSON T.J. (1979) Effects of potassium on buoyancy of banana fruit. *Experimental Agriculture* **15**, 173–176.
3 NEW S. & MARRIOTT J. (1983) Factors affecting the development of 'finger drop' in bananas after ripening. *Journal of Food Technology* **18**, 241–250 illus.
4 TROCHOULIAS T., WRIGHT W.E. & LOEBEL M.R. (1984) Use of insecticides to reduce corky scab injury in bananas. *Tropical Agriculture* **61**, 193–195, 246.
5 YOSHIOKA H., UEDA Y. & OGATA K. (1978) [Effect of elevated temperatures on ripening of banana fruit.] *Journal of the Japanese Society of Food Science and Technology* **25**, 607–611.

Guavas

The **guava**, *Psidium guajava* L., is native to South America and the West Indies but, after the voyages of discovery in the 16th century, it was taken to other parts of the tropics and subtropics. It seeds so easily as to be regarded in some places as a weed, but it is an important tree crop in India, the USA (Florida and Hawaii), South Africa, Brazil, the Dominican Republic, Haiti, Cuba, Guyana, the Philippines and New Zealand (6). Recent research work has been carried out in India (4,7,8), Malaysia (10), Australia (2,6,9), Hawaii (1), Mexico (5) and the Ivory Coast (3). Most exports are by air.

The fruit is spherical to pyriform, with a persistent calyx at the distal end. The thin smooth skin encloses a granular pulp which is white to pink in colour (6). In the centre the flesh is soft, and in it are embedded numerous small hard seeds. Ripe guavas have a distinctive aroma and are nutritionally valuable for their high vitamin C content (6).

It has been established that guava is a climacteric fruit (page 14) (1,2). Fruits picked when almost ripe have the best eating quality but cannot be stored. Fruits for export are harvested mature green (2), various maturity indices being used to determine the optimal picking date (10). A pre-harvest ethephon spray may be applied to promote uniformity. Care during harvesting is necessary to prevent damage to the skin and consequent development of post-harvest rots (5). Ripening by means of calcium carbide (7) is generally less satisfactory than the use of an ethephon dip. Other post-harvest treatments may include immersion in a heated fungicide dip (8,9) and coating with wax (4). Packing in perforated polyethylene bags is advantageous (8) and confers a storage life of a week or so at 20°C and up to 3 weeks at 5°C (2). At lower temperatures guavas are at risk from chilling injury.

257

---

1 AKAMINE E.K. & GOO T. (1979) Respiration and ethylene production in fruits of species and cultivars of *Psidium* and species of *Eugenia*. *Journal of the American Society for Horticultural Science* **104**, 632–635.
2 BROWN B.I. & WILLS R.B.H. (1983) Post-harvest changes in guava fruit of different maturity. *Scientia Horticulturae* **19**, 237–243.
3 FOUQUÉ A. (1979) Quelques observations sur les goyaviers. *Fruits* **34**, 767–770 illus.
4 KARTAR SINGH & CHAUHAN K.S. (1983) Effect of post-harvest application of Waxol-0-12 and $KMnO_4$ on guava fruits. *Punjab Horticultural Journal* **23** (1/2), 38–42.
5 LAKSHMINARAYANA S. & MORENO RIVERA M.A. (1978) Enfermedades y desordenes en la producción y mercadeo de la guayaba mexicana. *Chapingo. Nueva Época (Mexico)* No. 9, 27–33 illus.
6 MENZEL C.M. (1985) Guava: an exotic fruit with potential in Queensland. *Queensland Agricultural Journal* **111**, 93–98 illus.
7 PRASAD A. & SHUKLA J.P. (1979) Studies on the ripening and storage behaviour of guava fruits (*Psidium guajava* L.). *Indian Journal of Agricultural Research* **13**, 39–42.
8 TANDON D.K., ADSULE P.G. & KALRA S.K. (1984) Effect of certain post-harvest treatments on the shelf life of guava fruits. *Indian Journal of Horticulture* **41**, 88–92.
9 WILLS R.B.H., BROWN B.I. & SCOTT K.J. (1982) Control of ripe fruit rots of guavas by heated benomyl and guazatine dips. *Australian Journal of Experimental Agriculture and Animal Husbandry* **22**, 437–440.
10 YUSOF S. & MOHAMED S. (1987) Physico-chemical changes in guava (*Psidium guajava* L.) during development and maturation. *Journal of the Science of Food and Agriculture* **38**, 31–39.

DISEASES of guavas

ANTHRACNOSE, caused by *Glomerella cingulata* (**268**), is the most important disease of guavas. Most reports are from India, but it has also been recorded in Nigeria (1), Australia, the Philippines and Puerto Rico. Control begins in the orchard, with destruction of prunings and use of a preventive fungicide spray. Post-harvest measures include a fungicide dip, irradiation (page 20) and refrigeration. Some guava cultivars are moderately resistant to the disease (9).

ASPERGILLUS ROT is caused by several species of *Aspergillus*, including *A. niger* (1) and *A. flavus* (which produces a toxin in infected tissue). The disease may be controlled by careful handling to prevent physical injury, and by the use of a post-harvest fungicide dip.

BACTERIAL SOFT ROT, caused by *Erwinia* sp., is responsible for serious losses of guavas in Nigeria (1).

BOTRYODIPLODIA ROT is caused by *Botryodiplodia theobromae* (**197**). Guavas are susceptible to invasion at all stages of maturity, and there may be severe losses in the orchard from a dry stem-end rot of developing fruits. Post-harvest decay of ripe fruits usually takes the form of a soft, watery breakdown, resulting from infection via wounds or through the stem-end (1,8).

CANKER, caused by *Pestalotiopsis psidii* (Pat.) Mordue, has been recorded in Australia, Malaysia, India, Mozambique, Zambia, Nigeria, Venezuela, Ecuador and Puerto Rico. Infection occurs during moist weather (6). Each canker has a crater-like depression in the centre and a rough raised margin.

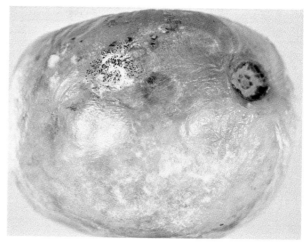

258 Canker

Distinctive black spores are produced on mature lesions (8). Although remaining fairly shallow, the cankers are unsightly and reduce market value. Control measures include orchard sprays, a post-harvest fungicide dip and low-temperature storage.

CHOANEPHORA ROT has been recorded in Nigeria (1). It is a soft rot, caused by *Choanephora cucurbitacearum* (**34**).

CURVULARIA ROT is a serious post-harvest disease in India, and is attributed to *Curvularia tuberculata* (4). Injuries permit infection. Circular honey-yellow spots develop on the skin, and their centres turn brown; secondary organisms generally follow in the wake of this fungus. Control is by careful handling and storage at 10°C (4).

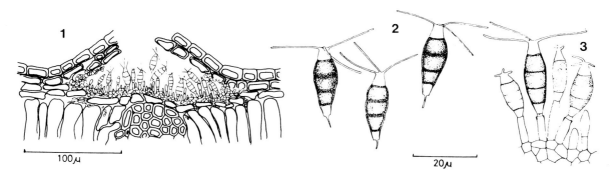

259 *Pestalotiopsis psidii* (from CMI Descr. No. 515, Mordue 1976) 1 acervulus 2 mature conidia 3 conidiogenous cells and immature conidia

260 Cylindrocladium rot

CYLINDROCLADIUM ROT, caused by *Cylindrocladium scoparium* Morgan, is one of the most prevalent types of decay in Malaysia. Infection occurs in the field, via insect injuries, and rotting continues after harvest.

DOTHIORELLA ROT, caused by the conidial state of *Botryosphaeria berengaria* (page 176), has been reported from Hawaii. The brownish lesions develop only on wounded fruits, and it is thought that insect punctures permit infection.

FUSARIUM ROT has been recorded on guavas in Malaysia, and can be serious in India, where the causal organism is *Fusarium solani* (page 150). Whitish spots develop, later darkening and becoming soft. A dry rot is reported from Nigeria, the causal fungus being either *F. equiseti* (**296**) or *F. oxysporum* (1).

MACROPHOMA ROT can cause serious post-harvest losses in India, where it has been identified as *Macrophoma allahabadensis* Kapoor & Tandon. Injured fruits are particularly liable to infection. Lesions are initially watersoaked and brownish. Affected areas become covered with mould, which may be orange-cinnamon to dark ivy-green at first, later dark brown to black and enveloping the entire fruit. Minute black bodies (pycnidia) form in the skin of infected fruits. Control is by careful handling and storage at 10°C.

MACROPHOMINA ROT, caused by *Macrophomina* sp. (page 273), has been recorded on guavas in Venezuela (2). Lesions are initially dark and circular; eventually the entire fruit undergoes a dry rot. Minute black bodies (pycnidia) are formed.

MUCOR ROT is reported from Hawaii, where it can affect mature green fruit even before harvest (3). Watersoaked lesions become covered with a yellowish mould bearing black spore-heads, and the decay is accompanied by a yeasty odour. The causal fungus, *Mucor hiemalis* (page 246), is essentially a wound parasite. Infection occurs when the skin is punctured during egg-laying by fruit flies contaminated with spores. Disease incidence is greatly reduced if fallen fruits (often infected) are promptly collected and destroyed (3).

PHOMA ROT occurs in India, and is attributed to *Phoma psidii* P. Henn (8). Infection usually occurs at the distal end in the region of the persistent calyx. Alternatively a lesion may develop on the side of the fruit. The affected area turns brown and becomes sunken, so that the water-soaked margin remains slightly elevated. There is sparse mould growth and an abundance of minute dark bodies (pycnidia). Some guava cultivars are moderately resistant to the disease.

PHOMOPSIS ROT is reported from India, Australia and Malaysia, and has been attributed to *Phomopsis destructum* Rao, Agrawal & Saksena (7) and *P. psidii* de Camara (8). Infection usually occurs at or near the stem-end or at the distal end in the region of the persistent calyx; a stylar-end ring rot has been described (5). Affected areas of the skin are water-soaked, becoming dark brown, soft and wrinkled. Numerous minute bodies (pycnidia) eventually develop, from which pinkish spores exude (7). Serious losses can occur in the orchard (5,7) and after harvest (8), and control measures include pruning and destruction of debris, together with preventive fungicide sprays.

PHYLLOSTICTA ROT is reported from Malaysia and India, where it affects guavas in the field and in transit. The fungus has been identified as *Guignardia psidii* Ullasa & Rawal (conidial state: *Phyllosticta psidiicola* (Petrak) van der Aa) and it induces sunken spots on ripening fruits. There may be a growth of

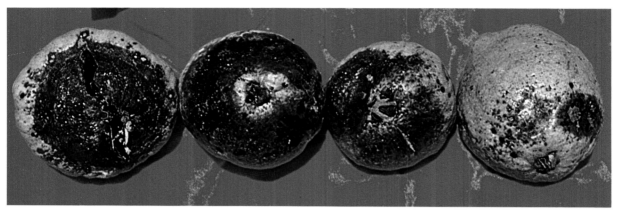

261 Phomopsis rot

dark mould showing concentric rings, and both sexual and asexual stages are produced in culture. In Hawaii, *G. musae* is associated with a physiological disorder of the blossom end.

PHYTOPHTHORA ROT is caused by *Phytophthora citricola* in Hawaii and *P. nicotianae* var. *parasitica* (page 64) in Cuba and India. It is a firm decay and, under humid conditions, the affected area becomes covered with a cottony white mould which spreads into adjacent fruits. The fungi are soil-borne; their spores are dispersed by rainsplash and can cause direct infection of uninjured fruits. Control is by spraying the trees and drenching the soil with a suitable fungicide.

RHIZOPUS ROT, caused by *Rhizopus stolonifer* (page 230), has been reported from India (8) and Hawaii. Both *R. stolonifer* and *R. oryzae* occur on guavas in Nigeria (1). Mature green or ripe fruit may be infected before harvest, especially under humid conditions in closely planted orchards. Wounding is a pre-requisite for infection, and in Hawaii fruit flies are thought to play an important role in transmission and infection. Lesions are soft and watery, and become covered with coarse mould strands bearing black spore heads (8).

SCLEROTIUM ROT, caused by *Corticium rolfsii* (**85, 87**), has been observed in India on market fruits and also on low-hanging fruits in the orchard (10). Injury is unnecessary for infection, and decay spreads rapidly during transit and storage at tropical temperatures. The persistent sclerotia are difficult to eradicate from the soil.

THIELAVIOPSIS ROT, caused by the conidial state of *Ceratocystis paradoxa* (page 156), occurs on guavas in India. The fungus induces a soft rot and the disease has caused extensive post-harvest losses.

---

1 ADISA V.A. (1985) Fruit rot diseases of guava (*Psidium guajava*) in Nigeria. *Indian Phytopathology* **38**, 427–430.
2 DÍAZ P., C. & RONDÓN A. (1971) Un tipo de *Macrophomina* patógeno en frutos de guayaba. *Agronomía Tropical* **21**, 111–118 illus.
3 ITO P.J., KUNIMOTO R. & KO W.H. (1979) Transmission of *Mucor* rot of guava fruits by three species of fruit flies. *Tropical Agriculture* **56**, 49–52.
4 KAPOOR I.J. (1983) Pathological and biochemical studies on *Curvularia* rot of guava. *Zeitschrift für Pflanzenkrankheiten und Pflanzenschutz* **90**, 591–598.
5 LIM T.K. & RAZAK A.R. (1986) Studies on a *Phomopsis* rot of bagged guava (*Psidium guajava*) fruits in Malaysia. *Fitopatologia Brasileira* **11**, 227–236 illus.
6 RAMASWAMY G.R., SOHI H.S. & GOVINDU H.C. (1984) Studies on spore germination in *Pestalotia psidii*, the causal organism of guava canker. *Indian Journal of Mycology and Plant Pathology* **14**, 289.
7 RAO D.P.C, AGRAWAL S.C. & SAKSENA S.B. (1976) *Phomopsis destructum* on *Psidium guajava* fruits in India. *Mycologia* **68**, 1132–1134 illus.
8 SRIVASTAVA M.P. & TANDON R.N. (1969) Postharvest diseases of guava in India. *Plant Disease Reporter* **53**, 206–208 illus.
9 TANDON I.N. & SINGH B.B. (1970) Studies on anthracnose of guava and its control. *Indian Phytopathology* **22**, 322–326.
10 ULLASA B.A. & RAWAL R.D. (1985) A new fruit rot of guava caused by *Sclerotium rolfsii*. *Current Science* **54**, 470–471 illus.

Litchis

Litchis or lychees (*Litchi chinensis* Sonn.) are native to southern China. The evergreen trees thrive in subtropical regions which have abundant summer rains and a dry frostless winter. Research has been carried out in India (1,7), China (3,4), Australia (9), South Africa (8,10), Madagascar (5), the USA (2) and Hawaii (6).

The fruit possesses a tough dry reddish peel or shell, and inside is a single seed surrounded by firm translucent white flesh. Litchis do not show an increase in respiration when exposed to ethylene (4), and appear to be non-climacteric. Hence they must be allowed to mature fully on the tree, the appropriate harvest date being calculated by counting days from flowering. Although the flesh of the litchi evolves very little ethylene, the peel produces substantial amounts (4). Related to this (4) is the phenomenon of **physiological browning**, the most important disorder of harvested litchis (1,2,10). Unless preventive measures are taken, litchis can lose their fresh red colour within a day or so of harvest (9). Internal quality is unaffected but market value is much reduced (2). Browning involves the formation of dark substances (polyphenols) which are produced in response to injury or desiccation (6). Control strategies include 'bagging' of developing fruits to prevent insect damage (8,9), prompt hydro-cooling of harvested fruits (5), post-harvest fumigation with sulphur dioxide (10) (though this is not always successful) (6), and reduction of moisture loss by the use of plastic film packaging (1,2,9). Recommended storage temperatures vary, but chilling injury may occur after a month at 5°C. The red colour is said to be retained best at 7° to 10°C (2), though lower temperatures have been used successfully. Storage for a week or so at ambient temperature (20° to 30°C) is feasible for packaged litchis, provided that the fruits have received prompt post-harvest treatment with a heated fungicide (9). Waxing may be beneficial (7) or ineffective (6).

Diseases are numerous, and control begins in the orchard with preventive measures against fruit-piercing insects (8,9). If the fruit-shell is punctured or if it cracks as a result of sun-scorch, then juices may exude, providing favourable conditions for mould growth (7,8).

262 Physiological browning

ASPERGILLUS ROT is caused by several species, including *Aspergillus flavus* and *A. niger* (page 160). The disease has been recorded in Australia (9), India (7) and South Africa (8).

PESTALOTIOPSIS ROT, caused by *Pestalotiopsis* sp. (page 123), occurs in India (7), and the author has observed it on litchis imported into the UK from South Africa.

PERONOPHYTHORA ROT, caused by *Peronophythora litchii* Chen ex Ko *et al.*, occurs in China and Taiwan, where it may reach epidemic proportions following rainy periods. The fungus, which is intermediate between *Peronospora* and *Phytophthora*, produces swimming spores (zoospores) capable of causing direct penetration of the intact peel (3).

SOUR ROT, caused by *Geotrichum candidum* (page 78), is reported from India and Australia.

YEASTY ROTS are characterised by the exudation of frothy juice having an odour of fermentation (8).

Other micro-organisms associated with decayed litchis include *Botryodiplodia theobromae* (page 81), *Colletotrichum gloeosporioides* (page 130), *Rhizopus oryzae* (page 230) and many others (7,8,9).

Two stages of pestalotiopsis rot

---

1 GAUR G.S., BAJPAI P.N. & AZAD C.S. (1978) Post harvest physiology of litchi fruits. 1. *Progressive Horticulture* **10**(3), 63–77.
2 HATTON T.T., REEDER W.F. & KAUFMAN J. (1966) Maintaining market quality of fresh lychees during storage and transit. *Marketing Research Report of the United States Department of Agriculture* No. 770, 9 pp.
3 HO H.H., LU J.Y. & GONG L.Y. (1984) Observations on asexual reproduction by *Peronophythora litchii*. *Mycologia* **76**, 745–747 illus.
4 JIANG J.P., SU M.X. & LEE P.M. (1986) [The production and physiological effects of ethylene during ontogeny and after harvest of litchi fruits.] *Acta Phytophysiologica Sinica* **12**, 95–103.
5 MOREUIL C. (1973) Quelques observations et essais sur le litchi. *Fruits* **28**, 637–640 illus.
6 NIP W.-K. (1988) Handling and preservation of lychee (*Litchi chinensis*, Sonn) with emphasis on colour retention. *Tropical Science* **28**, 5–11.
7 PRASAD S.S. & BILGRAMI R.S. (1973) Investigations on diseases of 'litchi'. 3. Fruit rots and their control by post-harvest treatments. *Indian Phytopathology* **26**, 523–527.
8 ROTH G. (1963) Postharvest decay of litchi fruit (*Litchi chinensis* Sonn.). *Technical Communication of the Department of Agricultural Technical Services, South Africa* **11**, 16 pp illus.
9 SCOTT K.J., BROWN B.I., CHAPLIN G.R., WILLCOX M.E. & BAIN J.M. (1982) The control of rotting and browning of litchi fruit by hot benomyl and plastic films. *Scientia Horticulturae* **16**, 253–262.
10 SWARTS D.H. (1985) Sulphur content of fumigated South African litchi fruit. *Information Bulletin of the Citrus and Subtropical Fruit Research Institute, South Africa* No. 155, 18–20.

Mangoes

The **mango**, *Mangifera indica* L., is one of the most highly prized of tropical fruits. At the same time, the 'king of fruits' is so abundant that it has also been called 'the apple of the Orient'. Mangoes have been cultivated in the Malay peninsula since ancient times and subsequently spread throughout the tropics. Nowhere, however, is mango of such overriding importance as in India; it has been grown in the sub-continent for over four thousand years and the number of cultivars is uncountable. Current world production is about 15 million tonnes, of which fully two thirds is grown in India. Other producers include, in order, Mexico, Pakistan, Brazil, Indonesia, China, Haiti, the Philippines, Madagascar and Tanzania. Export trades include Brazil to the UK, Burkina Faso to France, and India and Pakistan to the Arabian Gulf.

Mangoes grow on large, spreading, evergreen trees, the choicest fruits being derived from grafted, as opposed to seedling, trees (9). Irregular bearing is often a problem, a heavy crop being followed the next season by a light crop. The fruit is generally oval in shape, with a slight 'beak'. The skin is smooth and waxy, and the flesh encloses a large, rather flat stone containing a single seed.

Mangoes show a climacteric pattern of respiration (page 14) and, during ripening, the flesh (which may be somewhat fibrous) becomes soft and juicy with a rich aromatic flavour. The onset of ripening is accompanied by a five-fold increase in heat production, which has important repercussions on post-harvest handling (6). If mangoes are to be stored or exported, they should be harvested at the mature-green stage before the start of the climacteric (7). For most mango cultivars, fruits are judged to be mature once the 'shoulders' have risen above the stem-end (3). If mangoes are picked mature-green and promptly placed under refrigeration they may be kept for several weeks. Recommended storage temperatures are generally within the range of 10° to 15°C for green fruit, 7° or 8°C for ripe fruit. The optimal storage temperature is dictated by susceptibility to chilling injury (q.v.) which varies between cultivars. If harvested mangoes are left at tropical temperatures then ripening ensues within a few days (6).

A single picking inevitably comprises fruits of differing maturities (7). Uniform ripening is sometimes induced by treatment with ethylene (also produced by mangoes themselves). Either the gas is introduced into the ripening room (4) or, alternatively, the fruit is dipped in ethephon, a compound which produces ethylene. Less satisfactory results have been reported for calcium carbide, a compound which when moistened produces acetylene gas (1). Acetylene can stimulate the ripening process in early-picked fruit but, although an attractive skin colour develops, the flesh tends to taste flat.

With regard to diseases, control of anthracnose should begin in the orchard. Great care is needed during harvesting and handling, in order to reduce the likelihood not only of latex stain (from the cut stem) but also stem-end rots and lateral rots. Control of post-harvest decay can be achieved by hot water treatment (to inactivate deep-seated infections), incorporating a fungicide (2). Irradiation has also been used successfully (10). Waxing can delay ripening, though there have been several reports of the development of off-flavours. Design of packaging is important (1). The ideal pack is a case or carton containing a single layer of mangoes cushioned in soft material (2). Larger packs may be used, provided that there is good provision for ventilation of the contents (3). If the lid can be made with thick sides overhanging the inner by half the carton depth, then, in a block stow, small channels remain between the inners of adjacent cartons, allowing for airflow.

In view of the substantial increase in heat production which accompanies the onset of ripening (before any visual change is apparent), it is essential that fruit for overseas shipment be cooled as soon as possible, preferably within 24 hours of harvest (8). If more than a small proportion of mangoes have entered the climacteric by the time of loading into refrigerated space, there is a risk that heat production by the mangoes may impose too great a burden on the cooling machinery. If this happens, more and more fruits are triggered into ripening, in an atmosphere depleted of oxygen and rich in carbon dioxide and ethylene. The temperature rises further and ripening proceeds in abnormal fashion. The result is usually described as 'boiled' mangoes. A similar effect may result from attempts to store mangoes in atmospheres modified to reduce respiration rate (6). However, storage in controlled atmosphere, with careful regulation of temperature and carbon dioxide concentration, is reported to be effective in delaying ripening and reducing decay (5). Storage in perforated polyethylene bags is effective in minimising moisture loss, and an ethylene absorbent is beneficial.

Much work has been done on the mango, and conference proceedings include those of the 1st Symposium on Mango and Mango Culture (Delhi 1969) published in 1972 as *Acta Horticulturae* No. 24, 282 pp.; the 2nd Symposium (Bangalore 1985) summarised in *Fruits* **41**, 269–274; and the 1st Australian Mango Research Workshop (Cairns, Queensland 1984) published by CSIRO in 1986.

265

1. CAYGILL J.C., COOKE R.D., MOORE D.J., READ S.J. & PASSAM H.C. (1976) The mango (*Mangifera indica* L.). Harvesting and subsequent handling and processing: an annotated bibliography. *Report of the Tropical Products Institute (London)* G 107, 124 pp.
2. CHAPLIN G.R., LEE S.K., LEONG P.C. & WILLCOX M. (1982) Postharvest and marketing attributes of North Australian mangoes in Singapore and Sydney. *Singapore Journal of Primary Industries* **10**, 80–83.
3. CHEEMA G.S., KARMARKAR D.V. & JOSHI B.M. (1950) Investigations on the cold storage of mangoes. *Indian Journal of Agricultural Science* **20**, 259– 325 illus.
4. FUCHS Y., ZAUBERMAN G., YANKO U. & HOMSKY S. (1975) Ripening of mango fruits with ethylene. *Tropical Science* **17**, 211–216.
5. KANE O. & MARCELLIN P. (1979) Effets de l'atmosphère contrôlée sur la conservation des mangues (variétés Amélie et Julie). *Fruits* **34**, 123–129.
6. KRISHNAMURTHY S. & SUBRAMANYAM H. (1973) Pre- and post-harvest physiology of the mango fruit – a review. *Tropical Science* **15**, 167–193.
7. LAKSHMINARAYANA S. (1973) Respiration and ripening patterns in the life cycle of the mango fruit. *Journal of Horticultural Science* **48**, 227–233.
8. MATTERN F., PENNOCK W. & VALLES L., S. (1972) Supplying the New York market with high-quality Puerto Rican mangoes. *Journal of the Agricultural University of Puerto Rico* **56**, 1–10.
9. SINGH L.B. (1960) *The mango*. London: Leonard Hill, 438 pp illus.
10. SPALDING D.H. & REEDER W.F. (1986) Decay and acceptability of mangos treated with combinations of hot water, imazalil, and γ-radiation. *Plant Disease* **70**, 1149–1151.

See also:

HUBBERT C.A., JOHNSON G.I., MUIRHEAD I.F. & LEDGER S.N. (1987) Postharvest handling of mangoes – resource booklet. *Queensland Department of Primary Industries (Horticulture Branch) Publication* No. RQT87011, 27 pp.

PRINSLEY R.T. & TUCKER G. (Eds) (1987) *Mangoes – a review*. London: Commonwealth Science Council, 159 pp.

ANTHRACNOSE of mangoes caused by
Glomerella cingulata (Stonem.) Spauld. & v. Schrenk
Conidial state: *Colletotrichum gloeosporioides* (Penz.) Sacc.

OCCURRENCE Anthracnose is an important cause of loss in mangoes from all parts of the world, for example the USA (6,8), the West Indies (3,10), South Africa (4), India (9), Malaysia (5), the Philippines (1) and Australia (2,7).

SYMPTOMS Symptoms typically appear as the fruit ripens (10), and in countries where mangoes are a familiar commodity the development of black spots and streaks is taken as an indication that the fruit is approaching prime eating quality. On the other hand, in many markets the disfigurement can cause substantial reduction in value (4). The lesions are at first superficial but later become sunken and, under humid conditions, spore-masses are produced which change in colour from salmon-pink to dark brown (2). Eventually the flesh beneath the lesions undergoes a soft decay.

BIOLOGY The fungus may produce the sexual stage (perithecia giving rise to ascospores) as well as the asexual stage (acervuli giving rise to conidia). Perithecia have been found on dead branches and fallen leaves after periods of cold weather, but ascospores are not thought to play an important role in infection (2,3). In wet seasons acervuli are produced in abundance on all parts of the tree, and infection of fruits is chiefly by conidia washed down by rain (hence the characteristic 'tearstain' pattern which may eventually result); if moisture persists for several hours the fungus is capable of direct penetration of the intact skin of young fruits (10).

The infection characteristically remains quiescent, however, until the onset of ripening. Thus the crop may be in apparently perfect condition at the time of harvest, symptoms appearing only during marketing (1).

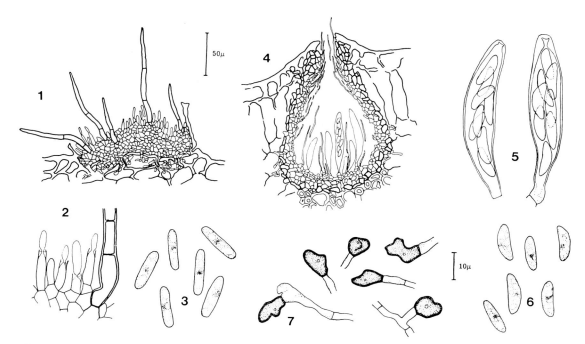

268 *Glomerella cingulata* (from CMI Descr. No. 315, Mordue 1971) 1 acervulus 2 conidiophores 3 conidia 4 perithecium 5 asci 6 ascospores 7 appressoria

CONTROL Anthracnose is controlled by both pre- and post-harvest measures. In wet seasons frequent orchard spraying may be necessary, from the time of flowering (since blossom infection can result in failure to set fruit) up till a few weeks before picking the fruit (9). The most usual post-harvest treatment involves immersion in hot water containing a fungicide; time and temperature are critical if anthracnose is to be controlled without heat injury to the fruit, and effective treatments vary with cultivar and locality (4,7,8). Following such a treatment, irradiation is sometimes used to give added protection (8).

---

1 DAQUIOAG V.R. & QUIMIO T.H. (1979) Latent infection in mango caused by *Colletotrichum gloeosporioides*. *Philippine Phytopathology* **15**, 35–46.
2 FITZELL R.D. & PEAK C.M. (1984) The epidemiology of anthracnose disease of mango: inoculum sources, spore production and dispersal. *Annals of Applied Biology* **104**, 53–59.
3 GORDILLO Q., L. & HERNANDEZ L., E. (1980) Forma sexual y medios de cultivo para *Colletotrichum gloeosporioides*, patógeno del mango en Cuba. *Ciencias de la Agricultura* **7**, 11–17.
4 JACOBS C.J., BRODRICK H.T., SWARTS H.D. & MULDER N.J. (1973) Control of post-harvest decay of mango fruit in South Africa. *Plant Disease Reporter* **57**, 173–176.
5 LIM T.K. & WAI O.C. (1986) Effects of selected fungicides *in vitro* on the mango anthracnose pathogen, *Colletotrichum gloeosporioides*. *Fitopatologia Brasileira* **11**, 67–74.
6 McMILLAN R.T. (1984) Control of mango anthracnose with foliar sprays. *Proceedings of Florida State Horticultural Society* **97**, 344–345.
7 MUIRHEAD I.F. (1976) Post-harvest control of mango anthracnose with benomyl and hot water. *Australian Journal of Experimental Agriculture and Animal Husbandry* **16**, 600–603 illus.
8 SPALDING D.H. & REEDER W.F. (1986) Decay and acceptability of mangos treated with combinations of hot water, imazalil, and γ-radiation. *Plant Disease* **70**, 1149–1151.
9 TANDON I.N. & SINGH B.B. (1968) Control of mango anthracnose by fungicides. *Indian Phytopathology* **21**, 212–216.
10 WARDLAW C.W. & LEONARD E.R. (1936) The storage of West Indian mangoes. *Memoir of the Low Temperature Research Station, Trinidad* No. 2, 47 pp.

BACTERIAL BLACK SPOT of mangoes caused by
Xanthomonas campestris pv. *mangiferaeindicae* (Patel, Moniz & Kulkarni) Dye

OCCURRENCE Bacterial black spot can be a serious disease of mangoes grown in hot humid climates, for example parts of South Africa (4,6), Australia (7), Taiwan (5), India (1,3,8,10) and South America (9).

SYMPTOMS Small dark lesions appear on the skin, sunken in the centre and raised at the borders. The pustules often form star-shaped cracks (5,7) and if the skin is raised then bacterial slime may be visible (2).

BIOLOGY Mango trees growing in poor soils are predisposed to the disease (7). During wet weather (6) bacteria are washed down the tree and are capable of infecting sound immature fruits via lenticels and stomata (10). It is only when the fruits ripen, however, that the typical spots develop (4). Infection of fruits also occurs through injuries resulting from abrasion by twigs (6).

CONTROL Primary infection of fruits may be reduced by spraying the trees with a bactericide immediately after fruit set (1,3,4), and post-harvest development of symptoms can be partially prevented by dipping the fruit after harvest (8). Long-term control measures involve the uprooting and burning of infected trees, and replacement with resistant varieties (8). In the meantime, work is in progress to devise a means of biological control, in which a bacterial antagonist is sprayed on to the trees.

---

1 BOSE P.C. & CHOKHA SINGH (1980) Chemical control of bacterial canker of mango. *Pesticides* **14**(8), 30–31.
2 DANIEL J.F., GARDAN L., LUISETTI J. & PRUNIER J.P. (1975) Identification de l'agent responsable de la maladie des taches noires de la mangue (*Mangifera indica*). *Fruits* **30**, 625–630 illus.
3 KISHUN R. & SOHI H.S. (1984) Control of bacterial canker of mango by chemicals. *Pesticides* **18**(11), 32–33.
4 KOTZÉ J.M., VILJOEN N.M. & STEYN P.L. (1976) Epidemiology of bacterial spot of mangoes. *Citrus & Subtropical Fruit Journal* No. 511, 5–7.
5 LIAO C.H. (1975) Studies on mango fruit spot. 2. Pathogenicity. *Bulletin of the Taiwan Agricultural Research Institute* No. 32, 62–66.
6 MANICOM B.Q. (1986) Factors affecting bacterial black spot of mango caused by *Xanthomonas campestris* pv. *mangiferaeindicae*. *Annals of Applied Biology* **109**, 129–135.
7 MOFFETT M.L., PETERSEN R.A. & WOOD B.A. (1979) Bacterial black spot of mango. *Australasian Plant Pathology* **8**, 54–56.
8 RAO A.P., RAO V.V.R. & PANDIT S.V. (1977) A note on mango canker in Andhra Pradesh. *Indian Journal of Mycology and Plant Pathology* **7**, 71.
9 ROBBS C.F., PONTE J.J. da & SALES M. da G. (1978) Nota sobre *Xanthomonas mangiferaeindicae* no nordeste do Brasil. *Fitopatologia Brasileira* **3**, 215–218.
10 SHEKHAWAT G.S., PATEL P.N. & RAJPAL SINGH (1979) Histology of bacterial canker affecting *Mangifera indica*. *Phytopathologische Zeitschrift* **95**, 12–23.

BLACK MOULD ROT of mangoes caused by
Aspergillus niger v. Tieghem

271

OCCURRENCE Aspergillus black mould rot causes occasional problems in the Philippines and elsewhere (3), but is of particular importance in India (1,2,4,5). The author has observed it on mangoes imported into the UK from Venezuela.

SYMPTOMS Greyish or pale brown spots may appear anywhere on the fruit surface, coalescing into dark brown or black lesions which are soft and sunken, and later covered with a sooty mass of black spores (**307**). The fungus may also cause a stem-end rot. On mature lesions there may be formation of spherical resting bodies (sclerotia) about 1 mm in diameter, initially white and later dark brown.

BIOLOGY The fungus persists in the soil on decaying vegetation, and its spores are blown about in the atmosphere. Fruits on the tree may be attacked and reduced to dried-up 'mummies' covered with black spores. Most infections occur during and after harvest, however, when the fungus penetrates wounds or the cut stem (4). Disease development is optimal at temperatures around 30°C.

CONTROL Black mould may be reduced by orchard fungicide sprays (5) and by post-harvest treatments with hot water containing a fungicide (1,2). It is important to maintain the bath at the correct temperature, since if the solution is too hot the fruit is injured and black mould is able to penetrate the lenticels. Careful handling is necessary at all stages to prevent mechanical damage to the fruit. Storage at temperatures between 10° and 15°C prevents development of black mould rot.

---

1 BHARGAVA S.N. & SINGH A.P. (1975) *Aspergillus* rot of mango: its control. *Indian Journal of Horticulture* **32**, 190.
2 PANDEY R.S., DWIVEDI D.K., SHUKLA D.N. & BHARGAVA S.N. (1980) Two new fungicides for the control of *Aspergillus* rot of mango. *National Academy of Science Letters* **3**, 263–264.
3 QUIMIO T.H. & QUIMIO A.J. (1974) A new fruit rot of mango in the Philippines. *Philippine Phytopathology* **10**, 85–87.
4 SRIVASTAVA O.P. (1968) Soft rot of Desi mango fruit and the pathogenicity of the isolated micro-organisms. *Journal of the Indian Botanical Society* **47**, 328–329 illus.
5 SUBRAMANYAM H., NARAYANA MOORTHY N.V., LAKSHMINARAYANA S. & DALAL V.B. (1972) Control of fungal spoilage in Alphonso mangoes by pre-harvest application of fungicides. *Acta Horticulturae* No. 24, 224–226.

BOTRYODIPLODIA ROT of mangoes caused by
Physalospora rhodina (Berk. & Curt.) Cooke
Conidial state: *Botryodiplodia theobromae* Pat.

272

OCCURRENCE   This serious disease has been reported from most countries where mangoes are grown, including India (3), the Philippines (2), Egypt (4), Puerto Rico (1) and Brazil (5).

SYMPTOMS   A dark brown to black lesion with a clearly defined margin may be produced anywhere on the fruit, although the disease often takes the form of a stem-end rot (q.v.). Decayed tissue is soft and moist, and in mature lesions the skin may be pimpled with minute black bodies (pycnidia) just visible to the naked eye.

BIOLOGY   The fungus occasionally produces the sexual stage (perithecia giving rise to ascospores) but exists chiefly in the asexual state (pycnidia giving rise to conidia) (1). The spore-bearing bodies form in the dead bark of mango trees, and fruits become contaminated with conidia washed down by rain. Infection occurs only through wounds or via the cut stem at harvest time. Rotting is extremely rapid at temperatures between 30° and 35°C but insignificant at 10°C (3,4).

CONTROL   The disease is controlled by subjecting harvested fruit to a hot fungicide treatment (5). It is also beneficial to ensure careful handling in order to reduce the incidence of wounding. A storage temperature of 10°C delays development of the disease.

---

1 ALVAREZ G., L.A. & LOPEZ G., J. (1971) Gummosis, dieback, and fruit rot disease of mango (*Mangifera indica* L.) caused by *Physalospora rhodina* (B. & C.) Cke. in Puerto Rico. *Journal of Agriculture of the University of Puerto Rico* **55**, 435–450 illus.
2 HALOS P.M. & DIVINAGRACIA G.G. (1970) Histopathology of mango fruits infected by *Diplodia natalensis*. *Philippine Phytopathology* **6**, 16–28.
3 PRASAD S.S. & SINHA A.K. (1981) *National Academy of Science Letters* **4**, 345–347.
4 RAGAB M.M., SABET K.A. & DAWOOD N.A. (1971) *Botryodiplodia theobromae* Pat., the cause of fruit-rot and die-back of mango in A.R.E. *Agricultural Research Review, Cairo* **49**(3), 81–97 illus.
5 SAMPAIO V.R., BARBIN D. & DEMÉTRIO C.G.B. (1981) Controle em póscolheito das podridões da manga. *Revista de Agricultura, Piracicaba* **56**, 173–182.

STEM-END ROTS of mangoes

OCCURRENCE Stem-end rots of mango are very common and may be caused by several fungi. In India (4), Sri Lanka (3), the Philippines (2) and the USA (5) the most important species is *Botryodiplodia theobromae* (q.v.). Species of *Phomopsis*, including *P. citri*, have been reported from India, Trinidad and the USA (5), and species of *Dothiorella* from Trinidad, Australia and Nigeria (1). *Colletotrichum gloeosporioides* (q.v.) may also cause a stem-end rot. The author has recorded both dothiorella rot and phomopsis rot on mangoes imported from Puerto Rico and Venezuela.

SYMPTOMS A soft dark brown lesion appears on the fruit, typically at the stem-end (2,3). (The same fungi may also cause lateral rots.) Mature lesions may show minute black bodies (pycnidia) projecting through the skin, and microscopic examination of these is necessary for identification of the causal fungus.

BIOLOGY All these fungi can persist in the orchard by colonising dead wood, usually in the asexual state (acervuli or pycnidia giving rise to conidia). The spores are washed down by rain and contaminate the fruits. Infection usually occurs at or shortly after harvesting, when the cut stem is invaded (4).

CONTROL Disease incidence can be reduced by prompt and careful handling, and by attention to hygiene. For example, it has been reported that when harvested mangoes were transported on a bed of banana leaves there was an increase in fruit infection by *B. theobromae* because the packing material was heavily contaminated with spores (2). Infection can be prevented by the use of clean containers, and by immediate application of a fungicidal paste to the cut stem of harvested fruit (4). Alternatively the fruits can be immersed in a hot water bath containing a fungicide. It is preferable to use a different chemical from those which may have been used as orchard sprays, in order to delay the appearance of tolerant fungal strains (5). Storage at 10°C delays the development of stem-end rot.

---

1 ALABI R.O. (1986) Disease of mango (*Mangifera indica*) fruits caused by *Dothiorella dominicana*. *International Biodeterioration* **22**, 21–25.
2 ALICBUSAN R.V. & SCHAFER L.A. (1958) Diplodia rot of mango. *Philippine Agriculturist* **42**, 319–322 illus.
3 FERNANDO M. (1937) A note on a soft rot of stored mangoes caused by *Botryodiplodia theobromae* Pat. *Tropical Agriculturist* **89**, 381–387 illus.
4 PATHAK V.N. & SRIVASTAVA D.N. (1969) Epidemiology and prevention of *Diplodia* stem-end rot of mango fruits. *Phytopathologische Zeitschrift* **65**, 164–175.
5 SPALDING D.H. (1982) Resistance of mango pathogens to fungicides used to control post-harvest diseases. *Plant Disease* **66**, 1185–1186.

OTHER DISEASES of mangoes

ALTERNARIA BLACK SPOT, caused by *Alternaria alternata* (**339, 340**), is of particular importance in Israel, where it is more common than anthracnose (9). The disease has also been recorded on mangoes in Egypt and in India. The fungus is capable of penetrating the lenticels of immature fruits but symptoms do not appear until after harvest (9). Control is achieved by pre-harvest sprays coupled, if necessary, with a post-harvest dip.

BACTERIAL ROT may be caused by species of *Erwinia*. Lesions are dark brown to black, sunken in the centre with raised borders, and sometimes exude gum. In Venezuela two species have been identified, *E. carotovora* (page 272) is greyish white in culture, while *E. herbicola* (Löhnis) Dye is brilliant yellow (5).

BLUE MOULD ROT, caused by *Penicillium cyclopium* Westling, has caused losses in India. It is reported that irradiation treatment fails to control the decay but a post-harvest fungicide dip is partially effective (8).

CHARCOAL ROT is caused by *Macrophomina phaseolina* (page 273) and has been recorded on mangoes in India (4). The fungus usually invades via the cut stem, and eventually the entire fruit darkens. Under humid conditions pale fluffy mould growth is visible on the lesion, and minute black bodies (pycnidia) form in the skin.

HENDERSONIA ROT is caused by the fungus *Hendersonia creberrima* Sydow & Butler and has been responsible for heavy losses in South African mangoes shipped under refrigeration to Europe (3). Pale brown lesions appear on the fruit and

275 Alternaria black spot

characteristically only a portion of each lesion darkens. The disease is controlled by post-harvest treatments with a heated fungicide.

MACROPHOMA ROT is caused by *Macrophoma mangiferae* Hingorani & Sharma and has been reported from India. Lesions are at first circular and watersoaked, later irregular and dark brown. Minute black bodies (pycnidia) eventually develop. Injury is not necessary for infection, and fruits are vulnerable at all stages of maturity. The disease is primarily a foliage blight and it is therefore advisable to burn fallen leaves which may be harbouring the fungus (6).

MUCOR ROT is a soft rot caused by a species of *Mucor*, and has caused losses in Bangladesh (7). The fungus may cause a lateral rot or, more usually, a stem-end rot. After a few days at temperatures above 20°C the entire fruit is decayed and evil-smelling.

PESTALOTIOPSIS ROT, recorded in India, is caused by *Pestalotiopsis mangiferae* (P. Henn.) Stey. Primary infection is via wounds, but decay can spread into sound healthy fruits in store (10). At temperatures around 10°C the fungus makes little progress, but at 25° to 30°C it can cause substantial losses.

PHYLLOSTICTA ROT is caused by the asexual stage of *Guignardia mangiferae* Roy. Irregular brown spots can develop even on unripe fruits in the orchard.

274 Hendersonia rot

PHYTOPHTHORA ROT, caused by *Phytophthora nicotianae* var. *parasitica* (page 64), has been recorded in Australia. In a humid atmosphere the dark lesions become covered with white spores.

POWDERY MILDEW is caused by the fungus *Oidium mangiferae* Berthet which is widespread in mango-producing countries. The disease is characterised by an apparently superficial white mould growth bearing abundant spores. Control is by application of orchard fungicide sprays.

RHIZOPUS ROT, caused by *Rhizopus oryzae* or by *R. stolonifer* (page 230), is a soft rot which develops rapidly at temperatures around 25°C. At places where the skin splits there is copious development of coarse white mould supporting a mass of black spore-heads. Infection takes place after harvest, typically via injuries. Packing material such as wood-wool may harbour these fungi. Control measures include careful handling, strict hygiene, and cool storage.

SCAB is caused by the fungus *Elsinoë mangiferae* Bitanc. & Jenkins, and results in a superficial blemish on the fruit, rather similar to scab of avocado (q.v.). At first the lesions are greyish-brown with dark irregular margins; later the centres may become cracked and corky (2). The disease is controlled by orchard fungicide sprays.

SOOTY BLOTCH is caused by *Gloeodes pomigena* (page 202) while **SOOTY MOULD** includes species of *Capnodium* and *Meliola*; all give rise to a dark blemish on the skin. If the mango tree has suffered an insect infestation resulting in deposits of honeydew secretions, these fungi are able to live on the honeydew and make superficial growth over the surface of the fruit. Control of the fungi is therefore achieved by controlling the insects (1).

276 Rhizopus rot

1 AHMED M., AHMED M., BALUCH M.A. & NAHEED R. (1981) Sooty mould on mango plants and its relationship with leafhoppers and climatic factors in Karachi – Pakistan during 1978–79. *Pakistan Journal of Scientific and Industrial Research* **24**, 140–144.
2 BITANCOURT A.A. & JENKINS A.F. (1946) A verrugose da mangueira. *Arquivos do Instituto Biológico, São Paulo* **17**, 205–228 illus.
3 BRODRICK H.T. & WESTHUIZEN G.C.A. (1976) *Hendersonia creberrima*, the cause of soft brown rot of mango in South Africa. *Phytophylactica* **8**, 13–15 illus.
4 DHAWAN S. (1982) Charcoal rot of mango. *Current Science* **51**, 140 illus.
5 GUEVARA M., Y., RONDÓN G., A. & SOLÓRZANO R. (1980) Bacteriosis del mango (*Mangifera indica* L.) en Venezuela. 1. Sintomatología e identificación. *Agronomía Tropical* **30**, 65–76 illus.
6 HINGORANI M.K., SHARMA O.P. & SOHI H.S. (1960) Studies on blight disease of mango caused by *Macrophoma mangiferae*. *Indian Phytopathology* **13**, 137–143 illus.
7 JOARDER G.K. & KHATUN M. (1975) Fungal spoilage of mangoes. *Bangladesh Journal of Scientific and Industrial Research* **10**, 233–236.
8 PALEJWALA V.A. & MODI V.V. (1985) Post harvest spoilage of mangoes by *Penicillium cyclopium* and its control by gamma-radiation and a fungicide. *Phytopathologische Zeitschrift* **112**, 63–68.
9 PRUSKY D., FUCHS Y. & YANKO U. (1983) Assessment of latent infections as a basis for control of postharvest disease of mango. *Plant Disease* **67**, 816–818 illus.
10 TANDON R.N., SISODIA U.S. & BILGRAMI K.S. (1955) Pathological studies of *Pestalotia mangiferae*. *Proceedings of the Indian Academy of Science* B **42**, 219–225 illus.

CHILLING INJURY of mangoes

277

OCCURRENCE This disorder is the main limiting factor in the storage life of mangoes. Reports on the critical temperature, below which there is a risk of injury, vary widely even for the same cultivar, from 7° C (5), through 10° C (1,4) to above 13° C (2,3). Investigations have been carried out on Indian cultivars e.g. Alphonso (4), and Bangalora (2), West Indian cultivars e.g. Julie (5), and Florida cultivars e.g. Kent (3).

SYMPTOMS Skin symptoms, which may not develop until the fruit is returned to higher temperature, include brown blotches (2), a greyish discoloration and/or pitting (1). Ripening occurs unevenly, and normal development of aroma and flavour is impaired. There is a marked increase in susceptibility to decay (4).

BIOLOGY Low temperature prevents normal metabolism of mango tissue, and the complex biochemical reactions associated with respiration proceed along alternative pathways (1). Susceptibility to chilling injury is greatest in fruits which are picked immature (5), and this could explain some of the discrepancies between the findings of different investigators. Mature green fruits (4) and ripening fruits (2) are also susceptible, whereas those which have already ripened can better withstand low temperatures (4). Mangoes which are cooled slowly to 'chilling temperatures' tend to show less severe symptoms than those cooled rapidly (2). Thus, increased resistance to chilling injury can be conferred by a programme of gradual adaptation to cold, involving a stepwise reduction in temperature spread over several days (4).

CONTROL Chilling injury being such a complex phenomenon, it is necessary to carry out local trials in order to determine the optimal storage temperature. Factors influencing susceptibility include cultivar, growing conditions, maturity when picked, and postharvest handling techniques. If definitive information is lacking, it is recommended that mangoes be stored at 13°C. Storage at 10°C may be feasible if they are first held at 20°C for 1 day, then at 15°C for 2 days, before being placed at 10°C (4). Cold-adapted mangoes, stored at 10°C and then ripened, can subsequently be held at even lower temperatures, such as 4° to 7°C (4).

---

1 FAROOQUI W.A., SATTAR A., DAUD K. & HUSSAIN M. (1985) Studies on the postharvest chilling sensitivity of mango fruit (*Mangifera indica* L.). *Proceedings of Florida State Horticultural Society* **98**, 220–221.
2 MUSA S.K. (1974) Preliminary investigations on the storage and ripening of 'Totapuri' mangoes in the Sudan. *Tropical Science* **16**, 65–73.
3 SAUCEDO V., C., ESPARZA T., F. & LAKSHMINARAYANA S. (1977) Effect of refrigerated temperatures on the incidence of chilling injury and ripening quality of mango fruit. *Proceedings of Florida State Horticultural Society* **90**, 205–210 illus.
4 THOMAS P. & OKE M.S. (1983) Improvements in quality and storage of 'Alphonso' mangoes by cold adaptation. *Scientia Horticulturae* **19**, 257–262.
5 THOMPSON A.K. (1971) The storage of mango fruits. *Tropical Agriculture* **48**, 63–70.

OTHER DISORDERS of mangoes

BLACK TIP is characterised by a hard dark area at the distal end of the fruit, and causes losses in many parts of India (5). The disorder is initiated during the early stages of fruit development, and the cause has been traced to the acidic fumes from nearby brick kilns. Mango cultivars differ in their sensitivity. If it is not feasible to reduce pollution, developing fruits can be protected by alkaline sprays which neutralise the fumes.

INTERNAL BREAKDOWN is a serious problem in the cultivar Alphonso grown in India. Large fruits are especially susceptible. Symptoms are visible only when the ripe fruit is cut in half; affected flesh is pale in colour and soft or spongy. Cavities may be present and there is an odour of fermentation. One type of internal breakdown can occur if harvested fruits are left in the sun.

JELLY SEED is described from South Africa, affecting the cultivars Zill and Sensation (3), and from Australia on several cultivars including Tommy Atkins. The condition exists at harvest time but develops further as the fruit ripens. If affected fruit is cut, watery translucent tissue is visible around the stone towards the stem-end, later spreading and conferring an over-ripe appearance. The only means of control is early harvesting (which may, however, result in immature fruit being offered for sale) or a change of cultivar.

SOFT NOSE has been reported from Florida on Kent, Haden and Sensation, and from the Canary Islands on several cultivars (1). Fruits harvested green are affected less than those harvested ripe (1). Symptoms start to develop as the fruit approaches maturity on the tree. The distal end softens while the shoulders of the fruit remain firm. In the most

278 Jelly seed

severe form of the disorder, affected tissue may become a spongy grey mass extending through much of the fruit. The tendency towards soft nose is aggravated by heavy nitrogen fertilisation and alleviated by high calcium levels.

SPONGY TISSUE is the name given to a serious disorder of Alphonso mangoes in India, characterised by the development of a spongy texture in part or all of the pulp. There is a sour odour and an off-taste. In Malaysia several local cultivars are prone to this disorder (4). Control is by picking the fruits early and hastening the ripening process by means of a post-harvest ethylene treatment (2).

STEM-END BREAKDOWN, described on numerous cultivars in Australia, is a disorder in which a distinct cavity forms between the stone and the stem-end.

Clearly it is probable that some of these disorders are one and the same.

---

1 GALÁN S., V., FERNANDEZ G., D. & CALVO R. (1984) Incidence of "soft-nose" on mangoes in the Canary Islands. *Proceedings of Florida State Horticultural Society* **97**, 358–360.
2 LAD B.L., GUNJATE R.J. & SALVI M.J. (1985) Effect of post harvest ethephon dipping of fruits on occurrence of spongy tissue disorder in Alphonso mango. *Indian Journal of Plant Physiology* **28**, 85–87.
3 LELYVELD L.J. van & SMITH J.H.E. (1979) Physiological factors in the maturation and ripening of mango (*Mangifera indica* L.) fruit in relation to the jelly-seed physiological disorder. *Journal of Horticultural Science* **54**, 283–287.
4 LIM T.K. & KHOO K.C. (1985) *Diseases and disorders of mango in Malaysia*. Kuala Lumpur: Tropical Press, 101 pp illus.
5 PAL R.N. & CHADHA K.L. (1980) Black-tip disorder of mango – a review. *Punjab Horticultural Journal* **20**, 112–121 illus.

See also:

WAINWRIGHT H. & BURBAGE M.B. (1989) Physiological disorders in mango (*Mangifera indica* L.) fruit – a review. *Journal of Horticultural Science* **64**, 125–135.

Papayas

The **papaya**, *Carica papaya* L., is also known as papaw and pawpaw. The plant originated in Central America and, in the 16th century and subsequently, was carried to all parts of the tropical world. Currently the ten leading producer countries are Brazil (6), Mexico, Indonesia, India (10), Zaire, the Philippines, China, Peru, Colombia and Mozambique, and total world production is about 3 million tonnes. There is an important export trade between Hawaii and the US mainland (1,2,4). Consignments are received in Europe from several countries, including South Africa, the West Indies (5) and Brazil.

The plant is a soft-wooded tree, possessing a slender, generally unbranched stem topped by a crown of large leaves. The papaya is usually dioecious (individual plants bear either male flowers or female flowers) but there are also hermaphrodite types (4). Pollination results in the development of globular, pyriform or elongated fruits which hang from the main stem. Botanically they are classed as 'berries' but may be up to 50 cm in length. The smooth flesh surrounds a central cavity containing a mucilaginous mass of seeds.

Papayas are climacteric fruits (page 14) and, for long-distance transport, must be harvested mature green (7). Judging the correct stage is difficult, and in any batch of harvested fruits there will be a range of maturities, some fruits perhaps incapable of normal ripening (8). Care is necessary during harvesting, to minimise injury and to prevent latex (which oozes out of the cut stem-end) from disfiguring the fruit.

Post-harvest treatments may include a hot water

spray or bath, a fungicide dip, and irradiation (1,2). The main fungal diseases are anthracnose and black rot (3,4,9,10). It may also be necessary to ensure that the treatments are sufficient to eradicate insect larvae in order to comply with quarantine regulations of the importing country. Packing the fruits in perforated polyethylene bags serves to minimise moisture loss (5); the cartons in which they are packed should possess suitable ventilation holes.

Ripening can be precipitated by exposure to ethylene gas (8). On the other hand, for maximum extension of storage life it is advantageous to remove ethylene produced by the fruit itself (7). The optimal storage temperature is generally given as 7°C for ripe and 10°C for green fruits, if chilling injury (q.v.) is to be avoided. Studies in Malaysia, however, indicated a preferred storage temperature of 20°C, since papayas kept for more than 7 days at 15°C and below failed to ripen (7).

---

1. AKAMINE E.K. & GOO T. (1977) Respiration, ethylene production, and shelf-life extension in irradiated papaya fruit after storage under simulated shipping countries. *Technical Bulletin of Hawaii Agricultural Experiment Station* No. 93, 12 pp.
2. ALVAREZ A.M. & NISHIJIMA W.T. (1987) Postharvest diseases of papaya. *Plant Disease* **71**, 681–686 illus.
3. FROSSARD P. (1969) Les maladies du papayer. Les maladies fongiques. *Fruits* **24**, 473–482.
4. HUNTER J.E. & BUDDENHAGEN I.W. (1972) Incidence, epidemiology and control of fruit diseases of papaya in Hawaii. *Tropical Agriculture* **49**, 61–71 illus.
5. LEE G.R., PROCTOR F. & THOMPSON A.K. (1973) Transport of papaya fruits from Trinidad to Britain. *Tropical Agriculture* **50**, 303–306.
6. MEDINA J.C., GARCIA J.L.M. & SALOMÓN E.A.G. (1980) Mamão: da cultura ao processamento e commercialização. *Frutas Tropicais* No. 7, 244 pp illus.
7. NAZEEB M. & BROUGHTON W.J. (1978) Storage conditions and ripening of papaya 'Bentong' and 'Taiping'. *Scientia Horticulturae* **9**, 265–277.
8. RODRÍGUEZ A.J., GUADALUPE R. & IGUINA de G., L.M. (1974) The ripening of local papaya cultivars under controlled conditions. *Journal of Agriculture of the University of Puerto Rico* **58**, 184–196.
9. SIMMONDS J.H. (1965) Papaw diseases. *Queensland Agricultural Journal* **91**, 666–677 illus.
10. SRIVASTAVA M.P. & TANDON R.N. (1971) Post-harvest diseases of papaya. *PANS* **17**, 51–54 illus.

See also:

CAPPELLINI R.A., CEPONIS M.J. & LIGHTNER G.W. (1988) Disorders in apricot and papaya shipments to the New York Market, 1972–1985. *Plant Disease* **72**, 366–368.

ANTHRACNOSE of papayas caused by
Glomerella cingulata (Stonem.) Spauld. & v. Schrenk
Conidial state: *Colletotrichum gloeosporioides* (Penz.) Sacc.

280

OCCURRENCE   Anthracnose is an important disease in most papaya-growing countries, including Hawaii (3,4,6), the Caribbean region (1), East Africa (10), West Africa (2), India (5), Taiwan (9), Singapore (7) and Australia (8).

SYMPTOMS   Symptoms appear only as the fruit ripens, and may not be apparent at the time of harvest. Brown sunken spots develop on the fruit surface, enlarging into water-soaked lesions which may support colonies of salmon-pink spores, sometimes in concentric rings (1,10). The flesh beneath is at first soft, but rotting is limited in extent and the perimeter of the lesion becomes hard and black as the fruit resists further decay. Sometimes there are multiple small lesions of the 'chocolate spot' type (6) which only occasionally develop into typical anthracnose lesions.

BIOLOGY   The fungus may produce the sexual stage (perithecia giving rise to ascospores) but much more common is the asexual stage (acervuli giving rise to conidia) (5). Spores are produced on the dying petioles of the lower leaves, and are released into the atmosphere by rainsplash. Air currents carry the spores to developing fruits, and if conditions remain moist for some hours the fungus develops a swelling (appressorium) from which an infection thread penetrates the skin (3). There is then a period of quiescence; further fungal development takes place only when the fruit ripens (1,8). The disease is most severe after hot wet weather (9).

CONTROL   Orchard hygiene is important, and fallen fruits and leaves should be collected and destroyed (2). The chief means of control is a carefully timed fungicide spray programme (7), followed by a post-harvest hot water treatment (9) or a fungicidal wax application (4). Hypobaric (low pressure) storage can be effective in delaying the process of infection without damaging the fruit.

---

1 BAKER R.E.D., CROWDY S.H. & McKEE R.K. (1940). A review of latent infection caused by *Colletotrichum gloeosporioides* and allied fungi. *Tropical Agriculture* **17**, 128–132.

2 BRUN J. (1952) Sur quelques parasites responsables des anthracnoses de la papaye. *Fruits* **7**, 271–272 illus.

3 CHAU K.F. & ALVAREZ A.M. (1983). A histological study of anthracnose on *Carica papaya*. *Phytopathology* **73**, 1113–1116.

4 COUEY H.M. & FARIAS G. (1979). Control of post-harvest decay of papaya. *HortScience* **14**, 719–721.

5 DASTUR J.F. (1920). *Glomerella cingulata* (Stoneman) Spauld. & v. Sch. and its conidial forms, *Gloeosporium piperatum* E. & E. and *Colletotrichum nigrum* E. & Hals., on chillies and *Carica papaya*. *Annals of Applied Biology* **6**, 245–268 illus.

6 DICKMAN M.B. & ALVAREZ A.M. (1983). Latent infection of papaya caused by *Colletotrichum gloeosporioides*. *Plant Disease* **67**, 748–750 illus.

7 LIM G. & TANG M.C. (1984) Anthracnose diseases of papaya fruits in Singapore. *International Journal of Tropical Plant Diseases* **2**, 191.

8 SIMMONDS J.H. (1941). Latent infection in tropical fruits discussed in relation to the part played by species of *Gloeosporium* and *Colletotrichum*. *Proceedings of the Royal Society of Queensland* **52**, 92–120.

9 TSAI W.H. (1969). Studies on ecology and physiology of papaya anthracnose and its control. *Journal of Taiwanese Agricultural Research* **18**; 51–57.

10 WALLACE G.B. & WALLACE M.M. (1948). Diseases of papaw and their control. *East African Agricultural Journal* **13**, 240–244 illus.

BLACK ROT of papayas caused by
Mycosphaerella caricae H. & P. Sydow
Conidial state: *Phoma caricae-papayae* (Tarr) Punith.

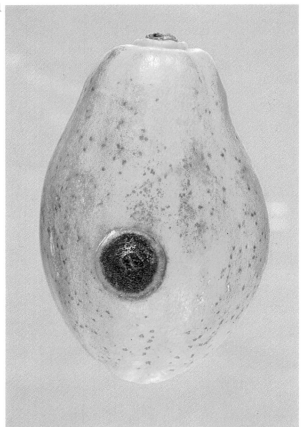

OCCURRENCE This disease is important in Hawaii (3,5), Brazil (1), Australia (9), India (4), East Africa (7) and South Africa (2).

SYMPTOMS A dry firm dark rot extends into the fruit from the stem-end (2, 6). Alternatively, small water-soaked spots may appear anywhere on the fruit, later becoming dark, sunken and irregular in shape (4).

BIOLOGY In India and Australia severe rotting can occur in uninjured developing fruits during the rainy season (4,9), whereas in Hawaii infection generally takes place via the cut stem at the time of harvest (8). The fungus produces both the sexual stage (perithecia giving rise to ascospores) and the asexual stage (pycnidia giving rise to conidia), and either may be responsible for infections in the orchard (3,10). Both types of spore-bearing bodies may develop on leaves but usually only pycnidia are to be found on fruits (8). The optimal growth temperature for the fungus is approximately 30°C (4).

CONTROL In wet growing areas it is necessary to apply fungicide sprays in the orchard from the time of fruit set (1,4,10). Post-harvest measures include a hot water bath or spray (3,5) which may be combined with fungicide (1) or irradiation treatment (2).

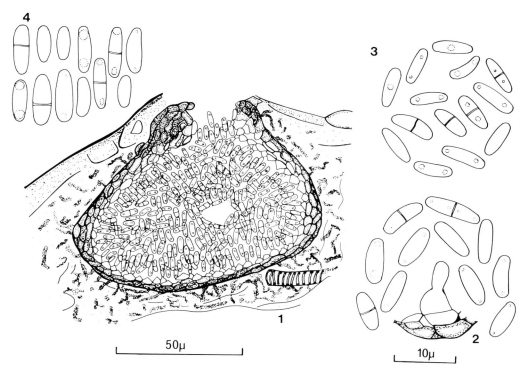

283 *Phoma caricae* (from CMI Descr. No. 634, Punithalingam 1979) 1 pycnidium 2 part of pycnidial wall, conidiogenous cell and conidia 3, 4 conidia

---

1 BOLKAN H.A., CUPERTINO F.P., DIANESE J.C. & TAKATSU A. (1976). Fungi associated with pre- and post-harvest fruit rots of papaya and their control in Central Brazil. *Plant Disease Reporter* **60**, 605–609.

2 BRODRICK H.T., THOMAS A.C., VISSER F. & BEYERS M. (1976). Studies on the use of gamma irradiation and hot water treatments for shelf life extension of papayas. *Plant Disease Reporter* **60**, 749–753 illus.

3 CHAU K.F. & ALVAREZ A.M. (1979). Role of *Mycosphaerella* ascospores in stem-end rot of papaya fruit. *Phytopathology* **69**, 500–503.

4 CHOWDHURY S. (1950). A fruit rot of papaya (*Carica papaya* L.) caused by *Ascochyta caricae* Pat. *Transactions of the British Mycological Society* **33**, 317–322.

5 COUEY H.M., ALVAREZ A.M. & NELSON M.G. (1984). Comparison of hot water spray and immersion treatments for control of post-harvest decay of papaya. *Plant Disease* **68**, 436–437.

6 HINE R.B., TRUJILLO E.E. & STANGHELLINI M.E. (1964). Stem-end rot and other fruit rots of papaya. *Hawaii Farm Science* **13**, 5–6.

7 HOPKINS J.C.F. (1938). A black rot of papaw fruit caused by *Phoma caricina* sp. nov. *Proceedings and Transactions of the Rhodesian Scientific Association* **35**, 128.

8 HUNTER J.E. & BUDDENHAGEN I.W. (1972). Incidence, epidemiology and control of fruit diseases of papaya in Hawaii. *Tropical Agriculture* **49**, 61–71 illus.

9 SIMMONDS J.H. (1965). Papaw diseases. *Queensland Agricultural Journal* **91**, 666–677 illus.

10 ULLASA B.A., SOHI H.S. & GANAPATHY K.M. (1974). *Ascochyta* leaf spot of papaya and its perfect state. *Indian Journal of Mycology and Plant Pathology* **4**, 218–219.

PHYTOPHTHORA ROT of papayas caused by
Phytophthora palmivora (Butler) Butler

OCCURRENCE This fungus causes a serious disease of papayas which has received particular attention in Hawaii (1,5,6,7), East Africa (10), India (9) and Taiwan (4). An epidemic of fruit and stem rot has been reported from Australia, after a cyclone (3).

SYMPTOMS Water-soaked spots appear on the fruit surface, soon to be covered with off-white mould supporting copious quantities of spores (5). Lesions may occur on any part of the fruit, including the stem-end (9).

BIOLOGY The fungus persists in the soil by means of thick-walled resting spores (chlamydospores) (5) and possibly also sexual spores (oospores). During wet weather the asexual stage assumes importance, and thin-walled lemon-shaped sporangia are dispersed by wind-blown rain (3,6); in wet conditions liberated zoospores are capable of causing direct infection of uninjured papaya fruits (8). If young fruits are infected they shrivel and fall to the ground, creating an abundant source of spores capable of initiating new infections in developing fruits (9,10). Recent infections may be undetectable at the time of harvest.

CONTROL The roots of papaya plants are also subject to attack, but invasion occurs only when they are young; thus, for new plantings in infested soil the disease may be controlled by filling the planting holes with virgin soil (7). Orchard fungicide sprays are essential to protect fruit in wet growing areas (1,4), and prompt hot water treatment is usually effective in preventing subsequent decay (2).

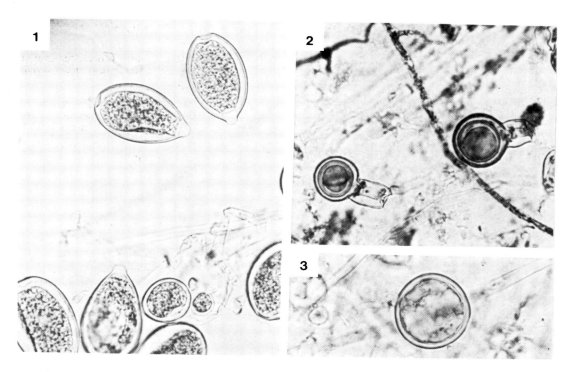

286 *Phytophthora palmivora* (from CMI Descr. No. 831, Stamps 1985) 1 sporangia 2 antheridia and oogonia 3 chlamydospore x 500

1 ALVAREZ A.M. & NELSON M.G. (1982). Control of *Phytophthora palmivora* in papaya orchards with weekly sprays of chlorothalonil. *Plant Disease* **66**, 37–39.
2 ARAGAKI M., KIMOTO W.S. & UCHIDA J.Y. (1981). Limitations of hot water treatment in the control of *Phytophthora* fruit rot of papaya. *Plant Disease* **65**, 744–745.
3 HAMILL S.D. (1987) Fruit rot of papaw caused by *Phytophthora palmivora* in Queensland. *Australasian Plant Pathology* **16**(1), 22.
4 HUANG T.H., CHENG D.W. & LEU L.S. (1976). *Phytophthora* fruit and root rot of papaya in Taiwan. *Plant Protection Bulletin, Taiwan* **18**, 293.
5 HUNTER J.E. & BUDDENHAGEN I.W. (1969). Field biology and control of *Phytophthora parasitica* on papaya (*Carica papaya*) in Hawaii. *Annals of Applied Biology* **63**, 53–60.
6 HUNTER J.E. & KUNIMOTO R.K. (1974). Dispersal of *Phytophthora palmivora* sporangia by wind-blown rain. *Phytopathology* **64**, 202–206.
7 KO W.H. (1982). Biological control of *Phytophthora* root rot of papaya with virgin soil. *Plant Disease* **66**, 446–448 illus.
8 PARRIS G.K. (1942). *Phytophthora parasitica* on papaya (*Carica papaya*) in Hawaii. *Phytopathology* **32**, 314–320.
9 SRIVASTAVA M.P. & TANDON R.N. (1971). Post-harvest diseases of papaya. *PANS* **17**, 51–54 illus.
10 WALLACE G.B. & WALLACE M.M. (1948). Diseases of papaw and their control. *East African Agricultural Journal* **13**, 240–244 illus.

RHIZOPUS ROT of papayas caused by
Rhizopus stolonifer (Ehrenb. ex Fr.) Lind

OCCURRENCE This fungus can cause a destructive rot in papayas from all producing areas, and losses have been reported from Hawaii (1,2), the Philippines (3), India (5), Pakistan (4) and Israel (2).

SYMPTOMS The fungus causes a soft watery rot, and under warm humid conditions there is copious development of a coarse white mould bearing black spore-heads visible to the naked eye.

BIOLOGY Spores of this fungus are ubiquitous, but sound uninjured papayas are resistant to invasion; initial infection is generally via injuries sustained during harvesting and handling (4). Infected fruit rapidly undergoes collapse, and leakage of juices results in heavy contamination and hence infection of neighbouring fruits (3).

CONTROL Careful harvesting and handling are necessary to prevent cuts and bruises to the skin (4). A post-harvest hot water dip is important (1), and its efficacy is increased by the addition of a suitable fungicide (3,4). Storage at 10°C serves to inhibit development of decay (5).

---

1 AKAMINE E.K. & ARISUMI T. (1953). Control of post-harvest storage decay of fruits of papaya (*Carica papaya* L.) with special reference to the effect of hot water. *Proceedings of the American Society for Horticultural Science* **61**, 270–274.
2 HUNTER J.E., BUDDENHAGEN I.W. & KOJIMA E.S. (1969). Efficacy of fungicides, hot water and gamma irradiation for control of post-harvest fruit rots of papaya. *Plant Disease Reporter* **53**, 279–284.
3 QUIMIO T.H., PORDESIMO A.N. & QUIMIO A.J. (1975). Control of papaya fruit rots by post-harvest dip in thiabendazole. *Philippine Agriculturist* **59**, 7–11.
4 SARWAR M. & KAMAL M. (1971) Studies on fruit rot of papaya caused by *Rhizopus oryzae*. *Pakistan Journal of Scientific and Industrial Research* **14**, 234–236 illus.
5 TANDON R.N. & MISHRA A.N. (1969). Fruit rot diseases of *Carica papaya* and *Musa paradisiaca* caused by *Rhizopus stolonifer*. *Indian Phytopathology* **22**, 334–341.

STEM-END ROTS of papayas

OCCURRENCE Stem-end rots probably occur wherever papayas are grown. In Hawaii the main cause of stem-end rotting is *Mycosphaerella* sp. (see black rot), but a *Phomopsis* sp. is occasionally responsible, and *Phytophthora palmivora* can also attack in this way (3). In the Philippines *Botryodiplodia theobromae* is associated with stem-end rots (4) and in India there are records relating to *B. theobromae* (2) and *Phomopsis caricae-papayae* (1). These two fungi also cause stem-end rot of papayas in the USA (5).

SYMPTOMS The skin around the stem-end becomes discoloured, and there may be development of mould and spores or spore-bearing bodies (pycnidia). Lesions caused by *B. theobromae* have a broad water-soaked margin and a rough black surface pimpled with pycnidia. Black rot is firmer with a narrow advancing margin; the lesion is black and wrinkled and the pycnidia are spherical and not joined together (3). *Phomopsis caricae-papayae* is characterised by a soft wet rot, a whitish mould growth and irregular black pycnidia, while *Phytophthora palmivora* produces numerous spores directly on an off-white mould.

BIOLOGY All these fungi are favoured by wet weather, which encourages liberation and dispersal of spores in the orchard. The newly cut stem-end is exposed to contamination during and shortly after harvest, and infection may occur before the wound has had time to heal.

CONTROL Spore production in the orchard can be reduced by fungicide sprays (3). With regard to harvesting techniques it has been noted that fruit is often prevented from rotting if part of the stem is left attached. Stem-end infections can be arrested by prompt immersion in a hot water bath, with careful control of time and temperature.

289

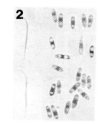

290 *Phomopsis caricae-papayae* (from CMI Descr. No. 827, Punithalingam 1985) 1 pycnidium 2 α- and β-conidia

---

1 DHINGRA O.D. & KHARE M.N. (1971). A new fruit rot of papaya. *Current Science* **40**, 612–613 illus.
2 GUPTA O. & NEMA K.G. (1979). Effect of different temperature and relative humidity on the development of fruit rots of papaya caused by *Botryodiplodia theobromae* and *Colletotrichum papayae*. *Indian Phytopathology* **32**, 106–107.
3 HUNTER J.E. & BUDDENHAGEN I.W. (1972). Incidence, epidemiology and control of fruit diseases of papaya in Hawaii. *Tropical Agriculture* **49**, 61–71 illus.
4 LANTICAN M.T. & QUIMIO T.H. (1976). Pathogenicity and cultural characteristics of *Botryodiplodia* spp. causing fruit rots. *Philippine Phytopathology* **12**, 66–74 illus.
5 STEVENS H.E. (1939). Papaya disease. *Proceedings of Florida State Horticultural Society* **52**, 57–63.

OTHER DISEASES of papayas

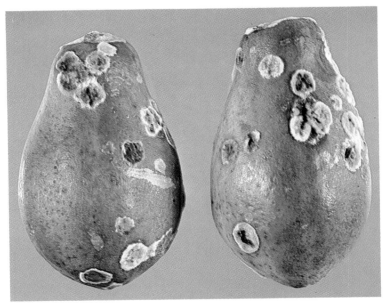

291 Alternaria rot

ALTERNARIA ROT, caused by *Alternaria alternata* (**340**), can be a problem in dry growing areas such as Israel and parts of Hawaii (1,2). The symptoms are round to ellipsoidal black lesions on the skin of the fruit, followed by rotting of the underlying tissue. If papayas can be marketed a few days after harvest, alternaria rot is not usually a problem. When cool storage is used to prolong post-harvest life, however, the fruits become increasingly susceptible to attack (1). Orchard fungicide sprays reduce the incidence of disease but cannot eliminate it.

ASPERGILLUS ROT has been recorded in India, the causal fungi being *Aspergillus flavus* and *A. niger* (**307**).

BLACK SPOT is caused by the fungus *Asperisporium caricae* (Speg.) Maubl. There is no decay, but market value is reduced by the blemish, the spots being up to 4 mm in diameter. The fungus is present in the Americas (10), and in Florida it is reported that most infection occurs during late winter and spring, although disease incidence there seems

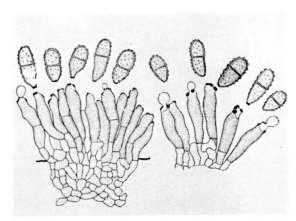

292 *Asperisporium caricae* (from CMI Descr. No. 347, Ellis & Holliday 1972) Conidia and conidiophores x 500

to have declined. In South Africa it is widespread and has recently received attention (3). There is also a recent report of its occurrence in Tanzania.

BLUE MOULD ROT is caused by *Penicillium* spp. It is not very common, but may occur after physiological damage such as chilling injury.

294 Blue mould rot

CERCOSPORA SPOT is caused by *Cercospora pappaea* Hansf. The black spots remain shallow and limited, constituting a blemish rather than a rot. Control is by means of fungicide sprays.

CLADOSPORIUM ROT is a dry rot caused by *Cladosporium herbarum* (page 226) and the disease is said to be common in Israeli papayas (2). In Hawaii the disease is known as **INTERNAL BLIGHT**, because the fungus invades the seed cavity, causing it to become dark and dry (5). Infection occurs via the flower-end of the fruit, and the eventual effect is that normal ripening is prevented. Orchard fungicide sprays can be used to reduce the incidence of infection (5).

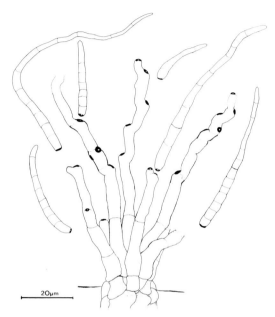

293 *Cercospora pappaea* (from CMI Descr. No. 919, Little 1987) Stroma, conidiophores, conidiogenous cells and conidia

295 Cladosporium rot

FUSARIUM ROT of papayas has been reported from Israel (2), India (9), the Philippines and Hawaii. Several species can cause decay, for example *Fusarium equiseti* (Corda) Sacc., *Fusarium oxysporum* Schlecht. and *Fusarium solani* (Mart.) Sacc. Young fruits, and mature but unripe fruits, are resistant to infection unless wounds are present. During storage intact fruits eventually become susceptible to direct penetration by the fungus. Hot water treatment can give good control (9).

296 *Fusarium equiseti* (from CMI Descr. No. 571, Booth 1978) Conidia x 750

GREASY SPOT, caused by *Corynespora cassiicola* (Berk. & Curt.) Wei, is an important disease of papayas in the Caribbean region. It is chiefly a leaf and stem disease but the fruit may also be attacked (8). The fungus is capable of direct infection of the intact skin, and produces circular to elliptical water-soaked spots, especially around the stem-end. Under humid conditions dark spores are produced on the lesions (**119**).

INTERNAL YELLOWING DISEASE is reported from Hawaii, where the causal organism has been identified as *Enterobacter cloacae* (Jordan) Hormaeche & Edwards. It is a potentially serious disease because infected fruits exhibit no external symptoms. Infected flesh is translucent with a bright yellow to lime-green discoloration and an offensive odour. There is evidence for insect transmission and infection of flowers (7); then follows a period of latency until the fruit ripens.

PINK MOULD ROT, caused by *Trichothecium roseum* (page 191), has been recorded on papayas in India.

PURPLE STAIN of papaya fruits is caused by a bacterium with the characteristics of *Erwinia herbicola* (Löhnis) Dye (6). On being cut open, infected fruits show a reddish-purple coloration of the vascular strands. Rotting occurs as the fruit ripens, and an offensive odour accompanies the decay. The bacterium also produces the purple pigment when cultured on artificial media.

STEMPHYLIUM ROT, caused by *Stemphylium botryosum* (page 85), is found occasionally in Israel (2). From Hawaii, where the causal fungus is *S. lycopersici* (Enjoji) Yamam., it is reported that hot water treatment can predispose papaya fruits to this disease (4). Careful temperature control and the use of a fungicide are recommended.

---

1. ALVAREZ A.M., HYLIN J.W. & OGATA J.N. (1977). Post-harvest diseases of papaya reduced by biweekly orchard sprays. *Plant Disease Reporter* **61**, 731–735.
2. BARKAI-GOLAN R. (1981). Post-harvest diseases of papaya. *Agricultural Research Organisation, Bet Dagan, Israel, Pamphlet* No. 197.
3. CHAMBERS K.R. & RIJKENBERG F.H.J. (1987) Culture of *Asperisporium caricae*, the papaya black spot organism. *Phytophylactica* **19**, 113.
4. GLAZENER J.A., COUEY H.M. & ALVAREZ A. (1984) Effect of postharvest treatments on *Stemphylium* rot of papaya. *Plant Disease* **68**, 986–988.
5. HINE R.B., HOLTZMANN O.V. & RAABE R.D. (1965). Diseases of papaya (*Carica papaya* L.) in Hawaii. *Bulletin of Hawaii Agricultural Experiment Station* No. 136, 26 pp illus.
6. NELSON M.N. & ALVAREZ A.M. (1980). Purple stain of *Carica papaya*. *Plant Disease* **64**, 93–95 illus.
7. NISHIJIMA K.A., COUEY H.M. & ALVAREZ A.M. (1987) Internal yellowing, a bacterial disease of papaya fruits caused by *Enterobacter cloacae*. *Plant Disease* **71**, 1029–1034 illus.
8. NORSE D. (1973). A stem and fruit disease of papaya in Barbados caused by *Corynespora cassiicola*. *Plant Disease Reporter* **57**, 404–406 illus.
9. PATHAK V.N., GOYAL J.P. & BHATNAGAR L.G. (1976). Effect of chemicals and hot water treatment on *Fusarium* and *Rhizopus* rots of papaya. *Indian Phytopathology* **29**, 210–211.
10. SALDAÑA M.I., MÁRQUEZ M. & RUÍZ P. (1985) Identificación de enfermedades fungosas del cultivo de la papaya (*Carica papaya* L.) en el estado de Tabasco. *Revista Mexicana de Fitopatología* **3**, 14–17.

CHILLING INJURY of papayas

OCCURRENCE The critical temperature, below which chilling injury may occur, is generally considered to be about 7°C for ripe fruits and 10°C for green fruits (3). Studies on unripe papayas in Malaysia, however, indicated that chilling injury can occur at temperatures as high as 15°C (1).

SYMPTOMS External signs are small spots, 1 to 2 mm in diameter, which are dark olive on green fruits and light brown on yellow fruits (3). They may coalesce to form scald-like areas on the skin. Mild bruises appear more pronounced, and ripening fruits develop a blotchy appearance (3). Other characteristics include breakdown of the pulp and increased susceptibility to decay (2,4), especially alternaria rot (q.v.).

BIOLOGY At temperatures below the critical value (which varies with cultivar), normal metabolism is disrupted (4). Injury may not be apparent until the fruit is returned to a higher temperature (2). Papayas, harvested at 'colour break' and stored for 4 days at 5°C, may develop skin discoloration shortly after return to ambient temperature (2). After longer cold storage there may be enhanced evolution of ethylene as a response to stress, and the fruits lose their ability to ripen normally (2).

CONTROL Unripe papayas may need to be held at 10°C (2), 12°C (3), 13°C (5) or 20°C (1) if chilling injury is to be avoided. However, damage can be reduced or prevented by 'pre-conditioning' fruit at 10° or 15°C for a few days prior to cool storage; this probably allows some ripening to occur, which reduces chill sensitivity (3).

297 Chilling injury followed by alternaria rot

---

1 BROUGHTON W.J., HASHIM A.W., SHEN T.C. & TAN I.K.P. (1977) Maturation of Malaysian fruits. 1. Storage conditions and ripening of papaya (*Carica papaya* L. cv. Sunrise Solo). *MARDI Research Bulletin* **5**, 59–72 illus.
2 CHAN H.T., SANXTER S. & COUEY H.M. (1985) Electrolyte leakage and ethylene production induced by chilling injury of papayas. *HortScience* **20**, 1070–1072.
3 CHEN N.-M. & PAULL R.E. (1986) Development and prevention of chilling injury in papaya fruit. *Journal of the American Society for Horticultural Science* **111**, 639–643.
4 EL-TOMI A.I., ABOU-AZIZ A.B., ABDEL-KADER A.S. & ABDUL-WAHAB F.K. (1974) The effect of chilling and non-chilling temperatures on the quality of papaya fruits. *Egyptian Journal of Horticulture* **1**, 179–185.
5 THOMPSON A.K. & LEE G.R. (1971) Factors affecting the storage behaviour of papaya fruit. *Journal of Horticultural Science* **46**, 511–516

Passion fruits

Passion fruits belong to the genus *Passiflora*, a perennial vine native to South America (2,7,8). *P. edulis* Sims forma *edulis* is the purple passion fruit or purple granadilla (meaning 'little pomegranate'). It thrives in the subtropics and in tropical highlands, for example Brazil (8), India (9,10), South Africa, Kenya and Uganda (3). A golden-yellow form, *P. edulis* forma *flavicarpa* Deg, is better adapted to the tropical lowlands and is grown in Caribbean countries, Guyana (4), Hawaii and the South Pacific Islands (5). Hybrids are grown in Australia and New Zealand (1,6). *P. quadrangularis* L. is known as the giant granadilla and requires a tropical climate such as that of Venezuela and the southern USA.

Purple passion fruits are picked according to development of skin colour whereas in the yellow variety the fruits are gathered from the ground. The hard rind or 'shell' may become wrinkled but this does not indicate deterioration in internal quality. Inside are numerous seeds surrounded by yellowish aromatic juicy pulp. The seeds are small enough to be eaten along with the pulp; there is also an important industry in passion fruit juice, for example in Brazil (8).

The fresh fruit, which has a climacteric pattern of respiration (page 14), produces much ethylene, the rate of evolution being several times greater than that of bananas and mangoes. Because of their susceptibility to chilling injury, the optimal storage temperature for passion fruits is within the range of 6° to 10°C. Packing in plastic film is recommended (2,9), and they may be kept for about a month.

BROWN SPOT is caused by *Alternaria alternata* (**339**) in the Pacific Islands (5); most other references are to *A. passiflorae* Simmonds, which occurs in Hawaii, East Africa and Australia (1,3,6). Circular, sunken, light brown spots develop on ripening fruits, the lesions often having a characteristic green border. Infection can occur by direct penetration even in the absence of water (2), but disease incidence is most severe during warm wet periods. Control is by pruning (and destroying the debris) and by spraying the vines regularly with fungicide.

SEPTORIA SPOT is caused by *Septoria passiflorae* Louw. Minute black bodies (pycnidia) develop in leaf lesions, and spores are splash-dispersed (6). Fruit infection occurs at any stage of growth; affected fruits ripen unevenly and are fit only for processing.

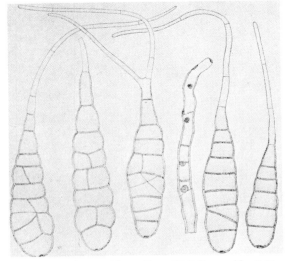

298 *Alternaria passiflorae* (from CMI Descr. No. 247, Ellis & Holliday 1970) Conidia and conidiophore x 500

299 Brown spot

Other fungi associated with decaying passion fruits include *Aspergillus niger* (page 160), *Cladosporium* sp. (page 226), *Colletotrichum gloeosporioides* (page 130), *Fusarium oxysporum* (page 150), *Penicillium expansum* (page 178), *Phytophthora nicotianae* var. *parasitica* (page 64) and *Rhizopus stolonifer* (page 230) (6,9,10). The bacterium *Pseudomonas syringae* pv. *passiflorae* (Reid) Young, Dye & Wilkie causes 'grease spot', while *Xanthomonas campestris* pv. *passiflorae* (Pereira) Dye is the cause of bacterial spot.

300 Cladosporium rot

---

1 ANON. (1952) Brown spot of passionfruit. *Agricultural Gazette of New South Wales* **63**, 373–374 illus.
2 COLLAZOS E., O., BAUTISTA G., A., MILLÁN M., B. & MAPURA M., B. (1984) Efecto de bolsas de polietileno en la conservación de maracuyá (*Passiflora edulis* var. *flavicarpa* Degener), curuba (*P. mollissima* HBK Bailey) y tomate (*Lycopersicon esculentum* Miller). *Acta Agronómica* **34**(2), 53–59.
3 EMECHEBE A.M. & MUKIIBI J. (1975) Fungicidal control of brown spot of passion fruit in Uganda. *Acta Horticulturae* No. 49, 281–289.
4 FOUQUÉ A. & FOUQUÉ R. (1980) Quelques notes sur la grenadille jaune *Passiflora edulis* Sims var. *flavicarpa*. *Fruits* **35**, 309–312 illus.
5 FULLERTON R.A. (1982) Brown spot of passion fruit on Niue Island caused by *Alternaria alternata*. *New Zealand Journal of Agricultural Research* **25**, 421–423 illus.
6 INCH A.J. (1978) Passion fruit diseases. *Queensland Agricultural Journal* **104**, 479–484 illus.
7 MARTIN F.W. & NAKASONE H.Y. (1970) The edible species of *Passiflora*. *Economic Botany* **24**, 333–343.
8 MEDINA J.C., GARCIA J.L.M. & LARA J.C.C. (1980) Maracujá – da cultura ao processamento e commercialização. *Frutas Tropicais* No. 9, 207 pp illus.
9 PRUTHI J.S., SRIVASTAVA H.C. & LAL G. (1958) Microbiological spoilage in purple passion fruit (*Passiflora edulis* Sims.) during storage. *Journal of Scientific and Industrial Research C* **17**, 129–131.
10 RAO N.N.R. & RAVISHANKAR H. (1982) *Colletotrichum* fruit canker of passion fruit. *Current Science* **51**, 375–376 illus.

Persimmons

301 Alternaria rot

The **persimmon** or **kaki** is the fruit of *Diospyros kaki* L., and is highly esteemed in Japan (3). It thrives in subtropical and warm temperate regions, including China, Australia (1), New Zealand (2), Hawaii, parts of the USA, Brazil, Israel (4), Egypt, Iran (5) and the south of France.

The fruit is yellow to reddish-orange with a thin skin, firm or soft orange flesh, and few or no seeds (1). It has a climacteric pattern of respiration (page 14) but ethylene production is low compared with that of other fruits. Some cultivars have an astringent taste owing to the presence of soluble tannins. There are various methods of removing astringency, for example by exposing harvested fruits to alcohol vapour or to high carbon dioxide concentrations (3). Some cultivars are non-astringent but can only be grown in the warmer regions (1).

Persimmons are harvested according to skin colour and sugar content; a pre-harvest spray with a growth-regulating chemical can be used to promote uniform maturity in the crop. Storage in polyethylene bags is recommended and, for those cultivars susceptible to chilling injury, a temperature of about $5°C$ is appropriate (5); storage life is approximately 1 month. Other cultivars may be stored for 3 months at $0°C$ (4).

ALTERNARIA ROT is caused by *Alternaria alternata* (page 172), which attacks developing fruits. Infections remain quiescent until after harvest, and black spots become apparent as the fruits ripen in store (4). Wound infection results in earlier appearance of symptoms.

Other causes of decay in persimmons include species of *Botrytis*, *Cladosporium*, *Colletotrichum*, *Mucor*, *Penicillium*, *Phoma* and *Rhizopus*.

CALYX SEPARATION is a physiological disorder which may affect certain cultivars; it has caused losses in New Zealand. Growing conditions are all-important, and excessive nitrogen fertilisation should be avoided (2). If the plants are thinned early in the season, this will enhance calyx growth and help to prevent the disorder.

---

1 GEORGE A.P. & NISSEN R.J. (1985) The persimmon as a subtropical fruit crop. *Queensland Agricultural Journal* **111**, 133–140 illus.

2 GLUCINA P.G. (1987) Calyx separation: a physiological disorder of persimmons. *Orchardist of New Zealand* **60**, 161–163 illus.

3 KAWASHIMA K., KAMIHISA Y. & KATABE K. (1984). [The control of pericarp blackening during ethanol treatment for the removal of astringency in Japanese persimmon.] *Journal of the Japanese Society for Horticultural Science* **53**, 290–297 illus.

4 PRUSKY D., BEN-ARIE R. & GUELFAT-REICH S. (1981) Etiology and histology of *Alternaria* rot of persimmon fruits. *Phytopathology* **71**, 1124–1128 illus.

5 SHAYBANY B., ROUHANI I. & AZARAKHSH H. (1978) Effect of storage temperature and container type on storage and shelf life of persimmon fruits. *Iranian Journal of Agricultural Research* **6**, 123–127.

Pineapples

The **pineapple**, *Ananas comosus* L. (Merr.), originated in South America, and the fruit's strange appearance (resembling that of a large pine-cone) provoked astonishment in the early explorers In the 16th century, and subsequently, the pineapple was introduced into Africa, Asia and Australasia, and it is now grown throughout the tropics (2,8). In the Azores it is cultivated in greenhouses. World production is approximately 10 million tonnes, the ten leading producer countries being the Philippines, Thailand, Brazil, India, the USA (including Hawaii), Vietnam, Indonesia, China (including Taiwan), Mexico and the Ivory Coast.

Much of the world's pineapple crop is grown exclusively for canning, but there are important export trades in the fresh fruit, for example from Hawaii to the US mainland (1), from the Philippines and Taiwan to Japan, from Cuba to the USSR, and from the Ivory Coast, Kenya and South Africa to various countries in Europe (8).

The structure which is loosely termed 'a pineapple fruit' is really a compound or multiple fruit, formed by the fusion of more than a hundred individual fruitlets (6). These develop from the inflorescence which arises on a single stalk from the centre of a rosette of spiky leaves. The flowers open in sequence, starting at the base. If left to bloom and fruit naturally, individual plants develop at different rates, necessitating serial harvesting (9). It is therefore of great advantage to synchronise flowering by the application of a growth-regulating chemical such as ethephon (which produces ethylene) or calcium carbide (which produces acetylene). Floral induction results in crop uniformity, which can be further enhanced by an application of ethephon to the maturing fruits a few days before the calculated date of harvest (8). Not only does this precipitate in all the pineapples the changes normally associated with 'ripening', but it also tends to promote these changes uniformly within each composite fruit.

Studies of a related plant (in which the fruitlets are separate rather than fused) confirmed that the pineapple is non-climacteric (page 14). It must therefore be allowed to develop most of its eating qualities whilst still attached (3). There is not necessarily a good correlation between external appearance and internal quality; for example in some climates the pineapple may remain green long after it is ripe inside. Specific gravity is a better index of maturity than is colour (10). Generally, however, pineapples for export are harvested after the base of the fruit has begun to turn yellow. Natural ethylene production is very low (3), but fruits degreened with ethephon, just before or just after harvest, should be regarded as potent producers of the gas. This makes them incompatible with ethylene-sensitive commodities such as bananas.

The main post-harvest diseases of pineapples are black rot (q.v.) and fruitlet core rot (q.v.), which call for control measures in plantation and packhouse (4,5,7,9). Care is necessary at all stages to prevent injury, and packaging has to be well-designed to prevent the sharp 'crown' leaves from piercing adjacent fruits. A serious physiological disorder is 'endogenous brown spot' or 'black heart', a type of chilling injury induced by unfavourable temperatures in the field or during storage and transport. Its expression is influenced by many factors (1,8). Notwithstanding the fact that this disorder sometimes occurs even at moderate temperatures, the usual carriage temperature for exported pineapples is 8° to 9°C, which has proved optimal for preservation of quality in most instances.

---

1 AKAMINE E.K. (1976) Problems in shipping fresh Hawaiian tropical and subtropical fruits. *Acta Horticulturae* No. 57, 151–161.
2 COLLINS J.L. (1960) *The pineapple: botany, cultivation and utilization.* London: Leonard Hill, 294 pp illus.
3 DULL G.G., YOUNG R.E. & BIALE J.B. (1967) Respiratory patterns in fruit of pineapple, *Ananas comosus*, detached at different stages of development. *Physiologia Plantarum* **20**, 1059–1065.
4 LIM W.H. (1985) Diseases and disorders of pineapples in peninsular Malaysia. MARDI Report No. 97, 53 pp illus.
5 MACHADO N.D. & VENTURA J.A. (1983) *Bibliografia internacional de doenças do abacaxizeiro.* Brasilia, Brazil: EMBRAPA, 311 pp.
6 OKIMOTO M.C. (1948) Anatomy and histology of the pineapple inflorescence and fruit. *Botanical Gazette* **110**, 217–231.
7 OXENHAM B.L. (1957) Diseases of the pineapple. *Queensland Agricultural Journal* **83**, 13–26 illus.
8 PY C., LACOEUILHE J.-J. & TEISSON C. (1984) *L'ananas, sa culture, ses produits.* Paris: Maisonneuve & Larose, 562 pp illus.
9 ROHRBACH K.G. & APT W.J. (1986) Nematode and disease problems of pineapple. *Plant Disease* **70**, 81–87 illus.
10 SMITH L.G. (1984) Pineapple specific gravity as an index of eating quality. *Tropical Agriculture* **61**, 196–199.

BLACK ROT of pineapples caused by
Ceratocystis paradoxa (Dade) Moreau
Conidial state: *Thielaviopsis paradoxa* (de Seynes) Höhnel

OCCURRENCE Black rot of pineapples is also known as soft rot, stem-end rot, water rot and water blister, and has been responsible for serious losses (5). The disease occurs in all the major producing countries (CMI Map 142), for example the Ivory Coast (5), India (6), the Philippines (9), Australia (8), Hawaii (3), Cuba (10) and Puerto Rico (2,7). It is the main cause of pineapple decay in Nigeria (1) but appears to be of secondary importance in South Africa.

SYMPTOMS Initially no symptoms are apparent unless the fruit is cut open, when a soft decay is visible either at the side or at the base. Affected tissue is water-soaked and dark yellow, later turning greyish black (2). Black spores may be produced within the flesh, but form most profusely if cut tissue is exposed to the atmosphere (4).

BIOLOGY The fungus survives on plant debris in the soil, in the form of thick-walled resting spores (chlamydospores). It also produces infective asexual spores (conidia) which are splashed on to the fruit by rain. The sexual stage (perithecia giving rise to ascospores) is occasionally recorded (7). Infection may occur before harvest, via insect punctures or growth cracks, or directly via the natural crevices between individual fruitlets (4). More usually, however, the fungus gains entry through the cut stem or through wounds sustained during handling (2). The optimal growth temperature is approximately 26°C and if decay starts at the stem-end then the entire fruit may be colonised within a few days (5).

CONTROL Pineapple propagation material can be treated with a systemic fungicide (3). It has been demonstrated that hygienic practices are beneficial in reducing losses due to this disease (8). Packing stations should be disinfected and culled fruits promptly destroyed. It is advisable to avoid packing injured, sunburned or wet fruit (2). Precautions are especially necessary if a wet period has followed a dry growing season when the crop might have been under stress and more prone to growth cracks. In order to prevent post-harvest infection the fruit must be treated with a fungicide. The method may involve total immersion (3,10) or the application of a spray or paste to the cut stem (2,6). It is important to treat the fruit promptly, within six hours of harvest (5). Refrigeration serves to check the development of symptoms, since at 8°C the fungus grows hardly at all. However, fruit known to be infected should not be shipped, because significant decay could ensue before attainment of the required carriage temperature (5).

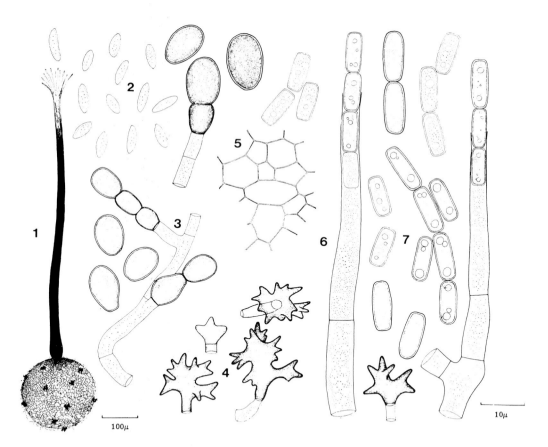

303 *Ceratocystis paradoxa* (from CMI Descr. No. 143, Morgan-Jones 1967) 1 perithecium 2 ascospores 3 chlamydospores 4 perithecial appendages 5 surface of perithecium wall 6 conidiophores 7 conidia

1 ADISA V.A. & FAJOLA A.O. (1982) Post-harvest fruit rots of pineapple (*Ananas comosus*) in Nigeria. *Fitopatologia Brasileira* **7**, 97–103.
2 BRATLEY C.O. & MASON A.S. (1939) Control of black rot of pineapples in transit. *Circular of the United States Department of Agriculture* No. 511, 12 pp illus.
3 CHO J.J., ROHRBACH K.G. & APT W.J. (1977) Induction and chemical control of rot caused by *Ceratocystis paradoxa* on pineapples. *Phytopathology* **67**, 700–703.
4 CHOWDHURY S. (1945) Ceratostomella diseases of pineapple. *Indian Journal of Agricultural Science* **15**, 135–139 illus.
5 FROSSARD P. (1978) Lutte contre la pourriture de l'ananas à *Thielaviopsis paradoxa*. Importance de la température de conservation et de la désinfection fongicide. *Fruits* **33**, 91–99 illus.
6 JAMALUDDIN, TANDON M.P. & TANDON R.N. (1975) *Ceratocystis* rot of pineapple and its chemical control. *Proceedings of the National Academy of Science, India B* **45**, 217–220.
7 LIU L.J. & RODRÍGUEZ M., A. (1973) Sexual compatibility, morphology, physiology, pathogenicity and *in vitro* sensitivity to fungicides of *Thielaviopsis paradoxa* infecting sugarcane and pineapple in Puerto Rico. *Journal of Agriculture of the University of Puerto Rico* **57**, 117–128 illus.
8 McKNIGHT T. (1941) Water blister disease of pineapples. *Queensland Agricultural Journal* **55**, 180–182.
9 ROLDAN E.F. (1925) The soft rot of pineapple in the Philippines and other countries. *Philippine Agriculturist* **13**, 397–405 illus.
10 ZALDÍVAR H. (1977) Lucha contra el deterioro de las frutas de piña durante su exportación. *Ciencias de la Agricultura* **1**, 70–78.

FRUITLET CORE ROT of pineapples caused by
***Gibberella fujikuroi* var. *subglutinans* Edwards**
Conidial state: *Fusarium moniliforme* var. *subglutinans* Wollenw. & Reinking
***Penicillium funiculosum* Thom**
***Erwinia ananas* Serrano**
***Pseudomonas ananas* Serrano**

OCCURRENCE Fruitlet core rot has also been described as fruitlet brown rot, fruitlet black rot, eye rot and black spot, and is an important cause of loss in pineapples intended for the fresh market or for canning. The disease has been reported from most producer countries, for example Australia (6), the Philippines (8), Malaysia (4,10), South Africa (2), the Ivory Coast (5), Brazil (1), Mexico (9), Guatemala (3) and Hawaii (7).

SYMPTOMS The disease is usually characterised by a light to dark brown soft rot of the axis of an individual fruitlet (page 155). In cross-section the rot appears as a brown spot in the centre of the fruitlet, whilst in longitudinal section the affected area is elongated and extends towards the heart of the fruit. In some circumstances diseased tissue is almost black, and wet or dry according to the humidity of the surroundings (2).

BIOLOGY Fruitlet core rot may be caused by several organisms (fungi and bacteria), either alone or in combination (2,3,9,10). They survive on debris in the soil and are disseminated by water and insects.

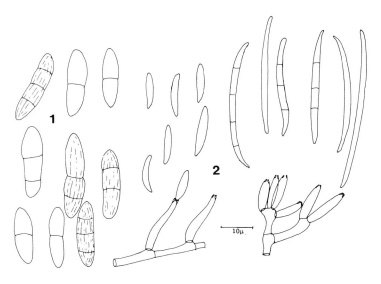

306 *Gibberella fujikuroi* var. *subglutinans* (from CMI Descr. No. 23, Booth & Waterston 1964) 1 ascospores 2 micro- and macroconidia and conidiophores

Rainfall during the first days of flowering is especially conducive to infection (3). Fungal spores and bacteria are carried into the floral cavities by watersplash, or by mites and mealybugs which damage the tissue and facilitate infection (1,3,5). Disease incidence is particularly high in years when a wet period follows a dry growing season, probably because the resulting growth cracks provide a means of entry (2). Infection can occur at any stage of fruit development but decay of the flesh is most rapid in ripe fruit (6).

CONTROL The chief means of reducing losses is to control the mites and mealybugs which spread the disease, and this is achieved by spraying the crop at intervals with an acaricide (5).

---

1 BOLKAN H.A., DIANESE J.C. & CUPERTINO F.P. (1979) Pineapple flowers as principal infection sites for *Fusarium moniliforme* var. *subglutinans*. *Plant Disease Reporter* **63**, 655–657.
2 EDMONSTONE-SAMMONS C.P. (1958) Some aspects of black spot in pineapple. *South African Journal of Agricultural Science* **1**, 111–120.
3 JOHANSSON N. (1934) A contribution to the knowledge of the etiology of fruitlet black rot disease of pineapple. *Svensk Botanisk Tidskrift* **28**, 384–404 illus.
4 LIM W.H. (1983) *Penicillium funiculosum* isolates associated with fruit blemishes of pineapple (cv. Masmerah) in peninsular Malaysia. *MARDI Research Bulletin* **11**, 179–186 illus.
5 MOURICHON X. (1983) Contribution à l'étude des taches noires (fruitlet core rot) et leathery pocket de l'ananas causés par *Penicillium funiculosum* Thom en Côte d'Ivoire. *Fruits* **38**, 601–609.
6 OXENHAM B.L. (1962) Etiology of fruitlet core rot of pineapple in Queensland. *Queensland Journal of Agricultural Science* **19**, 27–31 illus.
7 ROHRBACH K.G. & TANIGUCHI G. (1984) Effects of temperature, moisture, and stage of inflorescence development on infection of pineapple by *Penicillium funiculosum* and *Fusarium moniliforme* var. *subglutinans*. *Phytopathology* **74**, 995–1000.
8 SERRANO F.B. (1935) Control of bacterial fruitlet rots of the pineapple in the Philippines. *Philippine Journal of Science* **57**, 29–62.
9 SMITH M.A. & RAMSEY G.B. (1950) Bacterial fruitlet brown rot of Mexican pineapple. *Phytopathology* **40**, 1132–1135.
10 THOMPSON A. (1937) Pineapple fruit rots in Malaya. *Malayan Agricultural Journal* **25**, 407–420 illus.

OTHER DISEASES of pineapples

ASPERGILLUS ROT may be caused by various species of *Aspergillus*, for example *A. niger* (black mould) is sometimes found inside the flower cavities. *A. flavus* is reported to cause a dry rot in Nigerian pineapples (1).

307 Conidiophores and conidia of *Aspergillus niger*

BACTERIAL SOFT ROT, caused by *Erwinia chrysanthemi* Burkh. *et al*, is a serious disease in Malaysia. Infection occurs at flowering time but there are no symptoms until the fruit begins to ripen. Rotting then proceeds with great rapidity, and fruits can undergo complete collapse within a day or two, either before or after harvest (7).

BOTRYODIPLODIA ROT is recorded as a market disease in India (10), the causal organism being *Botryodiplodia theobromae* (**197**). It is a wound parasite and infection usually occurs via the cut stem. The fungus spreads up the axis and into the pulp, making the flesh appear dark brown and water-soaked. Minute black bodies (pycnidia) eventually develop in the rind. Decay can be held in check by a storage temperature of 10°C, but rapid development resumes at temperatures around 25°C (10).

INTERFRUITLET CORKING is caused by *Penicillium funiculosum* (page 158) and has been reported from South Africa and Hawaii (4,6). Individual flowers may be killed by the fungus, and affected fruits take on an unnatural glossy appearance. Beneath the skin there is cork formation between fruitlets, and this lowers the grade of pineapples (4).

LEATHERY POCKET, also recorded in South Africa (3) and Hawaii (4,6), is characterised by cork formation at a deeper level, in the ovaries at the base of individual flowers. The disease was formerly attributed to mite damage (3) but it has now been established that *Penicillium funiculosum* (page 158) is the primary cause (6). The fungus is probably disseminated by mites as they visit the flowers, and the disease is controlled by spraying the crop with an acaricide and a fungicide (3).

PINK DISEASE may be caused by any of several bacteria, including species of *Acetobacter* and *Enterobacter*. Losses have occurred in Hawaii (2,9), the Philippines (5), Australia and Mexico (8). Some strains induce a readily apparent brownish-pink discoloration of the pulp, causing losses in fresh pineapples. Other strains can infect without producing symptoms, and discoloration occurs only when the fruit is cooked, thus causing problems in the canning industry (9). Insects may be responsible for dissemination of the bacteria (8) but pink disease is severe only if flowering takes place during wet weather immediately following a long dry period.

Susceptibility to infection can be reduced by the use of potash fertiliser (which increases fruit acidity) (8). Cultivars show marked differences in susceptibility to the various strains of bacteria (9).

RHIZOPUS ROT may be caused by *Rhizopus oryzae* or *R. stolonifer* (page 230), as reported from Nigeria (1). Decay is rapid at tropical temperatures; infected fruits become soft and covered with fungal strands.

YEASTY ROTS are associated with ripe fruits. Yeast fungi are able to invade via wounds or sunburn injury and cause fermentation of the flesh, which turns bright yellow. Pockets of gas may be formed and juice bubbles out, eventually leaving the interior spongy and fibrous. Some yeasts do not produce gas but cause a glassy spoilage with a distinctive aroma (8). Losses from yeasty rots can be minimised by protecting fruits from sunburn and mechanical damage.

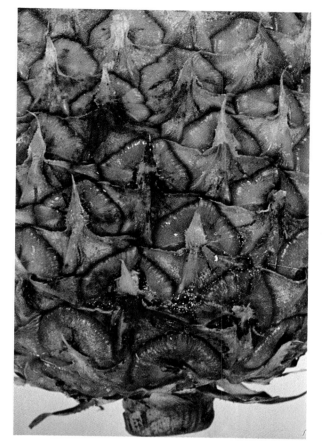

308 Yeasty rot

---

1 ADISA V. (1983) The effects of some environmental factors on the growth and pathogenicity of six pineapple fruit rot pathogens. *Fitopatologia Brasileira* **8**, 37–45.

2 GOSSELE F. & SWINGS J. (1986) Identification of *Acetobacter liquefaciens* as causal agent of pink-disease of pineapple fruit. *Journal of Phytopathology* **116**, 167–175.

3 GRICE D.S. le & MARR G.S. (1970) Studies on the control of pineapple fruit diseases. *Phytophylactica* **2**, 63–64.

4 HEPTON A. & ANDERSON E.J. (1968) Interfruitlet corking of pineapple fruit, a new disease in Hawaii. *Phytopathology* **58**, 74–78 illus.

5 KONTAXIS D.G. & HAYWARD A.C. (1978) The pathogen and symptomatology of pink disease of pineapple fruit in the Philippines. *Plant Disease Reporter* **62**, 446–450.

6 LIM T.K. & ROHRBACH K.G. (1980) Role of *Penicillium funiculosum* strains in the development of pineapple fruit diseases. *Phytopathology* **70**, 663–665.

7 LIM W.H. & LOWINGS P.H. (1979) Pineapple fruit collapse in peninsular Malaysia: symptoms and varietal susceptibility. *Plant Disease Reporter* **63**, 170–174 illus.

8 LINFORD M.B. (1952) Pineapple diseases and pests in Mexico. *FAO Plant Protection Bulletin* **1**, 21–25.

9 ROHRBACH K.G. & PFEIFFER J.B. (1976) The interaction of four bacteria causing pink disease of pineapple with several pineapple cultivars. *Phytopathology* **66**, 396–399 illus.

10 TANDON R.N. & BHARGAVA S.N. (1962) Botryodiplodia rot of pineapple (*Ananas comosus* Merr.). *Current Science* **31**, 344–345 illus.

CHILLING INJURY of pineapples

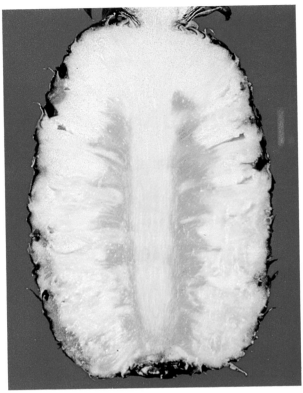

309

OCCURRENCE This important disorder leads to rejections in the cannery and heavy losses in pineapples sold fresh (3). Investigations have been carried out in Hawaii (2), South Africa (1), West Africa (4), Australia (5), Malaysia and Taiwan (3). Alternative names for the disorder include **endogenous brown spot**, **internal browning** (2,4) and, to describe the advanced stage, **blackheart** (1,3,5).

SYMPTOMS There are no external signs save a slight dullness of the crown leaves. The fruit should be cut in half longitudinally for examination. The earliest symptom is a watery (and later brownish) spot at the base of a fruitlet where it arises from the core tissue (2). As more lesions develop they coalesce, and in severely affected fruits much of the core and surrounding tissue is discoloured brown or even black (3).

BIOLOGY There are other contributory causes besides low temperature (2). Factors influencing the susceptibility of the pineapple crop to chilling injury include rainfall, shade, and the application of fertilisers and growth regulators (4). If there has been a spell of cool weather, then symptoms may develop a week or two later and be present at harvest (1). On the other hand, they may be induced or exacerbated by post-harvest refrigeration, the symptoms developing only after return to a higher temperature (2,4). The critical temperature, below which there is a risk of injury, is reported to be approximately 7°, 8°, 12°C or even higher (2).

CONTROL Susceptibility of the crop can be reduced by certain cultural measures such as irrigation during dry weather (4). Harvesting at optimal maturity is also important. Both incidence and severity of the disorder can be reduced by dipping the fruit in a wax; this treatment is equally effective whether applied before or immediately after the period at chilling temperature (2). Similarly, heat treatment (holding the fruit at a temperature above 30°C for 24 hours) suppresses symptoms whether applied before or after cold storage or refrigerated shipment (2). In contrast to most other commodities, pineapples held for extended periods at low temperatures appear to adapt. Thus, pineapples develop less internal browning (on removal to 25°C) after a 3 week period of refrigeration (at 0° to 8°C) than after a 1 week period of refrigeration (2,5). The recommended carriage temperature for pineapples is usually given as 8° to 9°C.

1 LELYVELD L.J. van & BRUYN J.A. de (1977) Polyphenols, ascorbic acid and related enzyme activities associated with black heart in Cayenne pineapple fruit. *Agrochemophysica* **9**, 1–6.
2 PAULL R.E. & ROHRBACH K.G. (1985) Symptom development of chilling injury in pineapple fruit. *Journal of the American Society for Horticultural Science* **110**, 100–105 illus.
3 SUN S.-K. (1971) A study of black heart disease of the pineapple fruits. *Plant Protection Bulletin, Taiwan* **13**, 39–47 illus.
4 TEISSON C., LACOEUILHE J.J. & COMBRES J.C. (1979) Le brunissement interne de l'ananas. 5. Recherches des moyens de lutte. *Fruits* **34**, 399–415.
5 WILLS R.B.H., HASSAN A. & SCOTT K.J. (1985) Effect of storage time at low temperature on the development of black heart in pineapples. *Tropical Agriculture* **62**, 199–200.

Pomegranates

The **pomegranate**, *Punica granatum* L., originated in Iran; it was grown in the Hanging Gardens of Babylon and was known in ancient Egypt. The crop is now cultivated in many subtropical regions, including Saudi Arabia, India (5), Pakistan, Afghanistan, Israel (1,3) and parts of the USA (2) and South America. The best fruits are produced in areas with cool winters and hot dry summers.

The plant is a large bush with striking orange-red flowers. The fruit has a thin leathery rind and the flower parts (sepals forming the calyx) persist as a stiff crown at the distal end (3). The interior of the fruit is divided into several compartments by walls of pith; there are numerous seeds, each surrounded by juicy pink flesh.

The pomegranate is non-climacteric and can be left on the tree for some time after reaching maturity (2,3). 'Bagging' the fruits reduces the incidence of pre-harvest blemishes and skin-cracking caused by rain (3). Spraying with a growth-regulating chemical is also effective against cracking.

Harvested fruits have a low rate of respiration and of ethylene production (2). They are prone to moisture loss and should be stored at high humidity (2); this may be achieved by the use of polyethylene-lined cartons (1). Pomegranates can be kept for about 3 months at 5°C (2) or 6°C (1). Storage at lower temperatures eventually results in chilling injury, characterised by discoloration and pitting of the rind, internal browning of the pith, paleness of the flesh, and increased susceptibility to decay (1,2). Storage at 10°C is satisfactory if a post-harvest fungicide is used (2).

SCALD is a browning of the rind which can be controlled by delaying the harvest, dipping harvested fruits in boiling water for 2 minutes, applying an antioxidant and/or storing the fruit in a low-oxygen atmosphere (1).

ASPERGILLUS ROT is caused by several species, including *Aspergillus flavus* and *A. niger* (**307**) (5). Following blossom-end infection, the

310 Aspergillus black mould rot

fungus permeates the entire pulp, transforming it into a dark mass. There may be no external indication except for a slightly abnormal skin colour.

CONIELLA ROT is caused by *Coniella granati* (Sacc.) Petrak & Sydow, which is similar to *Phoma* (4,5). Irregular black lesions develop, in which are numerous minute bodies (pycnidia). The disease has been recorded in the USA and in Kenya (4).

PESTALOTIOPSIS ROT is caused by *Pestalotiopsis versicolor* (Speg.) Stey (4). The lesions resemble tar spots and under moist conditions there is copious formation of characteristic spores.

Other causes of decay in pomegranate fruits include species of *Botrytis*, *Cladosporium*, *Penicillium*, *Phoma*, *Phomopsis* and *Rhizopus* (2,4,5). *Sphaceloma punicae* Bitanc. & Jenkins is the cause of anthracnose.

---

1 BEN-ARIE R. & OR E. (1986) The development and control of husk scald on 'Wonderful' pomegranate fruit during storage. *Journal of the American Society for Horticultural Science* **111**, 395–399.
2 ELYATEM S.M. & KADER A.A. (1984) Post-harvest physiology and storage behaviour of pomegranate fruits. *Scientia Horticulturae* **24**, 287–298.
3 SHULMAN Y., FAINBERSTEIN L. & LAVEE S. (1984) Pomegranate fruit development and maturation. *Journal of Horticultural Science* **59**, 265–274 illus.
4 SIBOE G.M., BIRGEN J.K. & SUBRAMONIAM V. (1982) Leaf-blotch and fruit-rot of pomegranate. *FAO Plant Protection Bulletin* **30**, 161–162 illus.
5 SONAWANE C.S., UTIKAR P.G. & SHINDE P.A. (1986). Post-harvest fungal flora of pomegranate. *Journal of Maharashtra Agricultural Universities* **11**, 107–108.

Sour sops and sweet sops

These are various species of *Annona*, native to South America but now grown in many parts of the tropics, for example India (6), Malaysia (3), Australia (4,7,9,10), Hawaii (5), the southern USA, Caribbean countries and parts of Africa (1). *A. muricata* L., the **sour sop** (5,8), thrives in the lowlands, *A. cherimola* Miller, the **cherimoya**, grows well at high altitudes such as in the Andes, while *A. squamosa* L., the **sweet sop** or **sugar apple** (1,3,6,9), is more adaptable. Hybrids of sweet sop and cherimoya are known as **atemoyas** (4,10). The name **custard apple** is used to describe *A. squamosa* and also its inferior relative *A. reticulata* L. or **bullock's heart**. Indeed, all sweet *Annona* species, including hybrids, are sometimes known as custard apples (4,7).

Annonas are composite fruits, formed by the fusion of many single-seeded fruitlets (5). Ripening does not occur on the tree, so fruits must be picked at the firm, mature stage and ripened after harvest (3). Ethylene production is very high, comparable to that of passion fruits (3,5). The climacteric respiration patterns of the atemoya indicate that it is closer to the cherimoya than to the sugar apple (4). Annonas are very susceptible to chilling injury and the optimal storage temperature may be as high as 15° to 20°C (10). However, lower temperatures can be used in conjunction with modified atmosphere storage; plastic film packaging has been recommended (6), preceded by a fungicide treatment. Atemoyas can be stored successfully for 2 weeks at 12° to 15°C. Storage life of all annonas is limited, however, and over-ripe fruits of sour sop and sweet sop tend to crack and split into individual segments. Commercial ripening with ethylene has the advantage of achieving uniform ripening in a batch of fruits (4). Several diseases of sweet sop have been described (1,2,9).

ANTHRACNOSE is a dry rot which has been recorded on harvested fruits in Egypt (1). Spore-bearing structures (acervuli) develop on the dark lesions and, in a humid atmosphere, give rise to pink spore-masses characteristic of *Colletotrichum gloeosporioides* (page 130). In Bangladesh the disease causes serious losses in the orchard. The sexual stage (perithecia giving rise to ascospores), *Glomerella cingulata*, has been found on badly rotted fruit. Good control is achieved by preventive fungicide sprays.

BLACK CANKER is caused by *Phomopsis anonacearum* Bondartseva-Monteverde, recorded in Australia during wet seasons (9). Purple spots develop on the fruit, most commonly at or near the distal end. As the lesions enlarge they remain shallow, but the surface becomes hard and cracked. Minute black bodies (pycnidia) develop, containing characteristic spores.

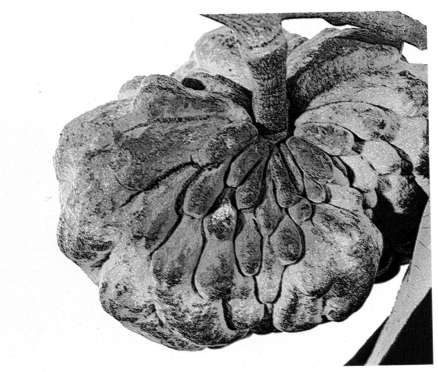

311 Black canker

BOTRYODIPLODIA ROT is caused by *Botryodiplodia theobromae* (page 81), and has been reported from Egypt (1), India, Mauritius (8) and Australia (9). Lesions are at first purple, later pimpled with black pycnidia. The flesh is rapidly invaded, becoming brown and corky (9) or soft (1), probably depending on environmental conditions and the presence or absence of secondary organisms.

GLIOCLADIUM ROT, caused by *Gliocladium roseum* (Link) Bainier, occurs in India. Watersoaked lesions turn brown and a soft rot develops.

PHYTOPHTHORA ROT is caused by *Phytophthora palmivora* (page 144) and has been recorded in Australia. Spores of the fungus are splashed up from the soil and can infect fruits of any age. Purple spots develop, and a brown decay extends deep into the fruit (9).

PINK MOULD ROT, caused by *Trichothecium roseum* (page 191), has been recorded in India on fruits in the orchard and market place.

Incidence of these diseases is best reduced by removal and destruction of dead wood and fallen fruit. Fungicide sprays are difficult to apply to these fruits, which are hidden by foliage in large trees (9).

312 Botryodiplodia rot

1 ABO-EL-DAHAB M.K. & EL-GOORANI M.A. (1971) Market and storage diseases of *Annona squamosa* L. fruits in U.A.R. (Egypt). *Phytopathologia Mediterranea* **10**, 107–109 illus.
2 ANON. (1981) *Annona* spp. *CAB Annotated Bibliography* No. C13, 104 refs. (1970–80).
3 BROUGHTON W.J. & TAN G. (1979) Storage conditions and ripening of the custard apple *Annona squamosa* L. *Scientia Horticulturae* **10**, 73–82.
4 BROWN B.I., WONG L.S., GEORGE A.P. & NISSEN R.J. (1988) Comparative studies on the postharvest physiology of fruit from different species of *Annona* (custard apple). *Journal of Horticultural Science* **63**, 521–528.
5 BRUINSMA J. & PAULL R.E. (1984) Respiration during postharvest development of soursop fruit, *Annona muricata* L. *Plant Physiology* **76**, 131–138.
6 CHAUDRY A.S., SINGH G.N. & SINGH A.R. (1985) Effect of wrapping materials and ripening media on the physico-chemical compositions of custard apple (*Annona squamosa* Linn.). *Indian Journal of Agricultural Research* **19**, 90–92.
7 GEORGE A.P., NISSEN R.J. & BROWN B.I. (1987) The custard apple. *Queensland Agricultural Journal* **113**, 287–297 illus.
8 LUTCHMEAH R.S. (1988) *Botryodiplodia theobromae* causing fruit rot of *Annona muricata* in Mauritius. *Plant Pathology* **37**, 152.
9 PURSS G.S. (1953) The fruit rots of the custard apple. *Queensland Journal of Agricultural Science* **10**, 247–265 illus.
10 WILLS R.B.H, POI A., GREENFIELD H. & RIGNEY C.J. (1984) Postharvest changes in fruit composition of *Annona atemoya* during ripening and effects of storage temperature on ripening. *HortScience* **19**, 96–97.

Other tropical and subtropical fruits

The **babaco** is a natural hybrid, *Carica* X *heilbornii* Badillo, closely related to the papaya. The fruits are characteristically 5-angled, and are yellow when ripe. They have received attention in New Zealand and South Africa as well as in their native South America. They are also grown in hot-houses on the Channel Islands. In view of their likely susceptibility to chilling injury, a storage temperature of about 10°C is probably appropriate for mature green fruits.

The **breadfruit**, *Artocarpus altilis* (Parkins.) Fosb., originated in the Malay archipelago and is now widely grown in the tropics (2). It is a rich source of carbohydrate and is the staple food in islands of the South Pacific. Breadfruit is considered acceptable only when fully mature but not yet soft, i.e. before conversion of starch to sugar. Owing to its high respiration rate, this climacteric fruit is very perishable at tropical temperatures. Chilling injury occurs at temperatures below 12°C (10). Storage at 13°C is satisfactory, and the use of plastic film packaging permits a storage life of 2 to 3 weeks (10). Decay by *Phytophthora palmivora* (page 144) has been reported from Western Samoa.

The **carambola**, *Averrhoa carambola* L., is also known as **star-fruit** because of its appearance in cross-section (2). It may be eaten as a vegetable when green or as a dessert fruit when ripe. After some controversy, a study of its response to ethylene established that the carambola is a climacteric fruit (Wang, 1988). Originating in Malaya, the plant is now grown in China, India, the Philippines, Hawaii, the southern USA and Brazil. Carambolas are susceptible to chilling injury, and storage temperatures between 5° and 10°C have been recommended; they may be kept for 3 to 4 weeks. Post-harvest disease organisms include *Alternaria alternata* (**339**) and *Cladosporium cladosporioides*. The author has observed decay due to *Botryodiplodia theobromae* in samples imported into the UK.

The **Cape gooseberry** or **Peruvian cherry**, *Physalis peruviana* L., a member of the tomato family, originated in the highlands of tropical America but is now also grown in South Africa and India. Each orange-yellow berry is at first totally enclosed by a papery brown calyx. Disease organisms include *Alternaria alternata* (page 172), *Cladosporium cladosporioides*, *Cochliobolus lunatus* Nelson & Haasis and *Fusarium equiseti* (Corda) Sacc.

Dates are fruits of the palm, *Phoenix dactylifera* L., which has been cultivated in the Middle East and North Africa for thousands of years. For many Arab peoples the date is the staple carbohydrate food. Present world production is about 2.5 million tonnes, of which approximately one tenth enters international trade, chiefly from Iraq, Saudi Arabia, Pakistan, Tunisia and Iran. There is also a date industry in the USA (California). Most exported dates are in processed form, having undergone drying, storage and partial rehydration (9). Susceptibility to physiological darkening, mould damage and mite infestation is determined by post-harvest handling practices and, in particular, by the moisture content of the dates. The presence of dead insect larvae usually indicates a failure to control

313 Botryodiplodia rot of carambola

314 Cladosporium rot of date

date moths in the grove. Small quantities of dates are despatched in the fresh state, and can be kept for a few weeks at 0°C or several days at ambient. Partially dried dates can be kept for a year at 0°C or lower, or for a few weeks at ambient temperature.

The **durian**, *Durio zibethinus* Murray, is native to the Malay archipelago and is little known outside South-east Asia. It is a large spiny fruit with an indescribable pervasive odour (repellent to many) but a smooth and luscious taste (2). An important cause of fruit rot is *Phytophthora palmivora* (page 144) which also attacks the roots, stems and leaves.

The **feijoa**, *Feijoa sellowiana* (O. Berg) O. Berg, originated in Brazil, and the bushy tree is a relative of the guava. If the crop is grown in a lime-free soil, feijoa fruits have a flavour reminiscent of pineapple and strawberry. They have received much attention in New Zealand (4) and are also grown in Italy and the USA, as well as in several countries of South America. Feijoas are climacteric fruits (page 14) which are picked mature green and ripened after harvest. They can be stored for about 1 month at 4°C (4). Lower temperatures have been used, in conjunction with a calcium dip treatment to maintain firmness and delay the onset of browning. Phytophthora fruit rot, caused by *Phytophthora cactorum* and *P. citricola*, has been reported from Sicily. Other diseases include anthracnose (page 130) and grey mould rot (page 258).

Figs, *Ficus carica* L., have been cultivated since ancient times. They are important in Mediterranean countries, especially Italy, Spain and Turkey, but will also grow in parts of the arid and semi-arid tropics. The predominant use is for the dried product, but a market also exists for fresh figs, which have a luscious taste when ripe. The structure of the fig is peculiar, owing to a unique process of development

315 Alternaria rot of fig

316 Blue mould rot of fig

from the inflorescence. The small flowers are normally invisible, being enclosed within a hollow fleshy structure (derived from stalk tissue) which has only a small opening. Pollination is achieved by a certain species of minute wasp, although there are some fig cultivars which are self-fertile. The swollen structure (syconium) which develops is a composite 'fruit' arising from vegetative tissue, and appears to be non-climacteric. Fresh figs may be kept for 1 to 3 weeks at 0°C. Disease organisms include *Alternaria alternata*, *Aspergillus niger*, *Botrytis cinerea*, and species of *Cladosporium*, *Penicillium* and *Rhizopus* (8). Since infection (or at least contamination) often occurs before harvest, orchard hygiene is important; dead wood together with fallen leaves and fruits should be removed and burned so that such debris cannot act as a source of spores (8). 'Souring' is a fermentation induced by yeasts and bacteria which are carried into the fig by fruit flies. Control of these insects is therefore necessary. 'Endosepsis' is a soft rot caused by *Fusarium moniliforme* Sheldon var. *fici* Caldis. Control measures are complicated by the fact that the fungus may be transmitted by the indispensable fig-wasp (see above).

The **jackfruit**, *Artocarpus heterophyllus* Lam., is the largest fruit in the world; specimens of 20 to 30 kg are not uncommon (2). It is a close relative of breadfruit (q.v.) and probably has similar storage characteristics.

The **kiwano**, *Cucumis metuliferus* (page 270), a member of the cucurbit family, is also known as a horned melon because of numerous protuberances on the fruit. When ripe, the fruit is yellow or orange externally but the flesh remains green. Initial trials on this crop have been carried out in New Zealand.

The **longan**, *Dimocarpus longan* Lour., is related to the litchi (q.v.) but has brown fruits. The main producers are Thailand and China, but the crop is also grown in Australia and Hawaii (6). The recommended storage temperature is 4°C (6).

The **loquat**, *Eriobotrya japonica* (Thunb.) Lindley, is of considerable importance in China and Japan, where it has been cultivated since ancient times. The crop is now grown in other parts of the sub-tropics (and tropical higlands), including northern India, many Mediterranean countries, parts of central and South America and the southern USA. Loquats are picked after they have turned from green to orange, since they are incapable of ripening after harvest. Cultivars vary with regard to keeping quality. The use of plastic film packaging permits successful storage for up to 3 weeks at 0°C (3). Choice of film is important, since unfavourable atmospheres can lead to flesh browning and superficial mould growth. Disease organisms include *Botryodiplodia theobromae* (**197**) and *Spilocaea eriobotryae* (Cav.) Hughes (loquat scab).

The **mangosteen**, *Garcinia mangostana* L., is native to the Malay archipelago. Its fruits have a thick purple rind which stains the fingers, but the taste of the white pulpy segments is renowned throughout South-east Asia (2). Mangosteens must be picked when almost fully ripe, and are prone to rind-hardening and decay by *Botryodiplodia theobromae* (**197**). However, carefully selected fruits may be held for 3 or 4 weeks at 10°C.

317 Botryodiplodia rot of mangosteen

The **naranjilla**, *Solanum quitoense* Lam., is a relative of the eggplant and the pepino (q.v.), and originated in the Andean region of South America. It is a large shrub with yellow-orange fruits.

The **pepino**, *Solanum muricatum* Aiton, is another Andean species which, however, grows at lower altitudes and has larger fruits than the naranjilla (q.v.). It is now cultivated in Australia, New Zealand and Japan, as well as in Peru and Chile. The fruits are round to elongate (5 to 20 cm), and generally yellow with purple stripes. The sweet juicy flesh is rich in vitamin C. Storage life is limited by susceptibility to chilling injury, and current exports are mostly by air.

The **prickly pear** or **tuna** is the fruit of a subtropical cactus, *Opuntia robusta* Mill., which is especially abundant in Mexico (5). The fruits, which may be red, orange, yellow or green, possess thick peel (with sharp spines) and juicy pulp containing numerous seeds. There is no climacteric rise in respiration after harvest (5). The optimal storage temperature is 8° to 10°C, according to cultivar, and fruits may be stored thus for about 2 weeks.

The **rambutan**, *Nephelium lappaceum* L., is a relative of the litchi (q.v.). It is native to the Malay archipelago and does not seem to thrive outside South-east Asia. The fruits are highly esteemed, being bright red with soft fleshy spines and firm, white, juicy flesh (2). They are very perishable, however, and soon appear dull with the onset of physiological browning. They are non-climacteric fruits (6). Susceptibility to chilling injury precludes storage at low temperature; 12°C has been recommended. In Thailand the main disease organisms are *Botryodiplodia theobromae* causing stem-end rot, and *Gliocephalotrichum bulbilium* causing brown spots both before and after harvest. Fungicide treatment and plastic film packaging would probably be beneficial.

Sapodillas are fruits of the chicle tree, *Manilkara achras* (Mill.) Fosb., which is native to central America. The crop is now grown throughout the lowland tropics and is much prized for the sweetness of its dull brown fruits. They show a climacteric rise in respiration and are especially susceptible to chilling injury during the ripening process. The optimal storage temperature for a 2-week period may be as high as 20°C, although a lower temperature may be imposed for a few days. Storage life can be extended by the use of a modified atmosphere (page 23) and removal of ethylene (1). Plastic film packaging also helps to minimise moisture loss. Disease organisms include *Phytophthora palmivora* (page 144) and species of *Pestalotiopsis* and *Phomopsis*.

The **tamarillo** or **tree-tomato**, *Cyphomandra betacea* (Cav.) Sendtner, is grown in tropical highlands such as the Andes. There are also commercial plantings in New Zealand and Australia, where disease problems include anthracnose caused by *Colletotrichum* spp. Other decay organisms are *Diaporthe phaseolorum* and *Phoma exigua*, and control involves field spraying and a post-harvest fungicide dip. Unlike the tomato, tamarillos are non-climacteric (7). They may be kept for 6 to 10 weeks in plastic film packaging at 3° to 4°C.

---

1 BROUGHTON W.J. & WONG H.C. (1979) Storage conditions and ripening of chiku fruits *Achras sapota* L. *Scientia Horticulturae* **10**, 377–385
2 CHIN H.F. & YONG H.S. (1981) *Malaysian fruits in colour*. Kuala Lumpur: Tropical Press, 129 pp illus.
3 GUELFAT-REICH S. (1970) Conservation de la nèfle du Japon (*Eriobotrya japonica*). *Fruits* **25**, 169–173.
4 KLEIN J.D. & THORP T.G. (1987) Feijoas: post-harvest handling and storage of fruit. *New Zealand Journal of Experimental Agriculture* **15**, 217–221 illus.
5 LAKSHMINARAYANA S. & ESTRELLA I.B. (1978) Postharvest respiratory behaviour of tuna (prickly pear) fruit (*Opuntia robusta* Mill.). *Journal of Horticultural Science* **53**, 327–330.
6 PAULL R.E. & CHEN N.J. (1987) Changes in longan and rambutan during postharvest storage. *HortScience* **22**, 1303–1304.
7 PRATT H.K. & REID M.S. (1976) The tamarillo: fruit growth and maturation, ripening, respiration, and the role of ethylene. *Journal of the Science of Food and Agriculture* **27**, 399–404.
8 RICCI P. (1972) Observations sur la pourriture des figues fraiches après récolte. *Annales Phytopathologiques* **4**, 109–118.
9 RYGG G.L. (1975) Date development, handling, and packing in the United States. *Agriculture Handbook of the United States Department of Agriculture* No. 482, 56 pp illus.
10 THOMPSON A.K., BEEN B.O. & PERKINS C. (1974) Storage of fresh breadfruit. *Tropical Agriculture* **51**, 407–415.

See also:

WANG T.-T. (1988) Ethylene in the postharvest physiology of tropical and subtropical fruit. *Technical Bulletin of the Food and Fertilizer Technology Center, Taiwan* No. 108, 11 pp.

CHAPTER 4
POME FRUITS

Apples, *Malus domestica* Borkh., and **pears**, *Pyrus communis* L., are the most important temperate-zone fruits. They originated in Asia Minor and spread east and west; they have been cultivated in Europe and western Asia since pre-historic times. In recent centuries they were established in North America, South America, Australia, New Zealand and South Africa, and are also grown in the cool highlands of some subtropical regions such as northern India and east Africa. Thousands of cultivars have been bred, of which a limited number are of commercial significance (5,10). Oriental or Asian pears are separate species. *Pyrus ussuriensis* var. *sinensis* is widely grown in China, while *Pyrus pyrifolia* (Burm. f.) Nakai (syn. *P. serotina* Rehder) is of great importance in Japan, where it is known as 'nashi'.

World production of apples is currently about 40 million tonnes, the ten leading producer countries being the USSR, the USA, China, France, Italy, Turkey, Argentina, West Germany, Spain and Japan. Equivalent data for pears are 10 million tonnes, mainly produced in China, Italy, the USA, Spain, Japan, France, the USSR, Turkey, West Germany and Argentina. There are important export trades (in one or both fruits) from France, Italy, Hungary, South Africa, New Zealand, the USA and, increasingly, Chile and Argentina.

Known as pome fruits, apples and pears have seeds in the form of 'pips' located in five cavities within the core tissue. The fruit having developed from an inferior ovary, the calyx remains are to be found at the distal end of the fruit (cf. citrus). A great deal of work has been done on the physiology of apples and pears, and much is known about the influence of orchard, cultural and climatic factors on subsequent storage potential (2,5,10). Pre-harvest sprays of a calcium compound and/or a growth-regulating chemical can have a marked effect on the subsequent storage life of apples and pears (6). With the notable exception of some of the oriental cultivars, apples and pears show a climacteric pattern of respiration (page 14). The time of harvest is critical, the optimal picking date for long storage being shortly before the anticipated rise in respiration rate (5,6). Maturity is judged by starch content, flesh firmness or other criteria, depending on cultivar and locality (5).

Post-harvest treatments may include drenching with various chemicals (1) against diseases such as blue mould rot and brown rot (3,4,7) or against disorders such as superficial scald and senescent breakdown (2,5,8). The optimal storage temperature varies with cultivar (Appendix 3), and must take into account the susceptibility to chilling injury (5). Apples are kept at temperatures slightly above $0°C$, while many types of pear benefit from storage at $-1°C$, provided that there is no risk of freezing (in view of this, the pulp temperature of the fruit itself must be monitored). Modification of the storage atmosphere can further extend storage life, and a substantial proportion of the apple crop (and an increasing proportion of the pear crop) is held in controlled atmosphere (CA) stores (5,6). Temperature, humidity, oxygen and carbon dioxide concentrations are carefully regulated, and increasing attention is now being given to the possibility of obtaining additional benefits by the removal of ethylene (6).

Since temperature control in most ships is less accurate than that in land stores, it is usual to stipulate a carriage temperature above rather than below $0°C$ for pears as well as for apples. The use of polyethylene liners in pear cartons makes it imperative to ensure that the fruit is properly pre-cooled before loading. The purpose of the liners is to maintain high humidity and a modified atmosphere, but they also act as insulators and impede heat exchange. Cold fruit respires at a low rate, and one fresh air change per hour should be more than sufficient to prevent a harmful build-up of carbon dioxide in the hold (page 206). With regard to compatibility with other commodities, it should be noted that apples and pears are potent producers of ethylene (6).

It is of great value to be able to predict storage potential and to plan marketing accordingly (2,5). Once removed from cold storage or refrigerated transport, apples have a shelf life of days or weeks before texture and taste deteriorate following ripening and senescence. Most types of pear ripen normally only at moderate temperatures, but their quality is better when ripened after cold storage than if ripened immediately after harvest (5). Furthermore, a cold treatment has the useful effect of synchronising ripening so that all the fruits ripen at the same rate when transferred to a higher temperature (5). Ripe pears remain edible for a very brief period (perhaps only 2 or 3 days) and are so soft as to be prone to bruising; hence a consignment which is yellow on arrival may have no commercial value. If pears are stored for too long, they lose the ability to ripen on removal to ambient temperature (5). Oriental pears vary greatly in their post-harvest characteristics, some cultivars having exceptional keeping qualities, even at moderate temperature.

Shelf-life of apples and pears can be extended by the use of films and coatings. The resulting modified atmosphere retards respiration, but care is needed in selecting an appropriate formulation if physiological disorders are to be avoided (9).

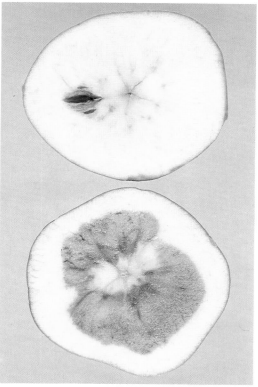

318 Brown heart of apple (carbon dioxide injury)

319 Brown core of oriental pear (natural senescence)

---

1 BOMPEIX G. (1985) Traitement des pommes après récolte contre les maladies de conservation. Données nouvelles et perspectives. *Arboriculture Fruitière* **32** (375), 43–46 illus.

2 BRAMLAGE W.J., WEIS S.A. & DRAKE M. (1985) Predicting the occurrence of poststorage disorders of 'McIntosh' apples from pre-harvest mineral analyses. *Journal of the American Society for Horticultural Science* **110**, 493–498.

3 CAPPELLINI R.A., CEPONIS M.J. & LIGHTNER G.W. (1987) Disorders in apple and pear shipments to the New York market, 1972–1984. *Plant Disease* **71**, 852–856.

4 EKSTEEN G.J. & COMBRINK J.C. (1987) *Manual for the identification of post-harvest disorders of pome and stone fruit.* Stellenbosch, South Africa: Fruit and Fruit Technology Research Institute, 25 pp illus.

5 FIDLER J.C., WILKINSON B.G., EDNEY K.L. & SHARPLES R.O. (1973) *The biology of apple and pear storage.* Farnham Royal, Slough, UK: Commonwealth Agricultural Bureaux, 235 pp illus.

6 LIU F.W. (1985) Factors influencing the effectiveness of low ethylene CA storage of apples. *Acta Horticulturae* No. 157, 113–119.

7 PIERSON C.F., CEPONIS M.J. & McCOLLOCH L.P. (1971) Market diseases of apples, pears and quinces. *Agriculture Handbook of the United States Department of Agriculture* No. 376, 112 pp illus.

8 PORRITT S.W., MEHERIUK M. & LIDSTER P.D. (1982) Postharvest disorders of apples and pears. *Agriculture Canada Publication* No. 1737, 66 pp illus.

9 SMITH S., GEESON J. & STOW J. (1987) Production of modified atmospheres in deciduous fruits by the use of films and coatings. *HortScience* **22**, 772–776.

10 ZWET T. van der & CHILDERS N.F. (Eds) (1982) *The pear – cultivars to marketing.* Gainesville, Florida: Horticultural Publications, 502 pp illus.

ALTERNARIA ROT of apples and pears caused by *Alternaria* spp.

320

OCCURRENCE *A. alternata* (Fr.) Keissler has been recorded on apples and pears in Israel (4) and Chile (10), on apples in Morocco (1) and India (7) and on pears in Canada (3) and the USA (5,6). *A. kikuchiana* S. Tanaka (8) is reported to cause serious damage to pears in Japan and Korea, and the author has isolated it from stems and fruits of pears imported into the UK from China. *A. mali* Roberts (5) occurs on apples in the USA (9) and in Korea (3). Some authorities consider both of these latter fungi to be types of *A. alternata* (**339, 340**).

SYMPTOMS Various symptoms can be produced. Dark grey mould may colonise the core cavity and proceed to rot the surrounding flesh (for 'mouldy core' and 'core rot' see page 182). Tissue which has suffered physical or physiological injury is often invaded, resulting in skin blemishes (3), small black corky lesions (9) or a shallow dark-coloured rot (1). Occasionally the stem undergoes a black decay, and even if the fruit remains unaffected its market value is reduced (6).

BIOLOGY These fungi can survive in the soil in the form of resting bodies (microsclerotia) or resting spores (chlamydospores) (1,8). Under favourable conditions asexual spores (conidia) are produced in abundance on leaf debris in the orchard (5) and are disseminated by wind and rain (9). They frequently lodge in dying flower parts and, in warm moist conditions (7), can cause an infection which is later manifest within the fruit. Some Asian pear cultivars are especially susceptible to infection by *A. kikuchiana*, which can invade sound young fruits via natural openings (stomata and lenticels), causing numerous black spots (8). The fungi also invade via wounds, possibly insect punctures (9), and may be associated with bitter pit, soft scald and sunscald lesions (q.v.).

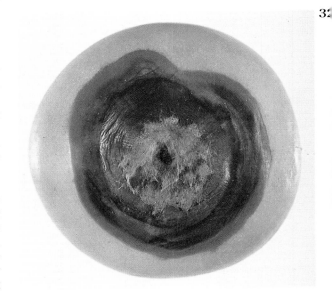

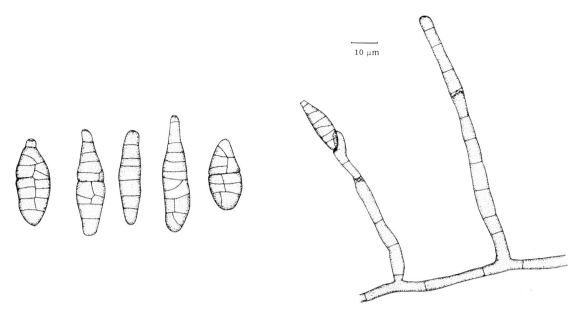

322 *Alternaria kikuchiana* (from CMI Descr. No. 954, David 1988) Conidia and conidiophores

CONTROL   There have been several reports of increased losses due to alternaria rot, following the use of systemic fungicides to control blue mould (6,10). Depending on the type of alternaria rot, possible means of control include a fungicide spray at blossom time (2), a post-harvest dip or drench (4,7) and precautions against physical injury and physiological disorders. In Japan there are efforts to breed more pear varieties with resistance to black spot, while in the UK statutory measures are enforced to prevent the introduction of oriental fungal strains into the British Isles.

---

1 AMRANI N. & NAJIM L. (1985) Contribution to a study of microscopical fungal flora of Morocco. 2. *Alternaria alternata*: microsclerotia and chlamydospores. *Cryptogamie, Mycologie* **6**, 265–270 illus.
2 LEE C.U. & KIM K.H. (1986) [Cross-tolerance of *Alternaria mali* to various fungicides.] *Korean Journal of Mycology* **14**, 71–78.
3 LOCKHART C.L. & FORSYTH F.R. (1974) *Alternaria alternata* storage decay of pears. *Canadian Plant Disease Survey* **54**, 101–102 illus.
4 PRUSKY D. & BEN-ARIE R. (1981) Control by imazalil of fruit storage rots caused by *Alternaria alternata*. *Annals of Applied Biology* **98**, 87–92.
5 ROBERTS J.W. (1924) Morphological characters of *Alternaria mali* Roberts. *Journal of Agricultural Research* **27**, 699–708 illus.
6 SITTON J.W. & PIERSON C.F. (1983) Interaction and control of *Alternaria* stem decay and blue mould in d'Anjou pears. *Plant Disease* **67**, 904–907 illus.
7 TAK S.K., VERMA O.P. & PATHAK V.N. (1986) Factors affecting conidial germination of *Alternaria alternata* (Fries) Keissler causing fruit rot of apple. *Korean Journal of Plant Protection* **24**, 129–133 illus.
8 TANAKA S. (1933) Studies on black spot disease of the Japanese pear (*Pirus serotina* Rehd.). *Memoir of the College of Agriculture, Kyoto University* No. 28, 31 pp illus.
9 TWEEDY B.G. & POWELL D. (1962) Cork rot of apples and its causal organism, a pathogenic strain of *Alternaria mali*. *Phytopathology* **52**, 1073–1079 illus.
10 VALDEBENITO S., R.M. & PINTO de T., A. (1972) Control de *Penicillium expansum*, *Botrytis cinerea* and *Alternaria alternata* en manzanas y peras. *Agricultura Técnica* **32**, 148–153.

BITTER ROT of apples and pears caused by
Glomerella cingulata **(Stonem.) Spauld. & v. Schrenk**
Conidial state: *Colletotrichum gloeosporioides* (Penz.) Sacc.
(syn. *Gloeosporium fructigenum* Berk.)

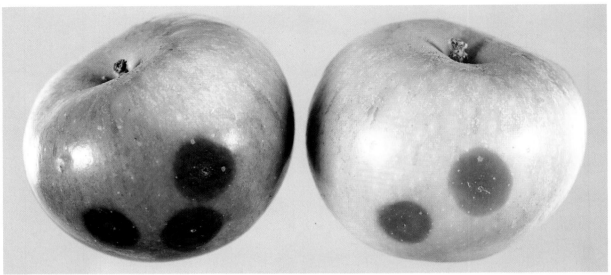

OCCURRENCE Heavy losses from bitter rot can occur in warm wet growing areas, and the disease is of particular importance in the south-eastern USA (4,10). Occasional epidemics have been reported from New Zealand (2) and Israel (5), and the disease also occurs in India (3) and in Poland (1). The term 'bitter rot' is sometimes used to include gloeosporium rot (q.v.) caused by related fungi (1).

SYMPTOMS The first symptoms are small light brown spots which later become slightly sunken. Minute dark structures (acervuli) form on the lesion, often in concentric circles. Under humid conditions the skin ruptures and spore-masses are extruded, turning from salmon-pink to dark brown (1). The rot beneath is characteristically cone-shaped, and a partially decayed apple has a bitter taste.

BIOLOGY Some strains of this fungus produce both the sexual stage (perithecia giving rise to ascospores) (**268**) and the asexual stage (acervuli giving rise to conidia), and both are important in infection (8,9) (cf. gloeosporium rot, q.v.). There are other strains which produce only the asexual state. The spore-bearing bodies develop in cankers on the tree; ascospores are liberated during rain showers and subsequently disseminated in air currents, while conidia are dispersed by rain water (9). Although the disease is often associated with ripening fruits (7), it has been established that apples are susceptible to infection at all stages of their development (4,10), the fungus being capable of direct penetration of the intact skin (2). The optimal temperature for infection is approximately 26°C, and epidemics are associated with wet growing seasons (2). Fruits may rot completely on the tree, forming shrivelled 'mummies' which survive the winter and, together with cankers, act as a source of spores the following year (1,2,6). Harvested fruits can become infected via injuries sustained during handling and transport (3).

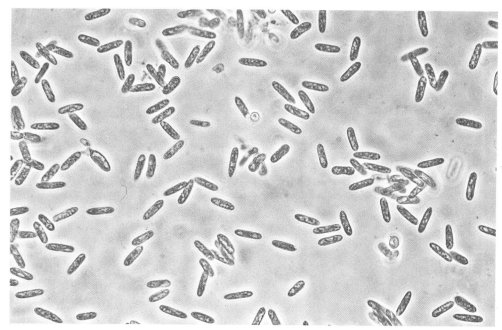

325 Conidia of *Colletotrichum gloeosporioides* (stained with cotton blue in lactophenol)

CONTROL The incidence of bitter rot may be reduced by pruning out dead wood and removing mummified fruits from the orchard (6). For complete control of the disease it is necessary to spray the trees with fungicide several times during the growing season. In irrigated orchards trickle irrigation is preferable to sprinklers, which are analogous to rain showers (5). Post-harvest fungicide treatment is also beneficial (3).

---

1 BORECKI Z. (1961) [Investigations on the bitter rot of apples caused by the fungi *Gloeosporium perennans* Zeller & Childs, *G. album* Osterw., and *G. fructigenum* Berk.] *Acta Agrobotanica* **10**(1), 53–97 illus.
2 BROOK P.J. (1977) *Glomerella cingulata* and bitter rot of apples. *New Zealand Journal of Agricultural Research* **20**, 547–555 illus.
3 CHAND J.N., KONDAL M.R. & AGGARWAL R.K. (1968) Epidemiology and control of bitter rot of apple caused by *Gloeosporium fructigenum* Berk. *Indian Phytopathology* **21**, 257–263
4 NOE J.P. & STARKEY T.E. (1982) Relationship of apple fruit maturity and inoculum concentration to infection by *Glomerella cingulata*. *Plant Disease* **66**, 379–381 illus.
5 PALTI J. & SHOHAM H. (1983) Trickle irrigation and crop disease management. *Plant Disease* **67**, 703–705.
6 ROBERTS J.W. (1915) Sources of the early infections of apple bitter-rot. *Journal of Agricultural Research* **4**, 59–64 illus.
7 SITTERLY W.R. & SHAY J.R. (1960) Physiological factors affecting the onset of susceptibility of apple fruit to rotting by fungus pathogens. *Phytopathology* **50**, 91–93.
8 STRUBLE F.G. & KEITT G.W. (1950) Variability and inheritance in *Glomerella cingulata* (Stonem.) S. & v. S. from apple. *American Journal of Botany* **39**, 110–119.
9 SUTTON T.B. & SHANE W.W. (1983) Epidemiology of the perfect stage of *Glomerella cingulata* on apples. *Phytopathology* **73**, 1179–1183.
10 TAYLOR J. (1971) Epidemiology and symptomatology of apple bitter rot. *Phytopathology* **61**, 1028–1029 illus.

BLACK ROT and WHITE ROT of apples and pears caused by
Botryosphaeria obtusa (Schw.) Shoem.
Conidial state: *Sphaeropsis malorum*
Botryosphaeria ribis Grossenb. & Duggar
Conidial state: *Dothiorella gregaria* Sacc.

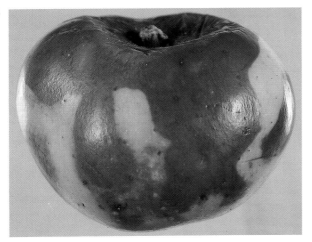

326

OCCURRENCE *B. obtusa* (syn. *Physalospora obtusa* (Schw.) Cooke) is responsible for black rot of apples and pears, which has been reported in the UK (4) and South Africa (3) but is of particular importance in the eastern USA (1,7). White rot is caused by *B. ribis*, which is synonymous with *B. berengeriana* de Not. (8) and is usually, but not always (8), considered to be synonymous with *B. dothidea* (Moug. ex Fr.) Ces. & de Not. White rot is especially important in the mid-west and southern USA (6). The disease has been recorded on apples in Australia, Korea (2) and Japan (8), and on pears in South Africa (3).

SYMPTOMS The first symtoms of both diseases are small brown spots on the skin of the fruit, often slightly sunken. In white rot the flesh becomes soft and the fruit may take on a bleached or cooked appearance, with drops of liquid in the rotted area. At temperatures around 23°C fruits may rot entirely within 10 days; lower temperatures are less favourable and tend to produce different symptoms, for example at 13°C the rot may be firm and brown, indistinguishable from the leathery type of lesion produced by the black rot organism. Because of the variability in symptoms produced by these species, identification necessitates culturing the causal fungus and examining its spores microscopically (6,9).

BIOLOGY Each of the fungi may produce both the sexual stage (perithecia giving rise to ascospores) and the asexual stage (pycnidia giving rise to conidia). The spore-bearing bodies are to be found throughout the year in dead branches and mummified fruits, and spore discharge is brought about by splashing rain (5,7,10). Conidia are waterborne while ascospores may be dispersed to greater distances in air currents. Fruits are invaded via lenticels, mechanical injuries (2) and insect punctures (4) and, in some cultivars, through the calyx (4) resulting in a blossom-end rot or core rot (q.v.). Young fruits are resistant to infection and decay, becoming susceptible as they mature (5). Rotted fruits drop from the tree, but fruits infected shortly before harvest appear healthy when placed in store. Both fungi are checked by storage temperatures around 0°C, but rotting is resumed at temperatures above 10°C. The decay does not usually spread from fruit to fruit.

CONTROL Measures include strict orchard hygiene and maintenance of good tree vigour. Weakening of the tree (by drought, for example) can result in infection and die-back of limbs, together with canker-formation (8,9). Dead wood and branches affected by fireblight (a bacterial disease) should be pruned out and destroyed, together with fruit mummies which would otherwise serve as a

327

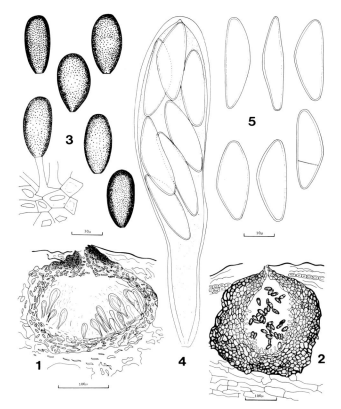

328 *Botryosphaeria obtusa* (from CMI Descr. No. 394, Punithalingam & Waller 1973) 1 pseudothecium 2 pycnidium 3 conidia 4 ascus 5 ascospores

source of fungal spores (4,5). In addition, the use of an orchard fungicide is necessary though, in view of the resistance of immature fruits, sufficient protection may be achieved by late-season spraying once the crop has attained a certain sugar content (2). Post-harvest fungicide treatments may be ineffective against infections which are well-established at the time of picking and packing.

---

1 BROWN E.A. & BRITTON K.O. (1986) *Botryosphaeria* diseases of apple and peach in the southeastern United States. *Plant Disease* **70**, 480–484 illus.
2 CHO W.D., KIM C.H. & KIM S.C. (1986) Pathogen physiology, epidemiology and varietal resistance in white rot of apple. *Korean Journal of Plant Protection* **25**, 63–70.
3 COMBRINK J.C., FOURIE J.F. & GROBBELAAR C.J. (1984) *Botryosphaeria* spp. on decayed deciduous fruits in South Africa. *Phytophylactica* **16**, 251–253 illus.
4 CROSSE J.E. & BENNETT M. (1951) Black rot and leaf spot of apple due to *Physalospora obtusa* (Schw.) Cooke. *Report of East Malling Research Station for 1950*, 137–138 illus.
5 DRAKE C.R. (1971) Source and longevity of apple fruit inoculum, *Botryosphaeria ribis* and *Physalospora obtusa*, under orchard conditions. *Plant Disease Reporter* **55**, 122–126.
6 FULKERSON J.F. (1960) *Botryosphaeria ribis* and its relation to a rot of apples. *Phytopathology* **50**, 394–398 illus.
7 HOLMES J. & RICH A.E. (1970) Factors affecting release and dissemination of *Physalospora obtusa* spores in a New Hampshire apple orchard. *Phytopathology* **60**, 1052–1054.
8 KOGANEZAWA H. & SAKUMA T. (1984) [Causal fungi of apple fruit rot.] *Bulletin of the Fruit Tree Research Station C (Morioka)* No. 11, 49–62 illus.
9 SHEAR C.L., STEVENS N.E. & WILCOX M.S. (1924) *Botryosphaeria* and *Physalospora* on currant and apple. *Journal of Agricultural Research* **28**, 589–598 illus.
10 SUTTON T.B. (1981) Production and dispersal of ascospores and conidia by *Physalospora obtusa* and *Botryosphaeria dothidea* in apple orchards. *Phytopathology* **71**, 584–589.

BLUE MOULD ROT of apples and pears caused by
Penicillium expansum Link

329

330

OCCURRENCE   Blue mould causes one of the most destructive rots of stored apples and pears. It is important in all producer countries, for example the USA (1,3,5), the UK (4), Poland (2), Italy (10), Israel (8), India (9) and Australia (7).

SYMPTOMS   The first symptoms are soft watery brown spots which undergo rapid enlargement at temperatures between 20° and 25°C. There is a distinct margin between soft rotted flesh and firm healthy tissue. In humid conditions masses of blue-green spores form on the surface of the lesion (1). The fungus can spread into neighbouring healthy fruits, forming 'nests' of decay.

BIOLOGY   *P. expansum* is generally considered to be a wound parasite, and infection commonly follows rough handling and washing procedures after harvest. Blue mould may also gain entry through *Gloeosporium*, *Mucor* or *Phytophthora* infection sites (q.v.) or via the lenticels (1). The resistance of lenticels to penetration has been related to fertiliser treatments and light intensity (4) but is chiefly determined by maturity, over-mature or long-stored fruit being very susceptible, especially when bruised or punctured. Fruit damaged by hail shortly before harvest shows an increased tendency to rot in store (10). Infection can occur even at 0°C and, although decay proceeds slowly at cold storage temperatures (3), rapid development ensues when the fruit is transferred to a warm environment (1). The mould produces a heat-resistant toxin (patulin), necessitating strict quality control of fruits used for processing.

CONTROL   Reduction in blue mould rot may be achieved by orchard fungicide sprays to control *Gloeosporium* infections which would otherwise provide a means of entry for blue mould. Calcium sprays (used to control bitter pit, q.v.) help to confer resistance to fungal infection. However, the chief means of controlling blue mould rot involves careful handling of fruit and strict hygiene in both orchard and packhouse (1). Squashed and rotting fruits should be destroyed before they can become a source of spores, and washing-water should be frequently changed and disinfected. A post-harvest fungicide dip or drench can be effective, especially when applied in conjunction with calcium treatment (5), but the time of application is critical and delays

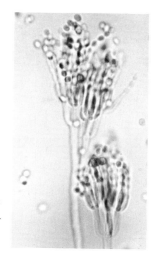

331 *Penicillium expansum* (from CMI Descr. No. 97, Onions 1966) Conidial heads x 500

of a few hours can result in large increases in decay. Use of the same or closely related chemicals before and after harvest is inadvisable because it can lead to the rapid development of tolerant strains of the fungus (7). It is preferable to use a range of fungicides with differing modes of action (8). Additional control may be achieved by subjecting the fruit to a hot-water bath immediately after harvest (9) and subsequently using controlled atmosphere storage. There have been investigations into the possibility of using biological control, in which the development of blue mould is inhibited by a non-invasive bacterium or yeast (6).

---

1 BAKER K.F. & HEALD F.D. (1934) An investigation of factors affecting the incidence of lenticel infection of apples by *Penicillium expansum*. *Bulletin of Washington Agricultural Experiment Station* No. 298, 48 pp.
2 BORECKA H. & BRYK H. (1982) [Role of the type and age of the inoculum in the development of *Penicillium expansum* (Link) Thom rot on stored apples.] *Acta Agrobotanica* **35**(2), 243–249.
3 BUCHANAN J.R., SOMMER N.F., FORTLAGE R.J., MAXIE E.C., MITCHELL F.G. & HSIEH D.P.H. (1974) Patulin from *P. expansum* in stone fruits and pears. *Journal of the American Society for Horticultural Science* **99**, 262–265.
4 COLHOUN J. (1962) Some factors influencing the resistance of apple fruits to fungal invasion. *Transactions of the British Mycological Society* **45**, 429–430.
5 CONWAY W.S. & SAMS C.E. (1987) The effects of postharvest infiltration of calcium, magnesium, or strontium on decay, firmness, respiration, and ethylene production in apples. *Journal of the American Society for Horticultural Science* **112**, 300–303.
6 JANISIEWICZ W.J. (1987) Postharvest biological control of blue mold on apples. *Phytopathology* **77**, 481–485 illus.
7 KOFFMANN W. & PENROSE L.J. (1987) Fungicides for the control of blue mould (*Penicillium* spp.) in pome fruits. *Scientia Horticulturae* **31**, 225–232.
8 PRUSKY D., BAZAK M. & BEN-ARIE R. (1985) Development, persistence, survival, and strategies for control of thiabendazole-resistant strains of *Penicillium expansum* on pome fruits. *Phytopathology* **75**, 877–882.
9 ROY M.K. (1975) Radiation, heat and chemical combines in the extension of shelf life of apples infected with blue mold rot (*Penicillium expansum*). *Plant Disease Reporter* **59**, 61–64.
10 TERZI A. & GORINI F.L. (1968) Studi sulla commercializzazione della pera 'Passa Crassana'. 5. Prove di lotta antimarciume post-raccolta. *Rivista della Ortoflorofrutticoltura Italiana* **52**, 617–621.

BROWN ROT of apples and pears caused by *Monilinia* spp.

Conidial state: *Monilia* spp.

OCCURRENCE *Monilinia fructigena* (Aderh. & Ruhl.) Honey, is widespread in Europe, Asia and South America (CMI Map 22), and causes serious losses in apple and pear fruits (3,10). It is uncommon in North America (2). *M. fructicola* (page 225) occurs in North and South America, South Africa, Japan, Australia and New Zealand (CMI Map 50), and is of minor importance on apples and pears (2,9). *M. laxa* (Aderh. & Ruhl.) Honey, occasionally causes fruit rots of apple and pear in Europe, Asia, and the Pacific coastal regions of North America (10) (CMI Map 44).

SYMPTOMS The first indication of fruit infection is the development of a small circular brown spot. Under humid conditions tufts of white mould break through the skin, often forming concentric circles around the point of infection. In the presence of light there is copious production of grey-brown spores (conidia). Occasionally infected fruits turn shiny black in store and, in the absence of light, there may be no spore formation. The fungus can spread into adjacent healthy fruit, but even in its advanced stage the rot tends to remain firm and dry (4,6).

BIOLOGY Fruits infected early in the growing season undergo rotting on the tree, turning into shrivelled 'mummies' which may remain attached or fall to the ground. The fungus thereby survives the winter and, when environmental conditions become favourable, spores are disseminated throughout the orchard. As well as producing conidia, *M. fructicola* (and, more rarely, *M. fructigena* and *M. laxa*) may also produce the sexual stage (apothecia giving rise to ascospores). The liberation of ascospores coincides with the emergence of young shoots and blossoms and, after infecting the flower parts, the fungus spreads into the developing fruit. Older fruits are more likely to be infected by conidia, dispersed by air currents, water splash or insects, and penetration is usually via wounds (7).

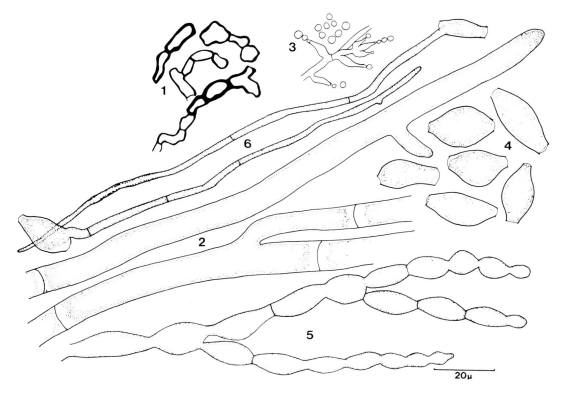

336 *Sclerotinia fructigena* (from CMI Descr. No. 617, Mordue 1979) 1 cells from sclerotium rind 2 hyphae (stained) 3 spermatia 4, 5, 6 mature, developing and germinating conidia

CONTROL   Orchard sprays are not very effective against the brown rot fungi, but a post-harvest fungicide dip or drench gives excellent control of secondary spread in store (1). Prevention of mechanical injury helps to reduce infection, and the use of refrigeration serves to slow down the rate of decay (5).

---

1 ALFORD D.V. & GWYNNE D.C. (1983) Pests and diseases of fruit and hops. In *Pest and disease control handbook* (Ed. by N. Scopes & M. Ledieu) pp. 295–374. Croydon, UK: British Crop Protection Council.
2 BATRA L.R. (1979) First authenticated North American record of *Monilinia fructigena*, with notes on related species. *Mycotaxon* **8**, 476–484.
3 BYRDE R.J.W. & WILLETTS H.J. (1977) *The brown rot fungi of fruit: their biology and control.* Oxford: Pergamon Press, 171 pp illus.
4 COLE M. & WOOD R.K.S. (1961) Types of rot, rate of rotting and analysis of pectic substances in apples rotted by fungi. *Annals of Botany* **25**, 417–434.
5 HALL R. (1972) Pathogenicity of *Monilinia fructicola*. 3. Factors influencing lesion expansion. *Phytopathologische Zeitschrift* **73**, 27–38.
6 HISLOP E.C., PAVER J.L. & KEON J.P.R. (1982) An acid protease produced by *Monilinia fructigena* in vitro and in infected apple fruits and its possible role in pathogenesis. *Journal of General Microbiology* **128**, 799–807.
7 MOORE M.H. (1950) Brown rot of apples: fungicide trials and studies of the relative importance of different wound-agents. *Journal of Horticultural Science* **25**, 225–234.
8 PRING R.J., BYRDE R.J.W. & WILLETTS H.J. (1981) An ultrastructural study of the infection of pear fruit by *Monilinia fructigena*. *Physiological Plant Pathology* **19**, 1–6.
9 SONODA R.M., OGAWA J.M. & MANJI B.T. (1982) Use of interactions of cultures to distinguish *Monilinia laxa* from *Monilinia fructicola*. *Plant Disease* **66**, 325–326.
10 WORMALD H. (1954) The brown rot diseases of fruit trees. *Technical Bulletin of the Ministry of Agriculture, Fisheries and Food, UK* No. 3, 113 pp. illus.

CORE ROT and MOULDY CORE of apples caused by various fungi

OCCURRENCE Fungal invasion of the core is particularly prevalent in apple cultivars having open calyces, such as Delicious, Gloster and Starking (8), and the problem has been reported from the USA (3), Canada (6), South Africa (4), New Zealand (1), India (9) and Hungary (7). The organism most commonly found in infected cores is *Alternaria alternata* (3,4,6), but many other fungi have been identified (5), including *Aspergillus niger* and *Fusarium* spp. (1,6,8), *Penicillium funiculosum* (4), *Pezicula malicorticis* (1), *Phoma* sp. (2), *Phomopsis mali* (10), *Pleospora herbarum* (4,7) and *Trichothecium roseum* (9). The black rot fungus (q.v.) can also attack in this way.

SYMPTOMS The seed cavities of the core region are disfigured by the presence of mould, which is often dark in colour (5). Sometimes the fungus progresses into the surrounding flesh, causing a dry rot or a soft wet rot, depending on the organism (1,4). Rotting of the core may commence whilst the apples are still on the tree (9) or alternatively only after several months of storage (10).

BIOLOGY These fungi characteristically invade the dead or dying flower parts at blossom time or later, and become established in the interior of the developing fruits, especially after late spring frosts (5).

CONTROL The incidence of mouldy core tends to increase following intensive spray programmes to control other fungi. Attempts to reduce mouldy core and core rot with orchard sprays have met with varying degrees of success (2,5).

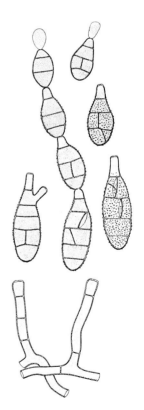

339 *Alternaria alternata* (CMI Drawing) Conidia and conidiophores

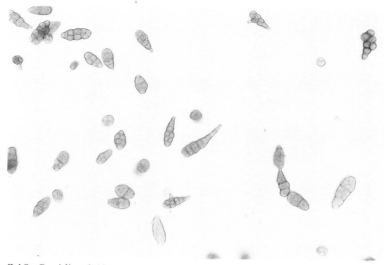

340 Conidia of *Alternaria alternata*

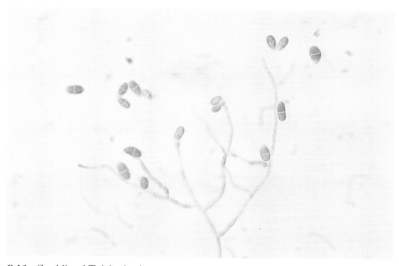

341 Conidia of *Trichothecium roseum* (stained with Trypan blue)

---

1. BRIEN R.M. (1937) The fungi associated with mouldy-core of apples. *New Zealand Journal of Agriculture* **54**, 283–286 illus.
2. BROWN E.A. & HENDRIX F.F. (1978) Effect of certain fungicides sprayed during apple bloom on fruit set and fruit rot. *Plant Disease Reporter* **62**, 739–741.
3. CEPONIS M.J., KAUFMAN J. & BUTTERFIELD J.E. (1969) Moldy carpels in Delicious apples on the Greater New York market. *Plant Disease Reporter* **53**, 136–138.
4. COMBRINK J.C., KOTZÉ J.M., WEHNER F.C. & GROBBELAAR C.J. (1985) Fungi associated with core rot of Starking apples in South Africa. *Phytophylactica* **17**, 81–83.
5. ELLIS M.A. & BARRAT J.G. (1983) Colonization of Delicious apple fruits by *Alternaria* spp. and effect of fungicide sprays on moldy-core. *Plant Disease* **67**, 150–152 illus.
6. HARRISON K.A. (1935) Mouldy core in Gravenstein apples. *Scientific Agriculture* **15**, 358–369.
7. IMRE K. (1983) [Ovary disease of Starking apple.] *Növényvédelem* **19**, 223–228.
8. MILLER P.N. (1959) Open calyx tubes as a factor contributing to carpel discoloration and decay of apples. *Phytopathology* **49**, 520–523.
9. RAINA G.L., BEDI P.S. & DUTT S. (1971) Occurrence of core rot of apple in nature in the Kulu valley of Himachal Pradesh, India. *Plant Disease Reporter* **55**, 283–284 illus.
10. ROSENBERGER D.A. & BURR T.J. (1982) Fruit decays of peach and apple caused by *Phomopsis mali*. *Plant Disease* **66**, 1073–1075 illus.

CYLINDROCARPON ROT of apples and pears caused by
Nectria galligena Bresad.
Conidial state: *Cylindrocarpon mali* (Allesch.) Wollenw.

342

OCCURRENCE The fungus causes European canker of apple and pear trees which occurs in all producer countries (CMI Map 38) but is of particular importance in the humid climates of northern Europe (10), the north-western USA, and New Zealand (2). Fruit-rotting may also occur, for example in Ireland (5,7), the UK, India and, after an unusually wet season, in California (6).

SYMPTOMS Lesions may occur at the calyx-end in the form of an 'eye-rot', or anywhere on the fruit surface. The rot is brown, sunken and with a definite margin between healthy and infected tissue (6). In the initial stages it resembles gloeosporium rot (q.v.) but can be distinguished by examination of the off-white spore-masses which form on mature lesions. Internally the rotted tissue is soft and often has a striated appearance (5,10).

BIOLOGY The fungus produces both the sexual stage (perithecia giving rise to ascospores) and the asexual stage (conidia), and both forms may be found in cankers on the tree. The time of spore discharge varies in different regions, being determined by the weather, in particular the incidence of rain. Conidia are dispersed by rain splash, while ascospores may be carried greater distances by air currents (10). Infection can occur through the calyx-end, the stem-end, through insect wounds or scab lesions, and the fruit may rot on the tree (5,6). Alternatively the fungus may enter a lenticel and remain quiescent until after harvest, when the fruit becomes increasingly susceptible to rotting (9). The fungus grows best at approximately 20°C and produces spores only in the light (1).

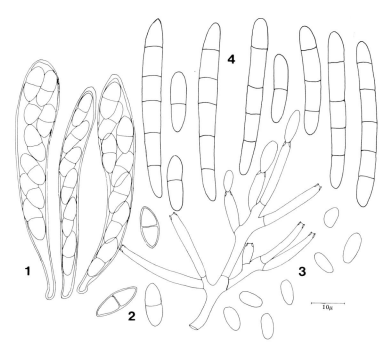

344 *Nectria galligena* (from CMI Descr. No. 147, Booth 1967) 1 asci 2 ascospores 3 microconidia and conidiophores 4 macroconidia

CONTROL The disease is controlled by eradicating or treating cankers before they can produce spores (3), and various fungicidal paints are available for this purpose (10). Young trees can be treated with a growth regulating chemical (8). It is beneficial to spray the orchard with fungicide at the time of leaf-fall, in order to prevent the fungus from becoming established in the leaf scars (2,4). Fruit-rotting can be controlled by orchard spraying coupled with a post-harvest fungicide dip (7).

---

1 AL-ZARARI A.J. (1978) Studies on *Nectria galligena* Bres. 1. The effect of temperature, light and media on growth and sporulation. *Pakistan Journal of Scientific and Industrial Research* **21**, 122–125.
2 BROOK P.J. & CLARKE A.D. (1975) Comparison of benomyl and captafol with Bordeaux mixture for control of European canker of apple. *New Zealand Journal of Experimental Agriculture* **3**, 271–272.
3 BYRDE R.J.W., CROWDY S.H. & ROACH F.A. (1952) Observations on apple canker. 5. Eradicant spraying and canker control. *Annals of Applied Biology* **39**, 581–587.
4 ENGLISH H., DUBIN H.J. & SCHICK F.J. (1979) Chemical control of European canker of apple. *Plant Disease Reporter* **63**, 998–1002.
5 KAVANAGH J.A. & GLYNN A.N. (1966) Brown rot of apples caused by *Nectria galligena* Bres. *Irish Journal of Agricultural Research* **5**, 143–144.
6 McCARTNEY W.O. (1967) An unusual occurrence of eye rot of apple in California due to *Nectria galligena*. *Plant Disease Reporter* **51**, 278–281 illus.
7 McDONNELL P.F. (1971) Control of *Nectria* decay of apple fruits by post-harvest chemical dip treatments. *Plant Disease Reporter* **55**, 771–773.
8 MEIER U. (1981) Untersuchungen zum Einfluss von Alar 85 auf die Pathogenese des Obstbaumkrebses (*Nectria galligena* Bres.). *Phytopathologische Zeitschrift* **101**, 43–50.
9 SAINDRENAN P. & BOMPEIX G. (1982) Production d'acide benzoïque chez les fruits immatures de cultivars communs du pommier, en réponse à l'infection par *Nectria galligena*. *Fruits* **37**, 249–257.
10 SWINBURNE T.R. (1975) European canker of apple (*Nectria galligena*). *Review of Plant Pathology* **54**, 787–799.

GLOEOSPORIUM ROT of apples and pears caused by
Pezicula alba Guthrie
Conidial state: *Gloeosporium album* Osterw.
Pezicula malicorticis (Jackson) Nannf.
Conidial state: *Gloeosporium perennans* Zeller & Childs

345

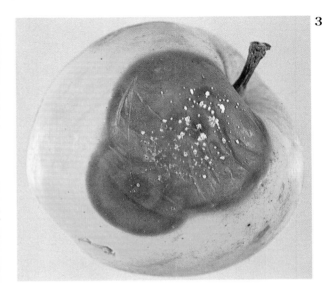

346

OCCURRENCE Gloeosporium rot is a major cause of loss in apples grown in cool humid regions (10); it is of lesser importance on pears, but occasional serious losses are reported (5). Both *Pezicula alba* and *P. malicorticis* are important in Europe (1,4,9), Australia and New Zealand (2), while in the USA the main species seems to be *P. malicorticis*, known there as *Neofabraea malicorticis* Jackson (5). *P. alba* has been reported from Canada (6) and South Africa (7). Both species are manifest only in mature stored fruits, hence the common name of 'ripe spot' (2).

SYMPTOMS *P. alba* produces a slow-growing circular firm brown rot which, under humid conditions, supports a white mould growth and waxy spore-masses. *P. malicorticis* tends to induce a zoned lesion, yellowish in the centre with a darker perimeter, hence the common names of 'bull's eye rot' or 'target spot' (5). In both diseases numerous lesions may develop on a single fruit; because of variability in symptoms the two species are distinguished by microscopic examination of the spores (9,10).

BIOLOGY These fungi may attack at blossom time (9) but characteristically gain entrance to sound developing fruits via the lenticels (1,9) and subsequently remain quiescent until some time after harvest (8). Infection of fruit results from the germination of asexual spores (conidia produced in pycnidia in cankers, diseased bark or dead leaves), and can occur at any time during the growing season when the weather is wet (5). The sexual stage (ascospores produced in apothecia) has been recorded for both fungi (4) but is rarely found and is not important in infection (9) (cf. bitter rot, q.v.).

CONTROL Complete eradication of cankers is impractical (2,9), and the main method of control is to use a fungicide. This may be applied in an orchard spray programme or, if legislation permits, as a post-harvest dip or drench (3). Refrigeration inhibits the onset of rotting, but eventually the fruit tissue loses its resistance. Infected fruit packed in polyethylene-lined cartons shows an increased tendency to decay (6).

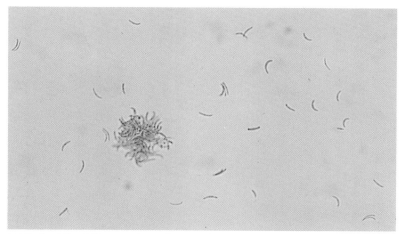

347 Conidia of *Gloeosporium album* (stained with cotton blue in lactophenol)

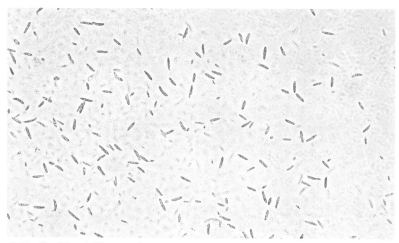

348 Conidia of *Gloeosporium perennans* (stained with cotton blue in lactophenol)

1 BOMPEIX G. (1978) The comparative development of *Pezicula alba* and *P. malicorticis* on apples and *in vitro* (air and controlled atmosphere). *Phytopathologische Zeitschrift* **91**, 97–109.

2 BROOK P.J. (1959) Experiments on the control of ripe spot on Sturmer apples. *New Zealand Journal of Agricultural Research* **2**, 463–474.

3 EDNEY K.L., BURCHILL R.T. & CHAMBERS D.A. (1977) The control of *Gloeosporium* storage rot of apple by reduced spray programmes. *Annals of Applied Biology* **87**, 51–56.

4 GUTHRIE E.J. (1959) The occurrence of *Pezicula alba* sp. nov. and *P. malicorticis*, the perfect stages of *Gloeosporium album* and *G. perennans* in England. *Transactions of the British Mycological Society* **42**, 502–506.

5 KIENHOLZ J.R. (1956) Control of bull's eye rot on apple and pear fruits. *Plant Disease Reporter* **40**, 872–877 illus.

6 LOCKHART C.L. (1967) Influence of controlled atmospheres on the growth of *Gloeosporium album in vitro*. *Canadian Journal of Plant Science* **47**, 649–651.

7 MATTHEE F.N. & WOLFSWINKLE L.D. (1963) *Gloeosporium album* – an apple disease new to the Western Cape. *Deciduous Fruit Grower* **13**, 145–146.

8 NOBLE J.P. & DRYSDALE R.B. (1983) The role of benzoic acid and phenolic compounds in latency in fruits of two apple cultivars infected with *Pezicula malicorticis* or *Nectria galligena*. *Physiological Plant Pathology* **23**, 207–216.

9 OLSSON K. (1965) A study of the biology of *Gloeosporium album* and *G. perennans* on apples. *Meddelanden Statens Växtskyddsanstalt (Stockholm)* **13** (104), 189–259 illus.

10 WILKINSON E.H. (1954) Fungal rots of apples with special reference to *Gloeosporium* spp. *Annals of Applied Biology* **41**, 354–358.

GREY MOULD ROT of apples and pears caused by
Botryotinia fuckeliana (de Bary) Whetzel
Conidial state: *Botrytis cinerea* Pers.

349

350

OCCURRENCE Grey mould rot can cause heavy losses and is particularly important in pears (1,3). It is found in all producer countries, for example the UK (10), Italy (5), Germany (4), Switzerland (2), Norway (9), Australia (1), South Africa (3) and the USA (6,7,8).

SYMPTOMS The disease can take several forms. In calyx-end rot the first symptom is a slight reddening of the skin surrounding the 'eye' of the fruit (6). A dark dry lesion appears which may progress no further but often develops into a soft brown rot affecting the entire fruit (3,10). Occasionally the initial infection is via the cut stem, but more commonly the fungus gains entry through skin breaks on any part of the fruit (8). Under humid conditions there is copious production of grey-brown spores. Tough black resting bodies (sclerotia), a few mm in size, may form eventually (**350**).

BIOLOGY The fungus survives as sclerotia in the soil and on plant debris, and under favourable conditions asexual spores (conidia) are formed. During wet and windy weather they can cause infection of dying blossoms (3,7). The fungus remains quiescent in the flower parts and only later invades the fruit. Direct infection may occur via wounds sustained during harvesting and handling (1). Rotting can proceed even at $-1°C$, and the decay spreads to adjacent healthy fruit, eventually causing extensive 'nesting' (7).

CONTROL Measures include clearing the orchard of dead plant material, spraying the trees with fungicide immediately after blossoming (3,4,7) and dipping the fruit after harvest (1,5). Use of the same or similar chemicals for pre- and post-harvest application should be avoided, as this practice can lead to the development of tolerant strains of the fungus and a consequent increase in losses. Simple disinfectants can help to reduce the number of spores available for infection, and gentle handling is important to minimise injury to the fruit (8). Attempts have been made to control the disease biologically by spraying the blossom with a fungus (*Trichoderma viride*) which is antagonistic to the grey mould (9).

---

1 BEATTIE B.B. & OUTHRED N.L. (1970) Benzimidazole derivatives as post-harvest fungicides to control rotting of pears, cherries and apricots. *Australian Journal of Experimental Agriculture and Animal Husbandry* **10**, 651–656.
2 BOLAY A. (1983) La protection phytosanitaire en arboriculture fruitière et en cultures de petits fruits. Maladies fongiques et bactériennes. *Revue Suisse de Viticulture, d'Arboriculture et d'Horticulture* **15**, 21–23.
3 COMBRINK J.C., GROBBELAAR C.J. & JONKER J. (1983) Preliminary investigation on calyx-end rot in pears. *Deciduous Fruit Grower* **33**, 367–370 illus.
4 MAPPES D. (1983) Die Kelchgrubenfäule des Apfels, ihre Ursache und Bekämpfung. *Mededelingen van de Faculteit Landbouwwetenschappen Rijksuniversiteit Gent* **48**, 581–590.
5 MORI P. & SOZZI A. (1980) Trattamenti per la prevenzione dei marciumi da conservazione delle pere. *Rivista della Ortoflorofrutticoltura Italiana* **64**, 623–629.
6 PALMITER D.H. (1951) A blossom-end rot of apples in New York caused by *Botrytis*. *Plant Disease Reporter* **35**, 435–436 illus.
7 SOMMER N.F., BUCHANAN J.R., FORTLAGE R.J. & BEARDEN B.E. (1985) Relation of floral infection to *Botrytis* blossom-end rot of pears in storage. *Plant Disease* **69**, 340–343 illus.
8 SPOTTS R.A. & PETERS B.B. (1982) Use of surfactants with chlorine to improve pear decay control. *Plant Disease* **66**, 725–727.
9 TRONSMO A. & YSTAAS J. (1980) Biological control of *Botrytis cinerea* on apple. *Plant Disease* **64**, 1009.
10 WILKINSON E.H. (1942) Dry-eye rot of apples caused by *Botrytis cinerea* Pers. *Journal of Pomology* **20**, 84–88.

MUCOR ROT of apples and pears caused by *Mucor piriformis* Fischer

352

OCCURRENCE This disease is of major importance in Canada and the USA, where pears in particular are affected (2,3,5). Mucor rot of apples occurs in the UK (4) and in South Africa (1).

SYMPTOMS Lesions may be at the stem-end (2), at the calyx-end, in the core region (4) or anywhere on the fruit surface (1). Affected skin is dark brown and parchment-like, and the flesh beneath is pale brown, soft and watery, with no odour (1,2). In a humid atmosphere there is profuse production of mould strands bearing black spore-heads (sporangia) (page 246).

BIOLOGY The fungus can survive unfavourable periods, such as a hot dry summer, in the sexual state (zygospores) (3). However, the asexual stage (sporangia giving rise to sporangiospores) is more abundant and is of prime importance in infection. Infested soil enters the packhouse on boxes used for harvest, and spores accumulate in the dump tanks (2). Injury to the skin of the fruit permits infection (1). Alternatively the fungus invades via the stem-end (2) or, in some apple cultivars, through the open calyx tube (4). Following infection there may be a period of quiescence lasting several weeks but, since decay proceeds even at 0°C, significant losses can occur during long-term storage (2).

CONTROL Orchard hygiene is important. It is advisable to collect and destroy fallen fruits before they can be colonised by the fungus; such a measure helps to minimise the build-up of spores in the soil (3). Harvesting in wet weather should be avoided if possible (2). Boxes and bins must be cleaned periodically to remove soil which would otherwise contaminate the treatment tanks (4). Chlorination (with frequent renewal) is beneficial, and the addition of a surfactant permits good contact between fungicidal liquid and fruit surface (5). Anti-scald preparations (page 214) should be correctly formulated, otherwise the fruit may suffer chemical injury and hence be predisposed to fungal invasion (4). Gentle handling is important, to minimise the incidence of wounds (1). In store, a relative humidity of about 95% is desirable, but a saturated atmosphere should be avoided.

---

1 COMBRINK J.C. & FOURIE J.F. (1984) Post-harvest decay of stone and pome fruit. *Deciduous Fruit Grower* **34**, 392–396 illus.
2 LOPATECKI L.E. & PETERS W. (1972) A rot of pears in cold storage caused by *Mucor piriformis*. *Canadian Journal of Plant Science* **52**, 875–879 illus.
3 MICHAELIDES T.J. & SPOTTS R.A. (1986) Mating types of *Mucor piriformis* isolated from soil and pear fruit in Oregon orchards. (On the life history of *Mucor piriformis*.) *Mycologia* **78**, 766–770 illus.
4 SHARPLES R.O. & HIMS M.J. (1986) Post-harvest diseases. Mucor core rot. *Report of East Malling Research Station for 1985*, 115–116.
5 SPOTTS R.A. & CERVANTES L.A. (1987) Effects of the nonionic surfactant Ag-98 on three decay fungi of Anjou pear. *Plant Disease* **71**, 240–242.

PINK MOULD ROT of apples and pears caused by
Trichothecium roseum Link

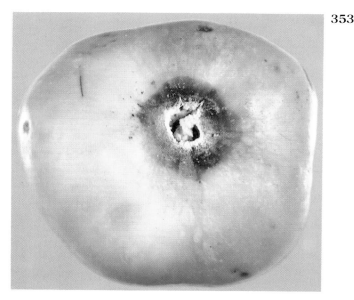

353

OCCURRENCE This disease has been recorded in Italy (5), Portugal (4), France, the UK, Belgium, Denmark, Germany (2), Czechoslovakia, Poland (1), the USSR, the USA and India (3).

SYMPTOMS Pink mould, which imparts a bitter taste to infected flesh, tends to follow in the wake of scab (q.v.) or black rot (q.v.). Alternatively it may cause a severe core rot (q.v.) (4,5). Characteristic pink spores are produced (**341**).

BIOLOGY The fungus is common in soil, and its spores (conidia) are probably disseminated by wind, water and insects. Infection occurs via wounds, insect galleries or lesions made by other fungi (4); there is one report of invasion through natural openings (lenticels) (1). Under humid conditions the fungus may penetrate the open 'eye' of the fruit and thence into the core and surrounding flesh (5).

Some apple and pear cultivars are prone to attack whilst still on the tree; after harvest, susceptibility to infection increases with length of storage period (1,2). Decay is rapid at temperatures between 20° and 25°C, but greatly inhibited below 10°C (1,4); consequently the disease is chiefly associated with storage at ambient temperature in warm climates (3).

CONTROL Care should be taken to prevent the occurrence of diseases or injuries which facilitate infection by pink mould (4). Control measures include the application of fungicide and insecticide sprays during the growing season, followed by careful harvesting and handling. Normal cold storage prevents development of pink mould (1); if fruit is to be stored at ambient temperature, post-harvest hot water treatment can inactivate deep-seated infections. A fungicide dip may also be used (3).

---

1 BORECKI Z. & PROFIC H. (1962) [The fungi *Trichothecium roseum* Link, *Cryptosporiopsis malicorticis* (Cord.) Nannf. and *Penicillium expansum* (Link) Thom as primary or secondary pathogens of apple.] *Acta Agrobotanica* **12**, 79–94 illus.
2 HATTINGEN R. (1958) Zur Frage der parasitären Natur von *Trichothecium roseum* Link bei Äpfeln und Birnen. *Jahresbericht des Bayerischen Landwirtschaftsrats (München)* **35**, 627–633.
3 KAUL J.L. (1986) Efficacy of biphenyl and sodium orthophenylphenate in controlling fungal rots of apple. *Indian Phytopathology* **39**, 282–286.
4 OLIVEIRA B. d' (1930) *Cephalothecium roseum* Corda subsidios para o estudio da podridão rósea do coração das maçãs. *Revista Agronómica, Lisboa* **18**(3), 49–63 and **18**(4), 9–67 illus.
5 PEYRONEL B. (1921) Il marciume amaro o marciume del cuore delle mele e delle pere. *Bolletino Mensile della R. Stazione di Patologia Vegetale* **2**, 23–27.

PHACIDIOPYCNIS ROT of apples and pears caused by
Potebniamyces pyri **(Berk. & Broome) Dennis**
Conidial state: *Phacidiopycnis malorum* Potebnia
(Syn. *Pyrenochaeta furfuracea* (Fr.) Rostr.)

354

355

OCCURRENCE This disease has been recorded in Belgium, France (1,2,9), Italy (5), Switzerland (6), Germany (4,10), Norway (8), Denmark and the USSR (7). The author has observed it on pears grown in the UK and on pears imported from Italy.

SYMPTOMS Lesions are almost invariably found only at the stem-end of the fruit (2,10) (cf. pyrenochaeta rot, q.v.), and the affected area is at first reddish-brown, later black and wrinkled (5). The firm leathery texture is somewhat similar to

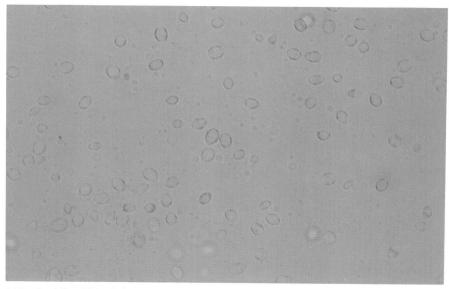

356 Conidia of *Phacidiopycnis malorum* (stained with cotton blue in lactophenol)

that of brown rot (q.v.) (6,9), but phacidiopycnis rot is distinguished by the formation of numerous characteristic greenish-grey bodies (pycnidia) up to 1 mm across and clearly visible to the naked eye (7). These structures darken and exude abundant spores (conidia) in the form of creamy droplets (1) or waxy coils (5,9).

BIOLOGY Under adverse growing conditions (following a hard frost, for example) the trees are predisposed to infection (1,3,9). The fungus becomes established in the bark, causing cankers on which may be found both the sexual state (apothecia giving rise to ascospores) (7,10) and the asexual state (pycnidia giving rise to micro- and macro-conidia) (4,5). Spores are disseminated by wind and rain, and cause new infections in weakened branches. The fungus may progress into fruit stalks and yet be invisible at harvest time (2); further development can occur even in cold store (5) and the disease is manifest as a stem-end rot (10).

CONTROL The chief means of control is careful pruning of cankered branches (9). All prunings should be burned lest they continue to provide a source of air-borne spores (1).

---

1 BARTHELET J. (1943) Recherches sur quelques parasites des arbres fruitiers. *Annales des Épiphyties (NS)* **9**, 27–45 illus.

2 BONDOUX P. (1967) Les maladies cryptogamiques des poires et des pommes au cours de l'entreposage. *Annales des Épiphyties* **18**, 509–550 illus.

3 BROOKS F.T. (1928) On the occurrence of *Phacidiella discolor* (Mout. & Sacc.) Potebnia in England. *Transactions of the British Mycological Society* **13**, 75–81 illus.

4 FICKE W. & PETER E. (1982) Über das Vorkommen in der DDR bisher nicht nachgewiesener Rindenbranderreger am Apfel. *Archiv für Phytopathologie und Pflanzenschutz* **18**, 31–38 illus.

5 GOVI G. (1952) Un marciume di frutti immagazzinati (*Phacidiopycnis furfuracea* (Rostr.) Jørst.). *Annali della Sperimentazione Agraria (NS)* **6**, 1121–1129 illus.

6 OSTERWALDER A. (1922) *Phacidiella discolor* (Mout. & Sacc.) A. Poteb. als Fäulnispilz beim Kernobst. *Landwirtschaftliches Jahrbuch der Schweiz* **36**, 852–853.

7 POTEBNIA A. (1912) Ein neuer Krebserreger des Apfelbaumes, *Phacidiella discolor* (Mout. & Sacc.) A. Pot., seine Morphologie und Entwicklungsgeschichte. *Zeitschrift für Pflanzenkrankheiten und Pflanzenschutz* **22**, 129–148 illus.

8 RAMSFJELL T. (1951) [Fungal diseases of apple fruit.] *Frukt og Bær* **4**, 79–94 illus.

9 VIENNOT-BOURGIN G. (1949) *Phacidiella* Potebnia. In *Les champignons parasites des plantes cultivées*, pp. 646–648 illus. Paris: Masson & Cie.

10 WOLLENWEBER H.W. (1937) Der schwarze Rindenbrand der Quitte. (Erreger: *Phacidiella discolor* (Mout. & Sacc.) Potebnia.) *Angewandte Botanik* **19**, 131–140 illus.

PHYTOPHTHORA ROT of apples and pears caused by
Phytophthora cactorum (Lebert & Cohn) Schroet.
Phytophthora syringae (Kleb.) Kleb.

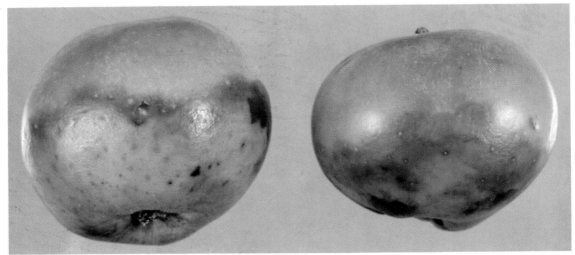

357

OCCURRENCE   In North America and Europe stored apples and pears occasionally suffer heavy losses due to *P. cactorum* or *P. syringae*, though these fungi are chiefly important as the agents of collar rot of the tree (1,5,6,7). In the UK, *P. syringae* has become a major cause of rotting in stored apples and pears (2,4,8). Fruit-rotting is also reported from Australia (9).

SYMPTOMS   The lesions are pale to mid-brown in apples and dark brown in pears, and the boundary between healthy and rotted flesh is indistinct. Affected tissue is characteristically firm (though it may become spongy at a later stage) and the vascular strands show a marked browning (6). If the skin of the fruit has been damaged, a white 'bloom' of mould growth may be visible on the exposed flesh, giving rise to both the asexual stage (sporangia) and the sexual stage (oospores) (3,10). On the other hand, there may be merely a 'marbling' of the skin, with no visible mould until the advent of secondary organisms (8).

BIOLOGY   *Phytophthora* species survive in the soil, and infection is closely related to orchard practices and the weather. If the ground below the trees is kept free of weeds (generally considered good orchard practice to conserve soil moisture), soil particles are splashed up by rain, and zoospores come into contact with low-hanging fruit. Wet weather around harvest time can lead to direct infection (1,2), generally via the natural openings (lenticels) (5). Harvested fruits left in bins in the orchard are subject to heavy contamination. Infected fruits later undergo rotting in store (9), and 'nests' of decay are formed as the fungus spreads to adjacent healthy fruits (2).

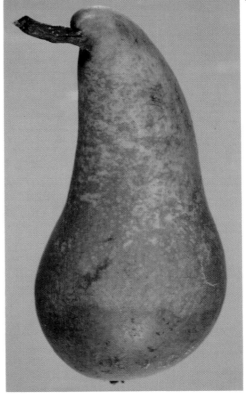

358

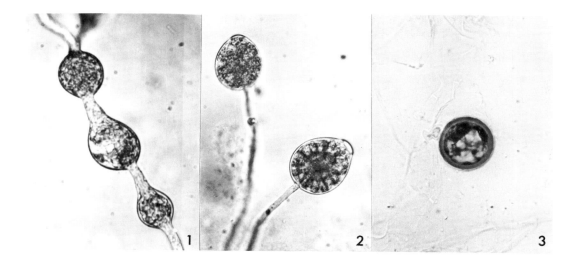

359 *Phytophthora syringae* (from CMI Descr. No. 32, Waterhouse & Waterston 1964) 1 hyphal swellings 2 sporangia, living material in water x 750 3 antheridium and oogonium (stained with cotton blue in lactophenol) x 500

CONTROL   The increased incidence of phytophthora rot may be related to the use of dwarfing rootstocks and pruning systems which encourage production of fruit near the ground. If the soil under the trees is kept bare it may be drenched with an appropriate fungicide. The fruit should be protected by orchard sprays and a post-harvest dip or drench (4,8). During wet seasons particular care should be taken to remove all harvested fruits from the orchard immediately after picking (2).

---

1 BOLAY A. (1977) *Phytophthora syringae*, agent d'une grave pourriture des pommes en conservation. *Revue Suisse de Viticulture, d'Arboriculture et d'Horticulture* **9**, 161–169
2 EDNEY K.L. (1978) The infection of apples by *Phytophthora syringae*. *Annals of Applied Biology* **88**, 31–36.
3 HARRIS D.C. & COLE D.M. (1982) Germination of *Phytophthora syringae* oospores. *Transactions of the British Mycological Society* **79**, 527–530 illus.
4 McCRACKEN A.R. (1984) Control of fungal rotting in gas-stored Bramley's Seedling apples. *Tests of Agrochemicals and Cultivars* No. 5 (*Annals of Applied Biology* **104**, Supplement), 26–27.
5 MOURICHON Z. & BOMPEIX G. (1979) Mise en évidence des voies de pénétration du *Phytophthora cactorum* (L. & C.) Schroeter dans les pommes 'Golden Delicious'. Étude réalisée en microscope électronique à balayage. *Fruits* **34**, 761–766.
6 ROSE D.H. & LINDGREN C.C. (1925) Phytophthora rot of pears and apples. *Journal of Agricultural Research* **30**, 463–468 illus.
7 ROSS R.G. & GOURLEY C.O. (1969) *Phytophthora syringae* fruit rot of apples in Nova Scotia. *Canadian Plant Disease Survey* **49**, 33–35.
8 UPSTONE M.E. (1978) *Phytophthora syringae* fruit rot of apples. *Plant Pathology* **27**, 24–30 illus.
9 WICKS T. & BUMBIERIS M. (1981) *Phytophthora* fruit rot of Delicious apples in South Australia. *Australasian Plant Pathology* **10**, 23–24.
10 WORMALD H. (1920) A *Phytophthora* rot of pears and apples. *Annals of Applied Biology* **6**, 89–100 illus.

POWDERY MILDEW of apples and pears caused by
Podosphaera leucotricha (Ell. & Ev.) Salmon
Conidial state: *Oidium* Link

360

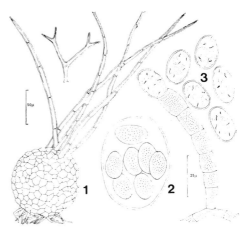

361 *Podosphaera leucotricha* (from CMI Descr. No. 158, Kapoor 1967) 1 cleistothecium 2 ascus and ascospores 3 conidia

OCCURRENCE Powdery mildew is chiefly known as a serious disease of buds and foliage, but it can also cause a blemish of apple and pear fruits (4,5). The disease is found in all growing areas and is particularly important in the UK (1,2,3) and parts of North America (4,5) (CMI Map 118).

SYMPTOMS The characteristic symptom on fruits is a smooth lace-like russeting of the skin which is non-progressive but which detracts from market appearance (5). A similar russeting may be caused by frost or by certain chemical sprays (1,4).

BIOLOGY The fungus produces both the sexual stage (cleistothecia giving rise to ascospores) and the asexual stage (conidia), and it is the latter which is of importance in infection (2). Fungal strands and asexual spores usually survive the winter in infected buds (3), though they may be killed by severe cold. In the spring when the buds unfold there is continuous spore production (2), and infection can occur even in dry conditions (5). The blossoms are vulnerable even before they open, and it is these early infections which eventually result in damaged fruits (5). After petal-fall the developing fruits become resistant (5).

CONTROL Mildew control strategy is based on removal of the primary source by pruning out or treating infected buds, followed by numerous protectant applications of fungicide during the growing season (1,3). It is essential to spray just before bloom, if latent fruit infection is to be prevented (4). Much effort has gone into devising spray schedules which are cost-effective and beneficial (3), with minimum adverse effect on fruit skin-finish (1).

---

1 BURCHILL R.T., FRICK E.L., COOK M.E. & SWAIT A.A.J. (1979) Fungitoxic and phytotoxic effect of some surface-active agents applied for the control of apple powdery mildew. *Annals of Applied Biology* **91**, 41–49.
2 BUTT D.J. & JEGER M.J. (1986) Components of spore production in apple powdery mildew (*Podosphaera leucotricha*). *Plant Pathology* **35**, 491–497 illus.
3 HUNTER L.D., BLAKE P.S. & SWAIT A.A.J. (1982) The use of fungicide-containing mixtures during the dormant season for safe and effective apple powdery mildew eradication. *Journal of Horticultural Science* **57**, 289–294.
4 SPOTTS R.A. (1984) Infection of Anjou pear fruit by *Podosphaera leucotricha*. *Plant Disease* **68**, 857–859 illus.
5 YODER K.S. & HICKEY K.D. (1983) Control of apple powdery mildew in the mid-Atlantic region. *Plant Disease* **67**, 245–248 illus.

RHIZOPUS ROT of apples and pears caused by
Rhizopus oryzae Went & Prinsen Geerligs
Rhizopus stolonifer (Ehrenb. ex Fr.) Lind

362

OCCURRENCE *R. oryzae* has been reported on apples in India (2,4) and Pakistan (3), while *R. stolonifer* has been recorded on apples in the USA (1) and on pears in Chile (5). It is conceivable that some other records are misidentifications of mucor rot (q.v.).

SYMPTOMS The first symptom is a soft brown spot which enlarges rapidly at temperatures around 25°C. If the skin covering the lesion is ruptured, juices leak out and there is a characteristic sour odour. Coarse white mould strands are formed on the exposed tissue and give rise to white globular spore-heads which later turn black (page 230) (2,3). A section through the fruit may show that the seeds are enveloped in mould.

BIOLOGY The asexual spores (sporangiospores) are ubiquitous, but generally cause infection only in fruits which are injured or over-ripe (3). Insect damage permits infection of unripe fruits on the tree; a pre-harvest stem-end rot has also been reported (2). Once established, these fungi can penetrate directly into adjacent healthy fruits. The optimal growth temperature is approximately 25°C for *R. stolonifer* and about 35°C for *R. oryzae* (2); both are inhibited by temperatures below 5°C and therefore do not appear during refrigerated storage (5). The sexual stage (zygospores) is rarely encountered, being formed only when two different strains of the fungus are present.

CONTROL Orchard sprays may be necessary to protect the fruit from insect attack and to reduce the incidence of primary infection (3). The fruit should be harvested at optimal maturity, handled with care and treated with a post-harvest fungicide (4). Refrigeration can arrest decay completely (5).

---

1 HEALD F.D. & RUEHLE G.D. (1931) The rots of Washington apples in cold storage. *Bulletin of Washington Agricultural Experiment Station* No. 253, 48 pp illus.
2 MEHTA P.R. (1939) A fruit rot of apples caused by a species of *Rhizopus*. *Indian Journal of Agricultural Science* **9**, 711–718 illus.
3 TANVEER M. & KAMAL M. (1966) Studies on rot of apple caused by *Rhizopus arrhizus*. 1. Symptoms, occurrence, associated fungi and pathogenicity. *West Pakistan Journal of Agricultural Research* **4**, 135–142 illus.
4 THAKUR D.P. & CHENULU V.V. (1970) Chemical control of soft rot of apple and mango fruits caused by *Rhizopus arrhizus*. *Indian Phytopathology* **23**, 58–61.
5 VALDEBENITO S., R.M. & PINTO de T., A. (1971) Identificación y patogenicidad de hongos aislados de pudriciones de peras Winter Nelis y Packham's Triumph almacenadas en "bins" en cámaras frigoríficas. *Agricultura Técnica* **31**, 49–51.

SCAB of apples and pears caused by
***Venturia inaequalis* (Cooke) Winter**
Conidial state: *Spilocaea pomi* Fr.
***Venturia pirina* Aderh.**
Conidial state: *Fusicladium pyrorum* (Lib.) Fuckel

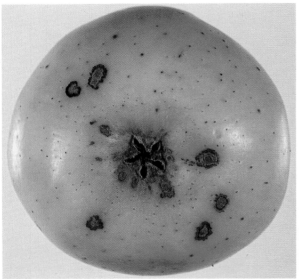

363

364

OCCURRENCE Scab is one of the most important diseases of apples and pears, particularly in cool humid climates such as the UK (3), Switzerland (9), Canada (6) and the northern USA (5,10). It has been recorded in most other growing areas, including India (2), South Africa (7), Israel (8) and Chile (4) (CMI Maps 120, 367).

SYMPTOMS The disease is primarily an orchard problem, but late-season infection can lead to the development of scab in store (10). Lesions consist of dark brown or black spots which become corky with age and sometimes support mould growth at the margins. The lesions are usually most numerous around the calyx-end and, if infection took place whilst the fruit was developing, the apple may be mis-shapen as a result.

BIOLOGY The fungus survives the winter in fallen apple leaves, where the sexual stage develops (perithecia giving rise to ascospores) (3,6). The spores are released during wet periods in early spring (4,5) and warm humid weather favours infection of both leaves and fruits (7). Asexual spores (conidia) are produced on the infected tissue (9) and are dispersed by rain splash; continuously wet weather can lead to heavy infection of mature fruits, and long storage should be avoided (7). Although the fungus does not spread into healthy fruit in store, slightly scabbed fruits will deteriorate further even at temperatures close to 0°C, the lesions increasing both in size and number (10).

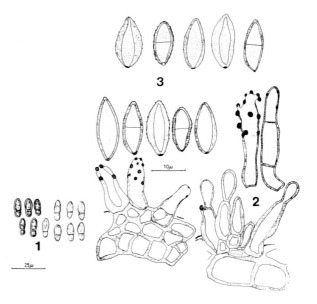

365 *Venturia pirina* (from CMI Descr. No. 404, Sivanesan & Waller 1974)   1 ascospores
2 conidiophores   3 conidia

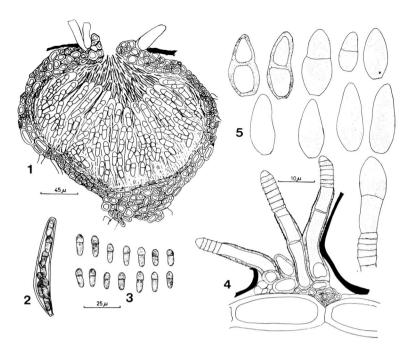

366 *Venturia inaequalis* (from CMI Descr. No. 401, Sivanesan & Waller 1974)
1 pseudothecium   2 ascus   3 ascospores   4 conidiophores   5 conidia

CONTROL   Fungicide spraying may be done according to a protective or a curative programme (7,8). Orchard inspections, coupled with a reliable weather forecasting service, are of assistance in timing the sprays correctly (2,5,7). It can also be beneficial to apply urea after harvest but before leaf-fall; the treated leaves decompose rapidly after falling from the tree, thus depleting the food source available to the fungus. There have been attempts to reduce scab by biological means (using various parasitic fungi) (1) and by breeding resistant apple cultivars (2).

---

1 BOUDREAU M.A. & ANDREWS J.H. (1987) Factors influencing antagonism of *Chaetomium globosum* to *Venturia inaequalis*: a case study in failed biocontrol. *Phytopathology* **77**, 1470–1475.
2 GUPTA G.K. (1984) Recent trends in forecasting and control of apple scab (*Venturia inaequalis*). *Pesticides* **19**(1), 19–31.
3 JEGER M.J. & BUTT D.J. (1983) Overwintering of *Venturia inaequalis* the causal agent of apple scab in relation to weather. *Annals of Applied Biology* **103**, 201–218.
4 LATORRE B.A., YANEZ P. & RAULD E. (1985) Factors affecting release of ascospores by the pear scab fungus (*Venturia pirina*). *Plant Disease* **69**, 213–216.
5 MacHARDY W.E. & GADOURY D.M. (1986) Patterns of ascospore discharge by *Venturia inaequalis*. *Phytopathology* **76**, 985–990.
6 St-ARNAUD M., COULOM L.J., NEUMANN P. & JACOB A. (1985) La maturation et l'éjection des ascospores du *Venturia inaequalis* à Frelighsburg (Québec) en relation avec la température et la pluie. *Phytoprotection* **66**, 153–161.
7 SCHWABE W.F.S. (1980) Epidemiology and control of apple scab in South Africa. *Phytophylactica* **12**, 219–222.
8 SHABI E., KOENRAADT H. & KATAN T. (1986) Further studies of the inheritance of benomyl resistance in *Venturia pirina* isolated from pear orchards in Israel. *Plant Pathology* **?5**, 310–313.
9 STADELMANN F.X. & SCHWINN F.J. (1982) Beitrag zur Biologie von *Venturia inaequalis* und *Venturia pirina*. *Zeitschrift für Pflanzenkrankheiten und Pflanzenschutz* **89**, 96–109.
10 TOMERLIN J.R. & JONES A.L. (1983) Development of apple scab on fruit in the orchard and during cold storage. *Plant Disease* **67**, 147–150.

SIDE ROT of apples and pears caused by
Phialophora malorum (Kidd & Beaum.) McColloch

367

OCCURRENCE The disease was first recorded in the UK, as a minor apple rot (2). It was more prevalent in the USA (3), but its incidence on apples appeared to decline. Side rot of pears, on the other hand, has come to be recognised as the cause of major storage losses in the Pacific north-west (1,4,5).

SYMPTOMS Typical lesions are dull brown, oval, slightly sunken but fairly shallow. Decayed flesh tends to be soft and moist, separating cleanly from adjacent healthy tissue. If the skin of the lesion is broken, however, the rot dries out and may become spongy (1). Cladosporium rot (q.v.) is indistinguishable and may occur in association (1,5).

BIOLOGY The fungus can survive in soil and in bark, and wet conditions favour spore dispersal. Fruits may be infected in the orchard (3), probably via the natural openings (lenticels) or through minute cracks in the skin (1). Infection can also occur during post-harvest handling. This is especially likely when fruit is transported through the packhouse in flotation tanks, which serve to minimise mechanical injury to pears but can become heavily charged with spores from contaminated fruit (5). Late-picked fruit seems to be particularly susceptible to infection (1,4).

CONTROL Partial control can be achieved by the use of pre-harvest sprays of a calcium compound and an appropriate fungicide. Some of the common fungicides, however, have no effect on this fungus (1,5). Certain salts used in the flotation baths are toxic to the spores, and the use of such chemicals helps to reduce the incidence of infection (5).

---

1 BERTRAND P.F., MacSWAN I.C., RACKHAM R.L. & MOORE B.J. (1977) An outbreak of side rot in Bosc pears in Oregon. *Plant Disease Reporter* **61**, 890–893 illus.
2 KIDD M.N. & BEAUMONT A. (1924) Apple rot fungi in storage. *Transactions of the British Mycological Society* **10**, 98–118 illus.
3 McCOLLOCH L.P. (1944) A study of the apple rot fungus *Phialophora malorum*. *Mycologia* **36**, 576–590 illus.
4 SPOTTS R.A. (1985) Effect of preharvest pear fruit maturity on decay resistance. *Plant Disease* **69**, 388–390.
5 SUGAR D. & POWERS K. (1986) Interactions among fungi causing postharvest decay of pear. *Plant Disease* **70**, 1132–1134.

OTHER DISEASES of apples and pears

ASPERGILLUS ROTS of apples and pears vary greatly in appearance, depending on the species of *Aspergillus*. Some cause firm and rather moist rots, others dry and leathery rots, while *A. niger* (page 160) produces a soft watery rot. Aspergillus rot is associated only with fruits stored under warm conditions.

BACTERIAL BLISTER SPOT of apples is caused by *Pseudomonas syringae* pv. *papulans* (Rose) Dhanvantari, and is of major importance on the cultivar Mutsu (Crispin) in north-eastern USA, Canada and Italy (1). There are usually multiple lesions (centred on lenticels), water-soaked at first, later purplish black. The bacteria survive the winter in buds and leaf scars, and attack apples during the early stages of fruit development; consequently it is necessary to spray with a bactericide immediately after petal fall.

BOTRYODIPLODIA ROT of apples and pears is caused by *Botryodiplodia theobromae* (page 81) which is found in warm regions such as the Philippines and India. Injured fruits are susceptible to infection (3). Rotting is maximal at about 35°C, slow at 20°C, and totally inhibited at temperatures below 10°C (3). However, a report from Israel describes a related fungus (*Diplodia* sp.) causing rapid decay of apples even at 15°C.

BROOKS FRUIT SPOT of apples is caused by *Mycosphaerella pomi* (Pass.) Lindau (conidial state: *Cylindrosporium pomi* Brooks). The symptoms are dark spots, later flecked with black, which are usually most abundant around the calyx-end. Severe infection can lead to cracking. The disease occurs in Australia and in the eastern USA (9) after wet growing seasons, and is controlled by fungicide sprays applied after petal fall.

CALYX-END ROT, caused by *Sclerotinia sclerotiorum* (page 66), has been recorded on stored apples and pears in New Zealand, and also on apples in Canada and the USA. Lesions are sunken, brown and leathery. They may remain restricted (5) or, alternatively, the fungus may invade the entire fruit, producing tough resting bodies (sclerotia) at breaks in the skin. As in 'dry-eye rot' caused by the grey mould fungus (q.v.), infection takes place at blossom time, and suggested means of control include spraying at late-bloom and improving orchard hygiene.

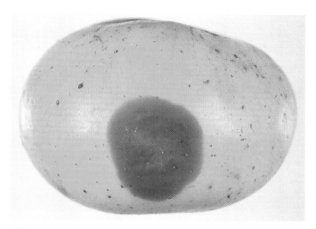

368 Fusarium rot

CLADOSPORIUM ROT of apples tends to follow certain physiological disorders, in particular the various types of scald. The causal fungus, *Cladosporium herbarum*, is a weak parasite in comparison with some other species of *Cladosporium* (page 226). Thus, while it can colonise moribund tissue affected by soft scald (q.v.), it generally does not progress into the surrounding healthy flesh. The rotted area, which is dark brown to black, remains shallow and limited. In pears cladosporium rot may be associated with side rot (q.v.).

FLYSPECK of apples is caused by *Zygophiala jamaicensis* Mason, and is occasionally found (often together with sooty blotch, q.v.) in fruit from cool humid growing areas. Symptoms are small, slightly raised black specks usually occurring in groups on the fruit surface as a disfiguring blemish. Control is by orchard fungicide sprays.

FUSARIUM ROT of apples and pears may be caused by any of several species of *Fusarium*, for example *F. avenaceum* (Fr.) Sacc., *F. culmorum* (W.G. Sm.) Sacc., *F. lateritium* Nees ex Fr., and *F. solani* (page 150). Infection usually takes place via a lenticel, and the rot is brown, soft and watery.

NIGROSPORA ROT of apples is caused by *Nigrospora sphaerica* (page 119), and has been reported from the eastern USA where it was formerly confused with black rot (q.v.). Initial symptoms may be small spots possessing haloes, or alternatively a rapid soft rot with brown to black lesions covered by a papery skin.

PHOMOPSIS ROT of apples is caused by *Diaporthe perniciosa* Marchal (Conidial state: *Phomopsis mali* Roberts). Records include the UK, Brazil and western USA (4). Infection usually occurs at the stem-end, but occasionally the fungus enters the calyx-end, eventually causing a core-rot (q.v.). Affected flesh is soft and watery. Uninjured fruits may be attacked but become susceptible to rotting only after reaching maturity. The optimal growth temperature for the fungus is approximately 25°C and slow growth occurs at 5°C (4). Control is by the use of fungicide sprays in the orchard, and prompt cooling of the harvested crop.

PYRENOCHAETA ROT of apples, caused by *Pyrenochaeta mali* M.A. Smith, has been reported from the USA (7). Lesions are circular, slightly sunken, tan to brownish-black, and centred on lenticels (cf. phacidiopycnis rot, q.v.). The rotted areas remain firm and eventually become covered with minute black bodies (pycnidia). The fungus grows best at about 25°C, and can also infect pears (7).

SCLEROTIUM ROT of pears has been recorded in India (8). The causal fungus is *Corticium rolfsii* (page 276) and the decay, which is soft and watery, occurs only under warm conditions.

SNOW-MOULD ROT of apples and pears has been recorded in western Canada as the cause of severe loss in cold-stored fruit, the fungus being *Coprinus psychromorbidus* Redhead & Traquair (10). Lesions are circular and sunken, with dark brown borders and paler centres, and may be numerous on the fruit. Infected flesh is firm and dry. After several months cold storage there is extensive development of white mould, and the decay spreads into adjacent healthy fruit. Control involves an orchard spray followed by post-harvest fungicide treatment.

SOOTY BLOTCH of apples and pears is caused by *Gloeodes pomigena* (Schw.) Colby, and is a non-progressive blemish often associated with flyspeck (q.v.). Cool wet weather is conducive to the development of dark granular smudges of various sizes which can be removed by scraping or rubbing. Control is by orchard fungicide sprays.

STEMPHYLIUM ROT of apples has been reported from several countries, including the UK, Sweden, Germany (2), the USA and New Zealand. There are various species, including *Stemphylium botryosum* (page 85) and *S. vesicarium* (Wallr.) Simmons. They tend to attack weakened tissue such as lesions resulting from sunscald (q.v.), and susceptibility to decay increases with storage time (2).

TRICHODERMA ROT of apples and pears is encountered occasionally, caused by *Trichoderma harzianum* or *Trichoderma viride* (page 82). In India a light brown malodorous rot of pears has been recorded, and in Australia a tan-brown rot with a coconut odour, affecting apples in cool store (6). Injury is necessary for infection of fruits (6), and under humid conditions a mass of green spores forms at the wound site. In the USA spores have been found in dump-tank water, and suggested control measures include hygiene in the packhouse and careful handling of fruit.

---

1 BAZZI C. & CALZOLARI A. (1983) Bacterial blister spot of 'Mutsu' apples in Italy. *Phytopathologia Mediterranea* **22**, 19–21 illus.
2 BEHR L. (1960) *Stemphylium botryosum* Wallr. als Erreger einer Lagerfaule am Apfel in Deutschland. *Phytopathologische Zeitschrift* **37**, 245–251 illus.
3 BHARGAVA S.N., SHUKLA D.N. & SINGH N. (1979) Some studies on *Botryodiplodia* rot of apple. *Proceedings of the National Academy of Science, India* B **49**, 43–45.
4 COMBRINK J.C., SOMMER N.F., TYLER R.H. & FORTLAGE R.J. (1976) Postharvest Phomopsis rot of apple fruits. *Plant Disease Reporter* **60**, 1060–1064 illus.
5 COVEY R.P. (1985) Calyx end rot (*Sclerotinia sclerotiorum*) of apples in Washington. *Plant Disease* **69**, 726 [Note].
6 PENROSE L.J., NICHOLLS M.R. & KOFFMANN W. (1984) Apple fruit rot caused by *Trichoderma harzianum*. *Australasian Plant Pathology* **13**(3), 46–47.
7 SMITH M.A. (1963) Apple rot caused by *Pyrenochaeta mali* n. sp. *Phytopathology* **53**, 589–591 illus.
8 SUMBALI G. & MEHROTRA R.S. (1980) Unrecorded sclerotial fruit rot of pears. *Indian Phytopathology* **33**, 476–477.
9 SUTTON T.B., BROWN E.M. & HAWTHORNE D.J. (1987) Biology and epidemiology of *Mycosphaerella pomi*, cause of Brooks fruit spot of apple. *Phytopathology* **77**, 431–437 illus.
10 TRAQUAIR J.A. (1987) Postharvest rot by *Coprinus psychromorbidus* on apples and pears in British Columbia. *Canadian Plant Disease Survey* **67**, 47–50 illus.

BITTER PIT of apples

OCCURRENCE This is one of the most important disorders of apples, and investigations have been carried out in many countries, including the USA, Canada, the UK (4), the Netherlands (2), South Africa (1), Australia (5) and New Zealand (3). Most cultivars are susceptible, including Cox's Orange Pippin, Golden Delicious, Granny Smith and Bramley's Seedling.

SYMPTOMS Bitter pit is sometimes discernible in apples on the tree, but usually develops only after harvest (4,5). The calyx-end of the fruit tends to be affected first. Pits are characteristically hard, sunken and discoloured, about 2 to 3 mm across. In addition there may be internal lesions; indeed, these may predominate (4). Affected tissue is dry, brown and spongy, with a slightly bitter taste.

BIOLOGY The causes are complex and are still being investigated (4) but, broadly, bitter pit is considered to be a nutritional disorder resulting from an imbalance of essential minerals. One hypothesis is as follows (1). During normal plant metabolism, organic acids are formed, and these are usually rendered harmless by reaction with calcium compounds absorbed through the roots and translocated to all parts of the plant, including the fruits. If the supply of calcium is insufficient, and if the proportions of, for example, magnesium or potassium are out of balance, then the organic acids accumulate at the ends of the conductive tissues (vascular bundles) in the flesh of the fruit (1). Eventually the affected cells are killed and turn brown. Calcium is an essential component of cell walls, and another hypothesis suggests that rapid redistribution of calcium to the core tissue results in localised deficiency and consequent cell death (4). Besides the nutrient status of the soil, orchard factors which predispose fruits to bitter pit include excessive tree vigour, excessive fruit thinning, and hot dry weather, especially if accompanied by wind.

369

Apples are more prone to the disorder if harvested immature, and large fruits from a light crop are particularly susceptible (2). All these factors are conducive to low concentrations of calcium in the fruit (4). The rate of development of bitter pit after harvest is determined by the conditions of storage (3,4).

CONTROL Integrated control is necessary because of the many factors involved (1). Soil analysis may indicate a necessity for applications of lime or other minerals (1,2). It may be advisable to modify pruning and thinning practices (2), and during fruit formation special care should be taken to prevent water stress (1,4). Leaf and fruit analyses can assist in prediction of bitter pit susceptibility (2). It may be necessary to spray the trees several times with a calcium compound (1,2) and, immediately after harvest, to drench the fruit with calcium chloride (5). Only mature fruit should be picked, and delaying the harvest affords a measure of control (cf. pear bitter pit, q.v.) (2). Storage conditions that delay fruit senescence are beneficial; controlled or modified atmospheres have an inhibitory effect on bitter pit and its appearance is delayed (3,4).

---

1 ANON. (1984) Bitter pit development and control in apples. *Deciduous Fruit Grower* **34**, 61–63.
2 BOON J. van der (1980) Prediction and control of bitter pit in apples. 2. Control by summer pruning, fruit thinning, delayed harvesting and soil calcium dressings. *Journal of Horticultural Science* **55**, 313–321.
3 HEWETT E.W. (1984) Bitter pit reduction in 'Cox's Orange Pippin' apples by controlled and modified atmosphere storage. *Scientia Horticulturae* **23**, 59–66.
4 PERRING M.A. (1986) Incidence of bitter pit in relation to the calcium content of apples: problems and paradoxes, a review. *Journal of the Science of Food and Agriculture* **37**, 591–606.
5 SCOTT K.J., O'LOUGHLIN J., ENGLAND B. & ROBERTS E.A. (1985) Effects of water rinses after calcium chloride dips, with and without additives, on the control of bitter pit of apples. *Australian Journal of Agricultural Research* **36**, 305–313.

BITTER PIT or CORKY SPOT of pears

370

OCCURRENCE   Bitter pit of pears is known as 'corky spot' in South Africa, where it occurs on the cultivar Packham's Triumph (5). In the USA (1,3,4) and Canada (2) the disorder is important on d'Anjou pears and may also be known as 'cork spot' or 'Anjou pit'. Incidence of the disorder varies from season to season but there appears to be no obvious correlation with the weather (2).

SYMPTOMS   Affected fruits may develop small sunken areas (predominantly near the calyx-end) in which the skin is darker green than elsewhere (2). The tissue beneath the pits is brown, dry and corky (4). Alternatively, or in addition, there may be corky lesions deep in the flesh, discernible only if the fruit is cut open (5). Symptoms may be apparent several weeks before harvest; experienced observers can recognise affected fruits by their slightly glassy external appearance and advanced maturity (5).

BIOLOGY   In common with apple bitter pit (q.v.) the disorder has complex origins but is associated with a deficiency of calcium (1,2,3,4,5). Predisposing factors include: a low level of calcium compounds in the soil, and hence insufficient uptake by the tree (1,5); excessive uptake of potassium compounds (which disturbs the K/Ca ratio) (1,5); water stress, which may be induced by high temperatures or inadequate irrigation (2,5). Incidence has also been positively correlated with increased irrigation (1). All these factors probably have an indirect influence on the ultimate concentration of calcium in the fruit (2). Pears with low calcium levels (less than 5 mg per 100 g fresh mass) not only tend to develop bitter pit, but also show enhanced rates of respiration and earlier maturity and senescence (3,4).

CONTROL   Mineral analyses of soil and leaves dictate the quantities of lime which must be applied to the soil (1,5). In addition, the trees may need to be sprayed with calcium chloride or calcium nitrate four or five times during fruit development (3,5). Methods of husbandry are also influential. Early pruning is important, the aim being to inhibit vigorous shoot growth and ensure a heavy crop (2,5). Attention should be given to the irrigation regime (1,5). Finally, the crop should be harvested promptly as soon as it reaches maturity (cf. apple bitter pit, q.v.) (5).

---

1 BRUN C.A., RAESE J.T. & STAHLY E.A. (1985) Seasonal response of 'Anjou' pear trees to different irrigation regimes. 2. Mineral composition of fruit and leaves, fruit disorders, and fruit set. *Journal of the American Society for Horticultural Science* **110**, 835–840.
2 MASON J.L. & WELSH M.F. (1970) Cork spot (pit) of 'Anjou' pear related to calcium concentration in fruit. *HortScience* **5**, 447 illus.
3 VAZ R.L. & RICHARDSON D.G. (1984). Effect of calcium on respiration rate, ethylene production and occurrence of cork spot in Anjou pears (*Pyrus communis* L.). *Acta Horticulturae* No. 157, 227–236.
4 WANG C.Y. & MELLENTHIN W.M. (1973) Chlorogenic acid levels, ethylene production and respiration of d'Anjou pears affected with cork spot. *HortScience* **8**, 180–181.
5 ZYL H.J. van (1978) Corky spot on Packham's Triumph pears. *Deciduous Fruit Grower* **28**, 115–117 illus.

BREAKDOWN of pears

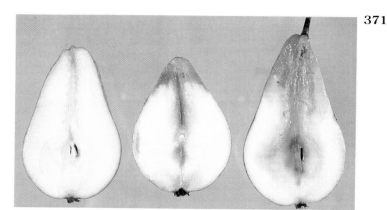

371

OCCURRENCE Breakdown may occur before or after ripening. Investigations have been carried out in all pear-growing countries, for example the UK (4), Canada (5), the USA (1), South Africa (2) and Australia (3). One of the most susceptible cultivars is Bartlett, otherwise known as William's Bon Chrétien (1,2,3).

SYMPTOMS External signs may include yellowing of the skin whilst the pears are still in store (2,5). (If premature ripening has occurred on the tree, a soft pale brown area surrounds the calyx-end.) Some cultivars, while remaining firm to the touch, develop numerous small brown spots (centred on the lenticels), and the entire skin may subsequently turn brown, sloughing off easily if rubbed. This disorder is known as **senescent scald** (2,3) (cf. superficial scald, q.v.) and is soon followed by general fungal decay. Early internal symptoms of breakdown are best seen if the fruit is cut in half longitudinally. There may be a darkening of the vascular strands in the 'neck' of the pear (**vascular browning**) (2) and/or a brownish watery alteration of the flesh, sometimes known as **internal breakdown**. The core region is especially susceptible and, in **core breakdown**, the central part of the fruit (containing the seeds) is readily detached from the surrounding tissue (1). Aroma and taste are seriously impaired.

BIOLOGY If pears are not consumed within a few days of normal ripening, they undergo rapid deterioration, characterised by a soft breakdown of the flesh (2). The length of time between optimal eating ripeness and senescence is determined by many factors, including growing conditions, date of harvest, rate of cooling, and duration and conditions of storage. Sometimes, in pears which have been picked late, or cooled slowly, or stored for too long, breakdown of skin and flesh occurs before ripening is able to proceed (1,5). Pears do not ripen normally at low temperatures (3), and eventually lose the ability to do so on removal to moderate temperatures (4).

CONTROL Harvesting at optimal maturity is of great importance if pears are to be stored for any length of time. It is essential to cool the fruit rapidly, immediately after harvest (5), and it is desirable to maintain the storage temperature as low as is feasible without risk of freezing (1,2,4). Storage potential should be assessed periodically by withdrawing samples of pears from store and holding at ambient temperature. If there is the slightest sign of abnormal ripening, the entire load should be marketed without delay.

---

1 BLANPIED G.D. (1975) Core breakdown of New York 'Bartlett' pears. *Journal of the American Society for Horticultural Science* **100**, 198–200 illus.
2 EKSTEEN G.J. & GINSBURG L. (1977) Respiration and senescence of pears in storage. *Deciduous Fruit Grower* **27**, 46–49, 59, illus.
3 HALL E.G. & SCOTT K.J. (1964) Cool storage of pears. *Agricultural Gazette of New South Wales* **75**, 1216–1221 illus.
4 KNEE M. (1973) Effects of storage treatments upon the ripening of Conference pears. *Journal of the Science of Food and Agriculture* **24**, 1137–1145.
5 PORRITT S.W. (1965) Effect of cooling rate on storage life of pears. *Canadian Journal of Plant Science* **45**, 90–97.

BROWN HEART of apples and pears

OCCURRENCE In the 1920s several shiploads of Australian apples were rendered practically worthless by this disorder, and the export trade was threatened (1). Even after the cause had been ascertained (3), sporadic losses continued into the 1930s (page 28). Brown heart is not common today, although it may occur in fruit packed in sealed polyethylene film (4,5). Investigations have been carried out in the UK (3), Australia (1,4,5), Canada and the USA (2). Cultivars with high susceptibility include Sturmer Pippin and Cox's Orange Pippin apples (1) and William's Bon Chrétien pears (5).

SYMPTOMS Affected fruits usually present a normal external appearance. When cut in half, however, they show patches of brown flesh, which may be distributed at random or may form a discoloured zone between the core and the skin (4,5). Once brown heart begins to develop, its incidence increases rapidly (5). Initially the injured tissue is firm and moist, but after some weeks' storage it becomes dry and spongy, and cavities may develop (3,4) (**318**).

BIOLOGY The primary cause of brown heart is an excessive concentration of carbon dioxide in the tissue (3). Susceptibility to injury does not, however, show a simple relationship to carbon dioxide concentration; it is also influenced by cultivar, orchard factors, time of harvest, speed of cooling and temperature of storage (1,2). Tasmanian Sturmer Pippin apples from a light crop may be injured by a concentration of 3% carbon dioxide, whereas many other cultivars can withstand up to 10% (1). Pears are generally less tolerant of carbon dioxide than are apples, and their susceptibility increases markedly with advancing maturity, whether due to late harvest, delayed cooling or prolonged storage (5). The heavy losses sustained in the first Australian shipments were the result of a lack of ventilation in the refrigerated holds (1,3). The early vessels (designed to carry frozen meat) had no provision for admitting fresh air, and the high respiration rate of the fruit, which was often loaded 'warm', quickly led to substantial concentrations of carbon dioxide. In more recent years the problem has re-appeared, owing to the use of polyethylene-lined cartons, in which carbon dioxide may accumulate beyond a critical level (2,4,5). A related disorder, invasive alcohol poisoning (q.v.), may occur at low oxygen concentrations (4).

CONTROL Although varietal differences in susceptibility to brown heart greatly influence recommended practices, it is generally advisable to ensure that apples and pears which are susceptible to the disorder should be cooled to storage temperature before carbon dioxide is allowed to accumulate in CA stores. Similarly, rapid establishment of low oxygen storage regimes for pears can result in the development of symptoms if the fruit is not cooled before the oxygen concentration is reduced. Carriage instructions for apple and pear shipments should stipulate fresh air ventilation to maintain the carbon dioxide concentration below 2%. Fruit which is to be shipped or stored in polyethylene-lined cartons should be pre-cooled before being packed (2). Harmful atmospheres may be avoided by careful choice of film gauge, by making perforations in the film, by leaving the liner unsealed, or by including a carbon dioxide absorbent in the pack (4,5).

For illustration, see **318**.

1 CARNE W.M. (1950) Brown heart of apples and its relation to our knowledge of apples and of ship carriage of perishable foods. *Journal of the Australian Institute of Agricultural Science* **16**, 59–64.
2 HANSEN E. & MELLENTHIN W.M. (1962) Factors influencing susceptibility of pears to carbon dioxide injury. *Proceedings of the American Society for Horticultural Science* **80**, 146–153.
3 KIDD F. & WEST C. (1923) Brown heart – a functional disease of apples and pears. *Special Report of the DSIR Food Investigation Board* No. 12, 54 pp illus.
4 MELVILLE F. (1963) Storage of Yates apples in polythene lined boxes. *Journal of the Department of Agriculture of Western Australia* **4**, 173–175 illus.
5 ROBERTS E.A., SCOTT K.J. & WILLS R.B.H. (1964) The effects of composition of the atmosphere and the length of storage on the development of brown heart in William's Bon Chretien pears held in polyethylene bags. *Australian Journal of Experimental Agriculture and Animal Husbandry* **4**, 371–375 illus.

BRUISING and OTHER INJURIES of apples and pears

OCCURRENCE Physical damage can occur whilst fruit is still on the tree, perhaps from hailstones or from contact with branches as they are swayed by the wind (2). Injuries are also sustained during harvesting, handling, packing and transport (1), and can take the form of **friction discoloration** (2,3), **external bruising** (5) or **internal bruising (brown spot)** (3), from either pressure or impact. Investigations have been carried out in several countries, including the USA (3,4), Canada (2), Australia, New Zealand (1) and the UK (5).

SYMPTOMS In hail-damaged fruit, sunken pits are present, mostly on the shoulders. Friction discoloration consists of brown areas on the skin. Bruises are flattened greyish lesions visible externally or, alternatively, brown spots in the flesh which are apparent only when the skin is removed (3). Affected internal tissue eventually dries out and becomes corky.

BIOLOGY Physical injury initiates certain biochemical reactions which result in the formation of brown substances (polyphenols) (4). Susceptibility to friction discoloration decreases with maturity (perhaps because of further development of the natural waxy cuticle), and large fruits are less affected than small fruits (4). If injured fruits are stored, the degree of discoloration increases with duration of storage (4). Susceptibility to bruising varies considerably with cultivar (5); it is probably influenced by orchard and climatic factors, and increases with delayed harvesting (1). Tissue is especially vulnerable when very turgid (1). If dropped, cold fruit sustains more injury than warm fruit (3), but for vibration damage the reverse is true. Whatever the type of injury, respiration rate is enhanced and, furthermore, there is an increased likelihood of infection and fungal decay, for example by blue mould (q.v.).

372

CONTROL Although wind-breaks can reduce limb-rub, there is no control for hail damage, and severely injured fruits may have to be diverted to a juice-factory. Much can be done, however, to reduce the incidence of surface abrasion and bruising of the flesh. Beneficial measures include training personnel to avoid excessive finger pressure during picking and to take care when filling and emptying field boxes. One recommendation for pears suggests a maximum drop of 4" (10 cm) on to a hard surface or 8" (20 cm) on to one another (3). Grading machinery should be well-designed and properly adjusted. Artificial waxing of fruit probably helps to reduce damage to the skin (2). In view of the severe vibration damage which can be caused by careless driving of road vehicles or by shunting of rail trucks, fruit should be well cushioned by appropriate packaging.

---

1 KLEIN J.D. (1987) Relationship of harvest date, storage conditions, and fruit characteristics to bruise susceptibility of apple. *Journal of the American Society for Horticultural Science* **112**, 113–118.
2 LOUGHEED E.C., LIDSTER P.D. & PROCTOR J.T.A. (1982) Friction discoloration of McIntosh apples from low-oxygen, controlled atmosphere storage. *Plant Disease* **66**, 1119–1120 illus.
3 MATTUS G.E., SCOTT L.E. & CLAYPOOL L.L. (1960) Brown spot bruises of Bartlett pears. *Proceedings of the American Society for Horticultural Science* **75**, 100–105 illus.
4 MELLENTHIN W.M. & YANG C.Y. (1974) Friction discoloration of 'd'Anjou' pears in relation to fruit size, maturity, storage and polyphenoloxidase. *HortScience* **9**, 592–593.
5 TOPPING A.J. & LUTON M.T. (1986) Cultivar differences in the bruising of English apples. *Journal of Horticultural Science* **61**, 9–13.

CHILLING INJURY or LOW TEMPERATURE BREAKDOWN of apples

373

OCCURRENCE   Cultivars susceptible to low temperature injury include Cox's Orange Pippin (3), Starking Delicious (2), McIntosh (4), Jonathan (5), Sturmer Pippin and Bramley's Seedling (1). Pears are not susceptible to this disorder. Investigations have been carried out in many countries, including the UK (1,3), the USA, Canada (4), Australia (5) and Japan (2).

SYMPTOMS   If affected fruits are cut open, parts of the flesh appear brown and moist (in contrast to the dry mealiness associated with senescent breakdown, q.v.). The core tissues are not at first affected, although later the vascular strands may become dark brown. Characteristic of this disorder is a narrow zone of normal tissue immediately beneath the skin. In advanced stages, however, this too turns brown and the skin becomes waterlogged and discoloured.

BIOLOGY   The critical temperature, below which injury can occur, varies with cultivar and growing conditions but is generally about 2° or 3°C. In common with the chilling injury of other fruits, the disorder increases with exposure time, and is especially severe if low temperature is imposed during the apples' climacteric rise (page 14) (1). Susceptibility is related to low levels of calcium (3,4) and phosphorus (4) in the fruit tissue. It tends to be associated with light crops, large fruit, cool weather during the latter part of the growing season, and high levels of humidity and carbon dioxide in store.

CONTROL   The primary means of control is to ensure that susceptible cultivars are stored at temperatures above the critical value. Because of climatic differences the critical temperature varies according to country; thus, for example, Cox's Orange Pippin apples are stored at 4°C in the UK but 1.5°C in Australia. Apples can better tolerate a low storage temperature if their mineral content is adequate, and this can be achieved by foliar sprays of phosphate compounds (4) and pre- and post-harvest treatments with calcium compounds (3). Some cultivars (e.g. Jonathan) may, however, be injured by such treatment (5). Harvest maturity is important, late-picked fruit being vulnerable because of the risk that cooling will coincide with the climacteric rise (1). Apples stored in controlled atmosphere may need to be kept at a temperature slightly higher than that suitable for the same cultivar stored in air.

---

1 KIDD F. & WEST C. (1934) The cause of low-temperature breakdown in apples. *Report of the DSIR Food Investigation Board for 1933*, 57–60.
2 KIMURA S., YASUDA T. & OKAMOTO T. (1985) Studies on chilling injury of apples. 7. Existence of acetaldehyde in mitochondria from the flesh of Starking Delicious after cold storage. *Bulletin of the Faculty of Agriculture, Hirosaki University* No. 43, 1–6.
3 PERRING M.A. (1985) Redistribution of minerals in apple fruit during storage: effects of late summer pruning, calcium sprays and low temperature breakdown. *Journal of the Science of Food and Agriculture* **36**, 333–342.
4 WEBSTER D.H. & LIDSTER P.D. (1986) Effects of phosphate sprays on McIntosh apple fruit and leaf composition, flesh firmness and susceptibility to low-temperature disorders. *Canadian Journal of Plant Science* **66**, 617–626.
5 WILLS R.B.H. & SCOTT K.J. (1981) Studies on the relationship between minerals and the development of storage breakdown in apples. *Australian Journal of Agricultural Research* **32**, 331–338.

CORE FLUSH or BROWN CORE of apples

OCCURRENCE This disorder has been recorded in several apple cultivars, for example Cox's Orange Pippin in the UK (1) and Granny Smith in Australia (2) and New Zealand (4). In these countries it is known as 'core flush'. In Canada (3) and the USA (5) the disorder is known as 'brown core' and is especially prevalent in the cultivar McIntosh.

SYMPTOMS This condition develops only after a period of storage. If an affected fruit is cut transversely, a pinkish or brownish discoloration of the core tissue is visible, either as a single diffuse circular area or as individual angular areas between the seed cavities (5). The discoloured flesh tends to be firm and moist. In McIntosh apples there may be an external symptom, viz. browning of the skin and flesh at the stalk-end, known as 'stem-cavity browning' (3).

BIOLOGY It has been variously suggested that core flush/brown core results from low temperature injury (5), carbon dioxide injury (4), or senescence (3). There appear to be several predisposing factors. In Australia fruit grown in warm districts is said to be more susceptible than fruit grown in cool districts, whereas in the USSR the disorder does not occur after hot growing seasons. In Canada the problem is said to be associated with cool wet or cloudy growing seasons. In some years and localities it is significantly worse in fruit from orchards which received heavy applications of nitrogenous fertiliser (3,5). Spraying of certain growth regulators (e.g. daminozide) can predispose the crop to core flush (1). Fruit picked too early (5) or too late (2,3) may be especially susceptible. Symptoms are aggravated by continuous low temperature storage and by high levels of carbon dioxide (4). Incidence of the disorder increases as the storage period is prolonged (2) and its development accelerates when affected fruit is brought out of cold store.

374

CONTROL Core flush/brown core and stem-cavity browning are best prevented by harvesting apples at optimal maturity (2) and storing them in controlled atmosphere, preferably with low oxygen and low carbon dioxide (2,3,5). Removal of ethylene from the storage atmosphere can be beneficial (2) but it may, on the other hand, compound the unfavourable effect of pre-harvest daminozide treatment (1). Trials have shown that the disorder may be held in check by the use of an antioxidant (page 214) (2), by the use of a moisture-absorbing chemical (4), by the practice of delayed cooling (2) or by intermittent warming during the storage period (4).

---

1 KNEE M. (1986) Opposing effects of daminozide and ethylene on induction of flesh softening in apple fruits. *Journal of Horticultural Science* **61**, 15–21.
2 LITTLE C.R., TAYLOR H.J. & McFARLANE F. (1985) Postharvest and storage factors affecting superficial scald and core flush of 'Granny Smith' apples. *HortScience* **20**, 1080–1082.
3 LOUGHEED E.C., MURR D.P. & MILLER S.R. (1978) Effect of diphenylamine upon storage scald, stem cavity browning and brown core of 'McIntosh' apples. *Plant Disease Reporter* **62**, 557–561.
4 SCOTT K.J. & WILLS R.B.H. (1976) Core flush of apples. 1. Effect of absorption of carbon dioxide, ethylene and water from the storage atmosphere. *Journal of Horticultural Science* **51**, 55–58.
5 SMOCK R.M. (1946) Some factors affecting the brown core disease of McIntosh apples. *Proceedings of the American Society for Horticultural Science* **47**, 67–74 illus.

FREEZING INJURY of apples and pears

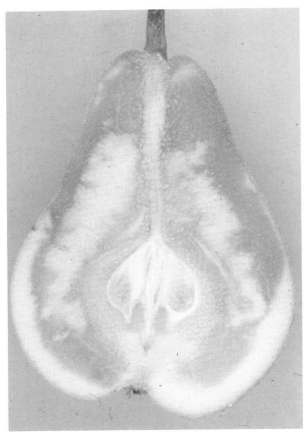

375

OCCURRENCE Fruits may suffer freezing injury whilst on the tree, in store or during transit (2). Exposure to freezing conditions tends to be more damaging to harvested fruits than to those still on the tree (5).

SYMPTOMS After slight freezing, the skin may show irregular areas of brown discoloration (2). Internal freezing injury can result in watersoaking, flesh-browning and/or discoloration of the vascular strands (2). Such symptoms are not diagnostic, however, since other disorders can produce a similar effect (4). Severely frozen flesh leaks on thawing, becoming soft and brown (1). After subsequent storage, cavities may develop in the injured tissue. Affected pears fail to ripen. Even in the absence of obvious symptoms, fruits which have been exposed to freezing temperatures are weakened and their storage life is curtailed (2). Apples soften and become mealy (1). In fruits which have been subjected to bruising whilst frozen, the bruises tend to be deep and characteristically cone-shaped (2).

BIOLOGY The freezing point of apples and pears varies with sugar content and other factors, but is likely to be in the region of $-1.5°C$. However, the flesh temperature may fall below this value without the onset of freezing. This phenomenon, known as supercooling, is a useful safety factor in the event of a temporary fall in the store temperature (3). A sudden vibration can, however, induce immediate freezing of supercooled fruit. Since freezing injury varies in intensity and in type of symptom, the extent of injury is probably related to the size and location of ice crystals formed within the flesh (1). Susceptibility to injury is determined not only by the degree and duration of freezing conditions but also by the physiological state of the fruit; thus susceptibility tends to increase as the season progresses (2).

CONTROL It is important to have accurate and well-positioned thermostats in fruit stores, so as to ensure that the air delivery temperature does not fall too low. Pears are especially vulnerable, being stored at temperatures only marginally above their freezing point. Care is also necessary during transport (4). Fruit which has been inadvertently frozen should not be handled until after thawing, otherwise bruising is likely to be severe (2).

---

1 BIR R.E. & BRAMLAGE W.J. (1973) Postharvest effects of freezing on apple. *Journal of the American Society for Horticultural Science* **98**, 218–221.
2 DIEHL H.C. & WRIGHT R.C. (1924) Freezing injury of apple. *Journal of Agricultural Research* **29**, 99–127 illus.
3 MARSHALL D.C. & PADFIELD C.A.S. (1962) The freezing point of pears. *Journal of Horticultural Science* **37**, 106–114.
4 ROSE D.H. & LUTZ J.M. (1933) Bruising and freezing of apples in storage and transit. *Technical Bulletin of the United States Department of Agriculture* No. 370, 14 pp illus.
5 SMOCK R.M. (1972) Freezing damage on attached and detached apple fruits. *HortScience* **7**, 174.

JONATHAN SPOT and LENTICEL SPOT of apples

376

OCCURRENCE Spot disorders of Jonathan apples have been investigated in several countries, including the USA (2,4), Australia (3), Japan (5) and Hungary (1). Some confusion exists in terminology (4,5) and it would appear that two disorders should be distinguished, 'Jonathan spot' occurring at random on the fruit surface and 'lenticel spot' being centred on the natural openings (lenticels) (4). Cultivars other than Jonathan may be affected, for example Rome Beauty and Stayman Winesap (3).

SYMPTOMS Brown to black spots develop on the skin, with a tendency to be darker and more numerous on the blushed side of the fruit (2). Jonathan spot is sometimes characterised by haloes surrounding the lesions. Lenticel spots may be slightly depressed and up to 5 mm in diameter, remaining fairly superficial unless invaded by weak fungi such as *Alternaria* spp. (q.v.) (2).

BIOLOGY The causes of these physiological disorders are still not fully understood, although characteristic spots have been induced artificially by ultra-violet light (1). This is consistent with the observation that spotting is more likely on fruits from exposed positions on the tree than on fruits which have been shaded during growth. Lenticel spots may appear before harvest (4), although development during the storage period is more characteristic. Predisposing factors include excessive use of nitrogen fertiliser and early harvesting (2,4,5). Jonathan spot appears only after harvest, and is especially likely on overmature fruit (4). Incidence of this disorder tends to be higher following insufficient use of nitrogen fertiliser, late harvest, delayed cooling or prolonged storage (3,4).

CONTROL After diagnosis of the type of disorder, it may be advisable to alter the fertiliser regime (3,4,5). Care should be taken to pick the crop at optimal maturity, cool the fruit promptly, and maintain as low a storage temperature as is feasible. Increased carbon dioxide levels help to delay the appearance of disorder, and it is therefore beneficial to use controlled atmosphere storage or to pack the fruit in polyethylene-lined cartons.

---

1 BALÁZS E. & TÓTH Á. (1974) Jonathan spot induced by ultraviolet light. *Acta Phytopathologica Academiae Scientiarum Hungaricae* **9**, 179–184.
2 BROOKS C. & COOLEY J.S. (1917) Effect of temperature, aeration and humidity on Jonathan spot and scald of apples in storage. *Journal of Agricultural Research* **11**, 287–317 illus.
3 MARTIN D., LEWIS T.L. & CERNY J. (1961) Jonathan spot – three factors related to incidence: fruit size, breakdown and seed numbers. *Australian Journal of Agricultural Research* **12**, 1039–1049.
4 RICHMOND A.E. & DEWEY D.H. (1969) Distinguishing characteristics of the Jonathan spot and lenticel spot disorders in the 'Jonathan' apple fruit. *Journal of the American Society for Horticultural Science* **94**, 245–248 illus.
5 TOMANA T. (1963) Histological and physiological studies on the cause of Jonathan spot in apples. *Bulletin of the Yamagata University (Agricultural Science)* **4**(2), 89–153 illus.

SENESCENT BREAKDOWN of apples

377

OCCURRENCE This disorder occurs in all cultivars, but they vary in susceptibility. Those studied include Cox's Orange Pippin in the UK (2), McIntosh in the USA (5), Spartan in Canada (3), Granny Smith in Australia (4) and in South Africa (1).

SYMPTOMS The first indication is a softening of the flesh, which subsequently becomes mealy and brown. Taste deteriorates also. In advanced stages the skin at the calyx-end (and later elsewhere) becomes dull and brown. Sometimes the skin ruptures and the flesh splits open.

BIOLOGY This is a disorder of 'old age', and eventually occurs in any cultivar which is stored for too long. However, it can develop during early storage in fruit which is harvested overmature (1), cooled too slowly, or held at an insufficiently low storage temperature. Susceptibility is associated with light crops and large fruit. Apples with a low calcium content are predisposed to the disorder. Symptoms are aggravated by a storage humidity above 90% RH, and their development accelerates when the fruit is removed from cold storage.

CONTROL It may be advisable to spray the crop with a calcium compound at various stages of fruit development (2). A pre-harvest analysis of mineral composition is of value in predicting potential storage life and shelf-life of particular crops (5). Cooling should be prompt and rapid; temperature and humidity in store should be maintained at optimal values. Controlled atmosphere storage can be very effective in delaying the onset of senescence (3,4).

---

1 GINSBURG L. (1961) Internal breakdown in South African apples grown in the Cape Province. *Bulletin de l'Institut International du Froid, Annexe 1961–1*, 93–104.
2 PERRING M.A., PEARSON K. & MARTIN K.J. (1985) The distribution of calcium in apples with senescent breakdown. *Journal of the Science of Food and Agriculture* **36**, 1035–1038.
3 PORRITT S.W. (1966) The effect of oxygen and low concentration of carbon dioxide on the quality of apples stored in controlled atmosphere. *Canadian Journal of Plant Science* **46**, 317–321.
4 STEVENSON C.D., WATKINS J.B. & BLAKE J.R. (1961) Controlled atmosphere storage of Queensland grown Granny Smith apples. *Queensland Journal of Agricultural Science* **18**, 463–475.
5 WEIS S.A., BRAMLAGE W.J. & DRAKE M. (1985) Comparison of four sampling methods for predicting poststorage senescent breakdown of 'McIntosh' apple fruit from preharvest mineral composition. *Journal of the American Society for Horticultural Science* **110**, 710–714.

SOFT SCALD of apples and pears

OCCURRENCE Investigations into this disorder have been carried out in the USA (1,2), the UK (4), Italy (3) and Australia (5). Susceptible apple cultivars include Jonathan (1,5), Rome Beauty (1), Ellison's Orange (4), Golden Delicious (3) and Red Delicious (2). Pears (3) are seldom affected, and the following information refers to apples. Soft scald (sometimes called 'ribbon scald' or 'deep scald') is quite distinct from superficial scald (q.v.).

SYMPTOMS Soft scald is characterised by discrete brown lesions which are smooth, slightly sunken, and generally confined to the equatorial parts of the apple (1). The flesh beneath is also affected, sometimes quite deeply; the tissue is initially pale brown, soft, spongy and moist, although later the smaller lesions may dry out somewhat (1). In advanced stages the lesions may show dark spots, as the injured tissue is invaded by opportunistic fungi such as *Cladosporium herbarum* (page 226) and *Alternaria alternata* (page 182).

BIOLOGY This disorder is essentially a form of low temperature injury, occurring in certain cultivars at temperatures below about 2° or 3°C (2). Symptoms develop only after a period of cold storage, and development seems to be especially likely if susceptible fruit is cooled whilst it is respiring at a high rate (4). The major predisposing factors are overmaturity at harvest and delay between picking and cooling (2), both of these factors having the effect of hastening the climacteric rise in respiration (page 14). Incidence of soft scald has been linked with dull cool wet summers, light crops, large fruit, and vigorous trees on heavy soils.

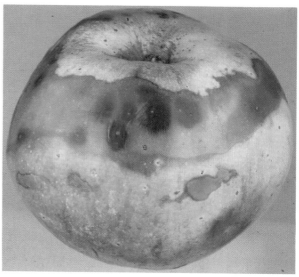

CONTROL Harvesting the crop at optimal maturity is especially important for cultivars prone to this disorder. If this can be achieved, the crop is more likely to be able to withstand immediate low temperature storage (2). Otherwise, special treatments may be required immediately after harvest. Applications of heat, carbon dioxide, or wax have proved beneficial in reducing the likelihood of soft scald development in apples (1). More recently treatments have included calcium chloride (3) and certain antioxidants combined with oils (5). An alternative strategy is to hold the harvested fruit at a temperature slightly above the critical value for the first few weeks of storage, and subsequently impose the normal cold storage regime (2).

---

1 BROOKS C. & HARLEY C.P. (1934) Soft scald and soggy breakdown of apples. *Journal of Agricultural Research* **49**, 55–69 illus.
2 GERHARDT F. & SAINSBURY G.F. (1952) Soft scald and its control in Delicious apples. *Proceedings of Washington State Horticultural Association* **48**, 97–100.
3 PRATELLA G.C., BERTOLINI P. & TONINI G. (1979) The effects of calcium chloride and growth regulators on the incidence of soft scald during the refrigerated storage of apples and pears. *Bulletin de l'Institut International du Froid* **59**, 1132 [Abstract].
4 TOMKINS R.G. (1968) Soft scald of Ellison's Orange apples. *Annual Report of the Ditton Laboratory 1967–68*, 19–21.
5 WILLS R.B.H. & SCOTT K.J. (1982) Use of dips containing diphenylamine and edible oils to reduce soft scald of apples. *HortScience* **17**, 964–965.

SUPERFICIAL SCALD of apples and pears

379

OCCURRENCE   This is an important disorder, especially of apples, and has been investigated in many countries, for example Australia (1,3), South Africa (5), Italy (4), the USA and Canada (2). Susceptible apple cultivars include Granny Smith (1) and Starking Delicious (2), while the main pear cultivars affected are d'Anjou and Packham's Triumph (3) (cf. senescent scald, q.v.).

SYMPTOMS   Diffuse brown blotches develop on the skin, randomly distributed on green apple cultivars but in red apples usually confined to the non-blushed side. Within the brown areas there may be slightly raised uninjured green spots, marking the natural openings (lenticels). In pears it is generally the neck region which is affected initially. The disorder is essentially a skin blemish, the interior of the fruit retaining its normal colour. In severe scald, however, the flesh immediately beneath the skin may turn brown, and such tissue is predisposed to invasion by fungi.

BIOLOGY   The causes of scald are complex, but are thought to be associated with the accumulation of volatile products generated by the fruit itself. A specific chemical (farnesene, identified as a natural constituent of fruit tissues) gives rise, on oxidation, to substances which are injurious to the fruit skin (1). Susceptibility to scald is governed by many factors, including cultivar, growing conditions, cultural practices and post-harvest environment (2). The disorder is more likely to occur in fruits with high nitrogen and low calcium content. It tends to be especially severe in a crop harvested after a hot dry period (3), and in the UK measurements of water deficit have permitted fairly accurate prediction of the occurrence of scald. Early picked fruits are most vulnerable, particularly the larger sizes. Symptoms are not apparent at harvest time, nor do they develop in cold storage until several months have passed. Removal to ambient temperature hastens symptom expression; fruits which appear perfect on leaving the cold store or refrigerated ship may become unmarketable within a week.

CONTROL   The grower should know by experience whether or not his crop might be liable to scald. Susceptible fruit should be harvested at optimal maturity and (if permitted by current legislation) promptly drenched with an antioxidant chemical such as diphenylamine (DPA) or ethoxyquin (1,2,3,4,5). Before such chemicals were developed, reliance was placed on oiled paper wraps which served to absorb volatiles (3). Fruit should then be immediately cooled as rapidly as possible and, in ordinary cold storage, adequate fresh air ventilation should be permitted during the storage period (1). Controlled atmosphere storage, in particular a low concentration of oxygen (5), can reduce the likelihood of scald development, though results vary widely with cultivar and locality (2). Techniques which reduce ethylene concentration below 1 part per million may also delay the development of superficial scald in certain apple varieties. The use of polyethylene liners, which can extend storage life by reducing moisture loss and gaseous exchange, carries with it the disadvantage of allowing harmful volatiles to accumulate (3).

---

1. ANET E.F.L.J. (1972) Superficial scald, a functional disorder of stored apples. 9. Effect of maturity and ventilation. *Journal of the Science of Food and Agriculture* **23**, 763–769.
2. CHEN P.M., OLSEN K.L. & MEHERIUK M. (1985) Effect of low-oxygen atmosphere on storage scald and quality preservation of 'Delicious' apples. *Journal of the American Society for Horticultural Science* **110**, 16–20.
3. HALL E.G., SCOTT K.J. & RILEY T.J. (1962) Control of superficial scald on Packham's Triumph pears. *Food Preservation Quarterly* **22**, 15–18.
4. PRATELLA G. (1983) Il riscaldo commune delle mele. *Informatore Fitopatologico* **33**(1), 17–25 illus.
5. TRUTER A.B. & EKSTEEN G.J. (1987) Storage potential of apples under controlled atmosphere storage at ultra-low oxygen concentrations. *Deciduous Fruit Grower* **37**, 141–145.

WATER CORE of apples and pears

OCCURRENCE Water core is particularly prevalent in apples grown in regions of high summer temperatures and intense sunlight, for example Australia, South Africa, and parts of the USA and South America (2). In more temperate climates it causes occasional problems, for example in the UK, Canada and Japan (5). Susceptible apple cultivars include Bramley's Seedling, Cox's Orange Pippin, Delicious, Jonathan and Winesap. Other varieties such as Golden Delicious, Granny Smith and McIntosh rarely exhibit this disorder. Water core may occasionally affect pears (5).

SYMPTOMS Symptoms develop before harvest, and can be observed if affected fruits are cut. Translucent or glassy areas of tissue are present, not necessarily in the core region, being sometimes located around the vascular bundles or scattered in the flesh towards the calyx-end of the fruit (2). Mild symptoms tend to disappear gradually after harvest, especially if the fruit is kept at ambient temperature for 3 or 4 days before cold storage. If water core is severe, however, it may be followed by an irreversible breakdown of the flesh (1).

BIOLOGY In all types of water core the characteristic water-soaking results from a leakage of cell sap into the intercellular spaces (which are normally filled with air), following an abnormally rapid conversion of starch to sugar. Predisposing factors include inadequate orchard nutrition, especially in respect of calcium, heavy or ill-timed pruning, and the alternation of hot bright days with cold nights (2). The most influential factor, however, is maturity, and delayed harvesting can result in a large increase in both incidence and severity of the disorder (2).

CONTROL The most practical way to minimise water core is to harvest fruit at the correct time, before it becomes overmature. The use of ethephon sprays (to promote early colouring) can induce extensive water core if the fruit is not harvested promptly. A non-destructive method of detection depends on the altered light transmittance in affected fruits (1). Furthermore, since these fruits have an increased density it would be feasible to segregate healthy fruits by flotation in an appropriate mixture of alcohol and water. Fruits showing severe water core should be marketed promptly; cold storage serves to delay the onset of breakdown, but the flesh will deteriorate rapidly when the produce is removed to ordinary temperatures.

---

1 BRAMLAGE W.J. & SHIPWAY M.R. (1967) Loss of watercore and development of internal breakdown during storage of 'Delicious' apples, as determined by repeated light transmittance measurements of intact apples. *Proceedings of the American Society for Horticultural Science* **90**, 475–483.
2 BROOKS C. & FISHER D.F. (1926) Water-core of apples. *Journal of Agricultural Research* **32**, 223–260 illus.
3 MARLOW G.C. & LOESCHER W.H. (1984) Watercore. *Horticultural Reviews* **6**, 189–251.
4 NARDIN C., SCIENZA A. & POMA T., C. (1983) The importance of mineral nutrition and $CaCl_2$ spraying in apple watercore prevention. *Acta Horticulturae* No. 138, 51–62.
5 YAMAKI S. & KAJIURA I. (1983) Change in the polysaccharides of cell wall, their constituent monosaccharides and some cell wall-degrading enzyme activities in the watercore fruit of Japanese pear (*Pyrus serotina* Rehder var. *culta* Rehder). *Journal of the Japanese Society for Horticulture* **52**, 250–255.

OTHER DISORDERS of apples and pears

381 Alcohol formation

ALCOHOL FORMATION can reach noticeable levels if the fruit is deprived of oxygen. Affected flesh has an odour of fermentation. In the early stages the condition is reversible and symptoms may disappear if the fruit is left in air (2). On the other hand, after prolonged exposure to anaerobic conditions, apples may develop 'invasive alcohol poisoning' in which large areas of the skin and flesh turn pinkish brown (10). Oxygen concentrations may fall below the threshold value in stores having faulty equipment, in unventilated containers (5) or within polyethylene-lined cartons during transport at above-optimal temperatures. Alternatively, the disorder may occur in apples which have been so heavily waxed that respiration is disrupted (1).

BLACK-END of pears has been studied in the USA and in Australia (7). The name describes a depressed blackened area of variable size at the calyx-end. The disorder is present at harvest and does not progress further, but it constitutes an unsightly blemish. Black-end is associated with trees grown on Japanese pear rootstocks, and tends to be prevalent in light crops. Water stress and high levels of boron in the fruit have also been implicated (7).

CHEMICAL INJURY can result from orchard sprays (calcium salts, for example) or from post-harvest treatments (such as an antioxidant used against superficial scald, q.v.). Such chemicals may cause injury as objectionable as the disorder they were designed to prevent. Skin lesions may be grey, brown or black, sometimes of a shape which indicates retention of liquid between adjacent fruits or on the floor of a bin. Care is necessary in preparation and application of chemicals, with due regard to the sensitivity of certain cultivars (1).

EXTERNAL CARBON DIOXIDE INJURY (as distinct from brown heart, q.v.) occurs in only a few apple cultivars, such as Bramley's Seedling (10) and McIntosh (6). Early picked fruit is especially susceptible. Lesions are sharply defined, sunken, becoming brown and finally almost black. Harmful concentrations of carbon dioxide can build up if the store is sealed before the fruit has been cooled (10). It is important to ensure that the temperature and composition of the storage atmosphere is appropriate to the cultivar (5).

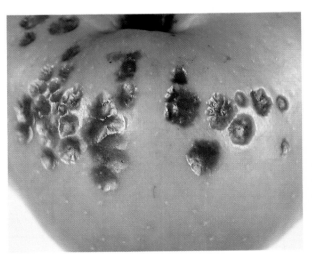

382 External carbon dioxide injury

JONATHAN BREAKDOWN occurs in certain apple cultivars, for example Jonathan and Delicious, and is characterised by a premature softening of the flesh. As in senescent breakdown (q.v.), predisposing factors include late picking, large fruit and low calcium content (5). In contrast, however, Jonathan breakdown is more prevalent at low than at moderate storage temperatures, and is worse if the fruit is cooled directly after harvest (10). A delay prior to storage allows a certain amount of moisture-loss from the fruit, which seems to render it less susceptible to breakdown (4).

McINTOSH BREAKDOWN, apparently unique to the cultivar, also involves softening of the flesh but often on one side of the fruit only. Affected apples are prone to bruise damage, resulting in further softening followed by local flesh browning. The texture becomes mealy throughout and the fruit may split (5). Predisposing factors include certain weather sequences during fruit development, the use of a growth-regulating chemical, low calcium-content of the fruit, late picking, and storage in polyethylene-lined boxes (9). It has been shown (9) that, if fruit is to be stored in controlled atmosphere, it is beneficial to hold it at 20° C for one or two days prior to storage.

RUSSETING is a roughening and discoloration of the skin which can be unsightly unless it is typical of the cultivar (e.g. Egremont Russet). It is caused by intensive cell division during the early stages of fruit development. Russeting is influenced by such factors as rootstock, tree vigour, soil type, nitrogen fertilisation, weather, chemical sprays and the presence of disease (e.g. powdery mildew, q.v.). Depending on cultivar and circumstances, timely spraying (starting after bloom) of a growth regulating chemical may prevent undue development of russet (8).

SPARTAN BREAKDOWN is characterised by a generalised softening and browning, which starts just beneath the skin and progresses throughout the fruit. It is caused by a deficiency of calcium, but has some of the characteristics of senescent breakdown, occurring in late-picked or long-stored apples of the cultivar Spartan. Large fruits are especially susceptible, and bruising can stimulate development of symptoms (5). The disorder worsens after the fruit is released from cold store. Control measures include pre- and post-harvest calcium treatments, followed by controlled atmosphere storage; excessive humidity should be avoided (3).

SUNBURN describes the bronzing of one side of an apple, following exposure to intense sunlight in the orchard. It is a non-progressive skin-blemish.

SUNSCALD is a more serious disorder of apples, typically caused when mature fruits are newly exposed to the sun as the branches bend under the weight of fruit (1). At harvest time the affected skin appears bleached, but during storage it darkens to almost black. In **delayed sunscald** there are no symptoms at harvest, but brown lesions develop in store. Alternaria rot (q.v.) often develops in the weakened tissue.

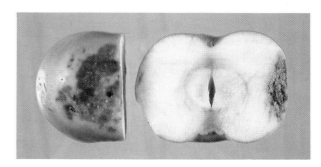

383 Sunscald

---

1 HALL E.G. & SCOTT K.J. (1977) *Storage and market diseases of fruit.* Melbourne: CSIRO, 52 pp illus.
2 NICHOLS W.C. & PATTERSON M.E. (1987) Ethanol accumulation and poststorage quality of 'Delicious' apples during short-term, low-O_2, CA storage. *HortScience* **22**, 89–92.
3 PORRITT S.W., LIDSTER P.D. & MEHERIUK M. (1975) Post-harvest factors associated with the occurrence of breakdown in Spartan apples. *Canadian Journal of Plant Science* **55**, 743–747.
4 SCOTT K.J. & ROBERTS E.A. (1968) The importance of weight loss in reducing breakdown of Jonathan apples. *Australian Journal of Experimental Agriculture and Animal Husbandry* **8**, 377–380.
5 SMOCK R.M. (1977) Nomenclature for internal storage disorders of apples. *HortScience* **12**, 306–308.
6 SMOCK R.M. & BLANPIED G.D. (1963) Some effects of temperature and rate of O_2 reduction on the quality of controlled atmosphere stored McIntosh apples. *Proceedings of the American Society for Horticultural Science* **83**, 135–138.
7 TAYLOR B.K., ENDE B. van den & JERIE P.H. (1987) Studies on black-end of pear in the Goulburn Valley, Australia. *Journal of Horticultural Science* **62**, 157–162.
8 TAYLOR D.R. & KNIGHT J.N. (1986) Russeting and cracking of apple fruit and their control with plant growth regulators. *Acta Horticulturae* No. 179, 819–820.
9 WALSH C.S. (1978) The effect of delayed storage, slow cooling and polyethylene box liners on 'McIntosh' breakdown. *HortScience* **13**, 534–536.
10 WILKINSON B.G. & FIDLER J.C. (1973) Physiological disorders. In *The biology of apple and pear storage* (Ed. by J.C. Fidler, B.G. Wilkinson, K.L. Edney & R.O. Sharples), pp. 63–131 illus. Farnham Royal, Slough, UK: Commonwealth Agricultural Bureaux.

CHAPTER 5
STONE FRUITS

The stone fruits considered in this chapter are species of *Prunus*, which is believed to have originated in central Asia. The **peach**, *P. persica* (L.) Batsch var. *persica*, has been grown in China for at least four thousand years and was introduced into the Graeco–Roman world from Persia. The closely related **nectarine**, *P. persica* var. *nucipersica* (Suckow) C. Schneider, is smooth-skinned with a rich flavour. It has been known in the east since ancient times and in Europe since the 16th century. Present world production (of peaches and nectarines combined) is of the order of 8 million tonnes, the leading producers being Italy and the USA, followed by Spain, Greece, China, France, the USSR, Turkey, Mexico and Japan.

The **apricot**, *P. armeniaca* L., has long been grown in China. It later became popular in Mediterranean countries, and in the 18th century was taken to North America and South Africa. World production is nearly 2 million tonnes, produced mainly in Turkey, Italy, the USSR, Spain, Greece, the USA, France, Morocco, Pakistan and China.

The **plum**, *P. domestica* L., exhibits great diversity and its fruits may be red, purple, green or golden. Some cultivars are suitable for drying; in the USA such fruits are called 'prunes' even when in the fresh state, whereas in the UK this term is reserved for the dried or rehydrated product. World production of plums is about 6 million tonnes, the leading producers being Romania, the USSR, Yugoslavia, the USA, China, West Germany, France, Hungary, Turkey and Italy. Much of the crop is processed, and the industry is of particular importance in countries of eastern Europe.

The **sweet cherry**, *P. avium* (L.) L., was domesticated in Asia Minor and was popular throughout the Roman Empire. Following a decline in the cultivation of cherries during the Middle Ages, they were again extensively planted in Europe, especially in Germany, from the 16th century onwards. Less is known of the **sour cherry**, *P. cerasus* L., but it is presently of considerable importance for processing into conserves and liqueurs. Leading producers of cherries include the USA, Romania, Germany and Italy.

Stone fruits intended for processing are sometimes harvested mechanically whereas fresh market fruit is picked by hand because of susceptibility to physical damage. Strategies to minimise injury at harvest time include lining buckets and bins with plastic film, improving the suspension of transport vehicles, and upgrading rough farm roads (8).

Peaches, nectarines, apricots and plums are climacteric fruits (page 14) and, provided they are sufficiently mature when picked, can undergo softening and sweetening after harvest. Cherries, on the other hand, appear to have a non-climacteric pattern of respiration and their texture and flavour do not alter significantly after harvest.

Post-harvest treatments have sometimes included exposure to high temperature (hot water or hot air) for a few minutes or for several hours, in attempts to kill insects or to control diseases and disorders. The use of such treatments is limited, however, because fruit is liable to be injured by high temperatures (8). More recently efforts have been directed towards refining biological control methods, which depend on the use of an antagonistic organism (e.g. certain bacteria) applied to the fruit surface.

Speed of handling is vital. Moisture is easily lost from cherries because of their small size and from peaches, nectarines and apricots because their skins lack a substantial waxy layer. Artificial wax is sometimes applied for reduction in moisture loss and for cosmetic effect. Peaches cannot be waxed immediately, being covered with 'fuzz' (hair-like projections called trichomes) which must first be brushed off. Brushing causes significant surface damage, and a waxed brushed peach tends to lose water more rapidly than a non-waxed non-brushed peach (8). Another function of artifical wax is to carry a fungicide (10), which would not otherwise adhere well to the smooth surface of a plum, for example. The fungicide should be a different type of compound from the one used as an orchard spray, in order to minimise the opportunity for development of insensitive fungal strains. All stone fruits are prone to serious post-harvest decay, the most important diseases being blue mould rot, brown rot, grey mould rot and rhizopus rot (2,4,6,7). Infection and development of disease organisms is minimised by prompt cooling; if hydro-cooling is used, the water must be properly chlorinated (unless a fungicide or an antagonist be incorporated) (10).

With regard to disorders, stone fruits are subject to various types of injury and breakdown (4), influenced by cultivar, orchard and climatic factors (5), harvesting and handling practices and storage environment. Chilling injury can occur at low storage temperatures, while ripening and decay (**384, 385**) are favoured by moderate to high temperatures (3). Special storage regimes have therefore been devised, such as delayed cold storage and/or intermittent warming, and such techniques may be appropriate in certain circumstances. The most practical method, however, is to cool the fruit immediately to 0°C and store for only a limited period so that disorders do not have time to develop. Typical storage periods are 1 to 2 weeks for cherries and apricots, 2 to 3 weeks for plums, and 3 to 6 weeks for peaches and nectarines; however, there is wide variation among cultivars. Controlled atmosphere storage can extend storage life (1,9), and cherries benefit from carbon dioxide treatment during transport to distant markets (8).

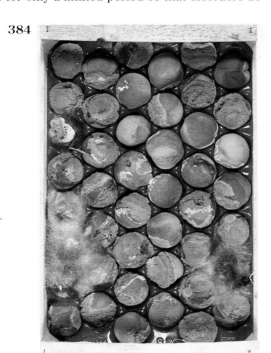

1 ANDRICH G. & FIORENTINI R. (1986) Effects of controlled atmosphere on the storage of new apricot cultivars. *Journal of the Science of Food and Agriculture* **37**, 1203–1208.
2 BORECKA H. & WOJTAS B. (1986) Deterioration of sweet cherry fruit during storage period. *Acta Agrobotanica* **39**(1), 23–28.
3 BROOKS C. & COOLEY J.S. (1928) Time–temperature relations in different types of peach rot infection. *Journal of Agricultural Research* **37**, 507–543.
4 CEPONIS M.J., CAPPELLINI R.A., WELLS J.M. & LIGHTNER G.W. (1987) Disorders in plum, peach, and nectarine shipments to the New York market, 1972–1985. *Plant Disease* **71**, 947–952.
5 FACTEAU T.J., ROWE K.E. & CHESTNUT N.E. (1987) Response of 'Bing' and 'Lambert' sweet cherry fruit to preharvest calcium chloride applications. *HortScience* **22**, 271–273.
6 FOURIE J.F. & HOLZ G. (1985) Postharvest fungal decay of stone fruit in the south-western Cape. *Phytophylactica* **17**, 175–177.
7 HARVEY J.M., SMITH W.L. & KAUFMAN J. (1972) Market diseases of stone fruits: cherries, peaches, nectarines, apricots and plums. *Agriculture Handbook of the United States Department of Agriculture* No. 414, 64 pp illus.
8 MITCHELL F.G. (1986) Protecting stone fruits during handling and storage. *Deciduous Fruit Grower* **36**, 199–204.
9 SOUTY M., CHAMBROY Y., JACQUEMIN G., REICH M. & BREUILS I. (1985) Stabilisation des abricots après récolte. *Fruits* **40**, 813–817.
10 WELLS J.M. & BENNETT A.H. (1976) Hydrocooling and hydraircooling with fungicides for reduction of postharvest decay of peaches. *Phytopathology* **66**, 801–805.

See also:

CHILDERS N.F. & SHERMAN W.B. (Eds) (1988) *The peach – world cultivars to marketing.* Gainesville, Florida: Horticultural Publications, 1000 pp illus.

ALTERNARIA ROT of peaches, nectarines, apricots, plums and cherries caused by
Alternaria alternata (Fr.) Keissler

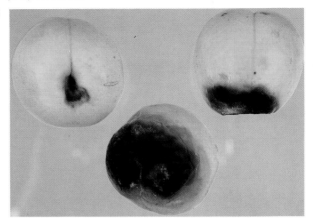

386

OCCURRENCE Alternaria rot is occasionally found in peaches, nectarines and apricots but is more common in plums and cherries. In Canada and the USA the incidence of alternaria rot is said to have increased in recent years (5), and the disease is also common in other countries such as India (1) and Argentina (2).

SYMPTOMS Lesions may occur on any part of the fruit and are characterised by the development of a dense mat of olive-green spores. In peaches, nectarines, apricots and plums lesions are firm, slightly sunken spots, although in some apricot cultivars symptoms do not progress beyond a reddish freckling of the skin where the fungus penetrated and then died back (3). In sweet cherries the affected tissue is a firm brown cone-shaped mass extending towards the stone, and the surface of the lesion may support a copious growth of fluffy white mould (4). The fungus is less able to invade the flesh of the sour cherry and rotting tends to be superficial; dark spores are produced in abundance but there is scant development of mould strands. Microscopic examination may be necessary to distinguish between alternaria rot and cladosporium rot (q.v.).

BIOLOGY The fungus is common in orchard debris, and its spores are disseminated in the air. Infection is usually associated with injury and in cherries this often takes the form of cracking or splitting of the skin whilst the fruit is still on the tree. Alternatively wounding may occur during harvesting and handling of the fruit, and the fungus invades via cuts in the skin. In apricots it has been observed that the fungus is able to penetrate sound fruit via natural openings (stomata) but that the resulting lesions remain limited (3).

CONTROL Calcium sprays can be used to check the tendency of cherries to crack or split after rain. For all stone fruits care is necessary to minimise mechanical damage during harvesting and handling. The fruit should be cooled as rapidly as possible after harvest. Controlled atmosphere storage can be used to delay the onset of rotting, the mould being inhibited by high concentrations of carbon dioxide. Post-harvest fungicides give additional protection (2).

1 BHARGAVA S.N., SHUKLA D.N. & NARENDRA SINGH (1978) Post-harvest diseases of cherry. *Proceedings of the National Academy of Science, India* B **48**, 168.
2 BOCKLET M.F., MASERA O. & AZAR de G., E. (1977) Cerezas Bing conservadas en frigorífico en bolsas de polietileno con distintos coadyuvantes. *Revista de la Facultad de Ciencias Agrarias, Universidad Nacional de Cuyo* **21** (1/2), 63–72.
3 LARSEN H.J., COVEY R.P. & FISCHER W.R. (1980) A red spot fruit blemish in apricots. *Phytopathology* **70**, 139–142.
4 LINDEGREN C.C. & ROSE D.H. (1924) Two hitherto unreported diseases of stone fruits. *Journal of Agricultural Research* **28**, 603–605.
5 McPHEE W.J. (1980) Some characteristics of *Alternaria alternata* strains resistant to iprodione. *Plant Disease* **64**, 847–849 illus.

ANTHRACNOSE or BITTER ROT of peaches, plums and cherries caused by
Glomerella cingulata (Stonem.) Spauld. & v. Schrenk
Conidial state: *Colletotrichum gloeosporioides* (Penz.) Sacc.

387

OCCURRENCE Anthracnose of sour cherries occurs in Poland (2), and plums are affected in India (1). Peach anthracnose has caused serious losses in Japan and the southern USA (3,5). Its current importance is difficult to estimate, because disease diagnosis is not always straightforward (4).

SYMPTOMS Lesions on peaches are circular, slightly sunken, firm and brown, and in the initial stages the symptoms are indistinguishable from those caused by species of *Botryodiplodia* and *Botryosphaeria* (q.v.) (4). Later, if conditions are humid, there may be a greyish mould growth and salmon-pink spore-masses arranged in concentric circles, characteristic of anthracnose (4). In severe infections the lesions may coalesce and cover a large part of the fruit (3).

BIOLOGY In the USA it was found that the fungus survived the winter in blue lupin (*Lupinus angustifolius* L.) which was planted as a cover crop (5). Some strains of the fungus produce the sexual state (perithecia giving rise to ascospores) and virtually all strains produce the asexual state (acervuli giving rise to conidia). It is the asexual spores which are important in infection, and they are dispersed by rain. The fungus is capable of attacking uninjured fruit (3) but infection takes place more readily via wounds caused by twig abrasion (4), and symptoms may be manifest whilst the fruit is still on the tree. If infection occurs shortly before harvest, however, the disease may not be detected in the packhouse. Further infections may occur via injuries sustained during harvesting and handling, and decay can spread into sound fruit in contact. The optimal growth temperature for the fungus is approximately 25°C, and it makes almost no progress below 5°C (3).

CONTROL In the USA peach anthracnose appeared to decline in importance during the 1950s, possibly because of the destruction of the lupin crop by unusually cold weather, followed by a decision to refrain from planting lupins in the vicinity of peach orchards. In Poland sour cherries are protected by fungicide sprays after petal fall (2).

---

1 MAITI S., SAHAMBI H.S. & GHOSH S.P. (1979) A new fruit rot on plum (*Prunus domestica*). *Indian Journal of Mycology and Plant Pathology* **9**, 249.
2 OLSZAK M. & PIOTROWSKI S. (1985) Studies on the control of tart cherry bitter rot caused by *Colletotrichum gloeosporioides* (Penz.) Sacc. (*Glomerella cingulata* (Stonem.) Spauld. et Schrenk). *Fruit Science Reports* **12**(4), 155–161.
3 RAMSEY G.B., SMITH M.A. & HEIBERG B.C. (1951) Anthracnose of peaches. *Phytopathology* **41**, 447–455 illus.
4 RITTENBURG L.H. & HENDRIX F.F. (1983) Peach fruit rots caused by *Botryosphaeria* spp. and *Glomerella cingulata*. *Plant Disease* **67**, 449–450.
5 WEIMER J.L. & DUNEGAN J.C. (1949) Identity of anthracnose of lupine and peach caused by *Glomerella cingulata*. *Plant Disease Reporter* **33**, 416–418.

BACTERIAL SPOT of peaches, nectarines, apricots, plums and cherries caused by
Xanthomonas campestris pv. *pruni* (Smith) Dye

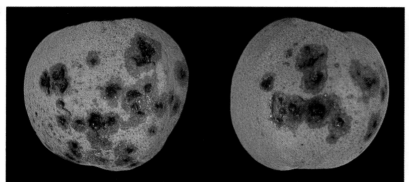
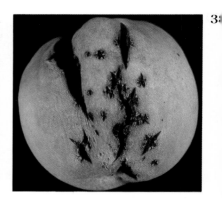

OCCURRENCE Bacterial spot is a serious disease in several countries, for example the eastern and central USA (2,4), Canada (1), South Africa (3), Australia and New Zealand (5). It is unknown in the UK but occurs in parts of southern Europe, South America and Japan (CMI Map 340, CMI Descr. 50).

SYMPTOMS In peaches, nectarines, apricots and plums the first symptoms are pale brown circular spots which later become sunken, dark in the centre and watersoaked at the edge. A yellow exudate consisting of a mass of bacteria may ooze from the lesions. Spots may occur anywhere on the fruit surface and sometimes coalesce, leading to extensive disfigurement, with deep cracks and sometimes gum-formation (2). On cherries the spots tend to be concentrated towards the distal end of the fruit, and some cherries may be so deformed as to have flattened sides.

BIOLOGY The bacterium survives the winter in bark cankers (5). During wet periods in the spring bacterial ooze is dispersed by wind, rain and dew, and infects the leaves via natural openings (stomata), causing a leaf spot and eventual defoliation (4). Leaf blight can also result from autumn infection of unhealed leaf scars. In warm wet conditions bacteria exuded from leaf spots are washed on to the fruits and are capable of penetration via the stomata, at any stage during fruit development (2). If young fruits are infected they are likely to become deformed and may not even be harvested. Fruits infected when mature suffer from superficial disfigurement, against which post-harvest treatments are ineffective.

CONTROL Since the organism can be transmitted on pruning shears (3), hygienic practices are essential. Maintaining good tree vigour is important, as this is associated with resistance to the disease (5). Further control may be achieved with carefully timed bactericide or antibiotic sprays (1). Chemical control is costly, however, and not always effective, and the breeding of more resistant cultivars is an important long-term objective (4).

---

1 DHANVANTARI B.N., DIRKS V.A. & BROWN R.J. (1978) Effectiveness of antibiotics for control of bacterial spot of peach in southwestern Ontario. *Canadian Journal of Plant Science* **58**, 953–959.
2 DUNEGAN J.C. (1932) The bacterial spot disease of the peach and other stone fruits. *Technical Bulletin of the United States Department of Agriculture* No. 273, 53 pp illus.
3 GOODMAN C.A. & HATTINGH M.J. (1988) Mechanical transmission of *Xanthomonas campestris* pv. *pruni* in plum nursery trees. *Plant Disease* **72**, 643.
4 WERNER D.J., RITCHIE D.F., CAIN D.W. & ZEHR E.I. (1986) Susceptibility of peaches and nectarines, plant introductions, and other *Prunus* species to bacterial spot. *HortScience* **21**, 127–130.
5 YOUNG J.M. (1987) Orchard management and bacterial diseases of stone fruit. *New Zealand Journal of Experimental Agriculture* **15**, 257–266.

BLUE MOULD ROT of peaches, nectarines, apricots, plums and cherries caused by
Penicillium expansum Link

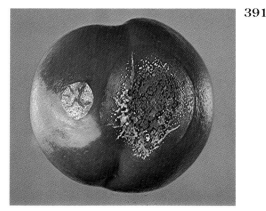

OCCURRENCE Blue mould rot is a common post-harvest disease of stone fruits in all producer countries, including the USA (1,2,4,5) and Chile (3). It is especially important on plums (1) and cherries (2,5), occurring to a lesser extent on apricots, peaches and nectarines (3,4).

SYMPTOMS Lesions are circular, soft and pale brown, becoming covered with a white mould which produces blue-green spores (conidia). On peaches, nectarines and apricots there is usually a dense mass of spores. On plums and cherries the spore-bearing mould strands tend to be clumped together in tufts which are arranged in concentric circles around the point of infection. There is a characteristic musty odour.

BIOLOGY The fungus survives on dead plant material and its spores are common in the atmosphere. It can only attack injured or overmature fruit. Cherries may suffer from cracking after rain, and all types of stone fruit are subject to injury during harvesting and handling. A clear relationship has been established between the severity of bruising in plums and the incidence of blue mould rot (1).

CONTROL The primary means of control is prevention of physical injury to the skin of the fruit (1). It is also important to minimise the number of spores available for infection by destroying culled mouldy fruits which may be lying around the packing station. Post-harvest treatments which are effective against blue mould rot include washing, hydro-cooling and the use of disinfectants and fungicides. Care is needed when using hot-water treatments for the control of brown rot (q.v.), lest physiological injury to the fruit open the way for blue mould (4). Prompt cooling is desirable, and controlled atmosphere storage can be beneficial (4). Carbon dioxide treatment during transit has been used successfully to minimise blue mould rot in cherries (2).

---

1 CEPONIS M.J. & FRIEDMAN B.A. (1957) Effect of bruising injury and storage temperature upon decay and discoloration of fresh, Idaho-grown Italian prunes on the New York city market. *Plant Disease Reporter* **41**, 491–492.
2 ENGLISH H. & GERHARDT F. (1942) Effect of carbon dioxide and temperature on the decay of sweet cherries under simulated transit conditions. *Proceedings of the American Society for Horticultural Science* **40**, 172–176.
3 MORALES M., A.R. (1984) Manejo de post-cosecha de nectarinos y duraznos. *Revista Fruticola* **5**(2), 56–59 illus.
4 SMITH W.L. & ANDERSON R.E. (1975) Decay control of peaches and nectarines during and after controlled atmosphere and air storage. *Journal of the American Society for Horticultural Science* **100**, 84–86.
5 YARWOOD C.E. & HARVEY H.T. (1952) Reduction of cherry decay by washing. *Plant Disease Reporter* **36**, 389.

BROWN ROT of peaches, nectarines, apricots, plums and cherries caused by
Monilinia **spp.**
Conidial state: *Monilia* spp.

392

393

OCCURRENCE In many countries where stone fruits are grown, brown rot is of major importance. *M. fructicola* (Wint.) Honey, appears to be restricted to the Americas (1,2,7,10), South Africa, Japan (9), Australia (4,5) and New Zealand (8) (CMI Map 50). *M. fructigena* (**336**) is widespread in Europe and Asia and has also been recorded in parts of North Africa and South America (9). *M. laxa* (page 180) is found in most countries which produce stone fruits (3); in the USA this species is generally confined to the west coast (6,10).

SYMPTOMS The first sign of disease is a small watersoaked spot which rapidly becomes brown. Affected skin is somewhat tough and, especially in peaches, may turn black. Spores (conidia) are produced in abundance, often in pustules arranged in concentric circles; spores of *Monilia cinerea* (the conidial state of *Monilinia laxa*) appear greyish in mass; so also do those of *M. fructicola*, whereas those of *M. fructigena* are buff.

BIOLOGY The brown rot fungi survive the winter on mummified fruits and blighted twigs in the orchard (5). Wet weather during the spring months results in the production and dissemination of spores. *M. fructicola* frequently produces the sexual stage (ascospores discharged from apothecia) (1); this is less common in *M. fructigena* and *M. laxa* (9). All three species produce abundant asexual spores which are disseminated by rainsplash (8), wind (10) and insects. *M. fructicola* and *M. laxa* (especially the latter) may infect blossoms and thence progress into the young fruits; alternatively, spores produced on blighted flowering shoots may later cause infection of nearby developing fruits by penetrating the stomata (8). Infections of immature fruits characteristically remain quiescent, and the fungus becomes active only when the fruits ripen. Infection of mature fruits may take place via natural openings (stomata), or directly through the skin or, in peaches, via the hair-sockets (4). Much more commonly, however, the brown rot fungi invade stone fruits either through pre-harvest wounds caused by insects or adverse weather, or through injuries resulting from harvesting and handling operations. Subsequently brown rot can spread rapidly through a batch of harvested fruit.

CONTROL It is important to practise good orchard hygiene in order to reduce the quantity of spores available for infection the following spring; this entails the removal of mummified fruits and blighted twigs, or the application of fungicides which suppress spore-production (5). Subsequently, at least in some locations, it is necessary to spray at the time of blossoming in order to protect the flowers, and it may also be worthwhile to make a further fungicide application shortly before harvest (10). The use of a range of unrelated chemicals helps to delay the emergence of fungal strains tolerant of a particular type of fungicide (5). Regular monitoring is desirable, to permit early detection of insensitivity (6). Control of insects is important if puncture wounds are to be minimised, and suitable irrigation can help to prevent growth splits (particularly prevalent in cherries). Care during harvesting and handling serves to reduce physical damage, thereby decreasing the likelihood of post-harvest infection. Stone fruits may be immersed in a hot-water fungicide bath to de-activate

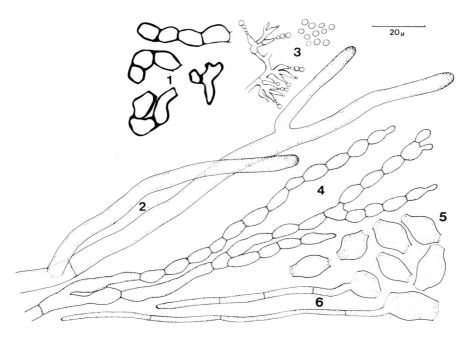

394 *Sclerotinia fructicola* (from CMI Descr. No. 616, Mordue 1979) 1 cells from sclerotium rind 2 hyphae (stained with cotton blue in lactophenol) 3 spermatia 4,5,6 developing, mature, and germinating conidia

recent infections and prevent subsequent rotting; time and temperature control are critical if the treatment is to be effective without injuring the fruit. Prompt cooling is then required, and if this is achieved by iced water (hydro-cooling) then it is advisable to disinfect the water with chlorine, unless, of course, a specific bacterium is being used as an antagonist (biological control) (7). An alternative or additional treatment involves the application of wax containing a fungicide (7). Finally, although the brown rot fungi can make some growth even at 0°C, the use of refrigeration minimises the development of decay (7). Cultivars differ in susceptibility to brown rot, some showing partial resistance (2).

---

1 BIGGS A.R. & NORTHOVER J. (1985) Inoculum sources for *Monilinia fructicola* in Ontario peach orchards. *Canadian Journal of Plant Pathology* **7**, 302–307.
2 FELICIANO A., FELICIANO A.J. & OGAWA J.M. (1987) *Monilinia fructicola* resistance in the peach cultivar Bolinha. *Phytopathology* **77**, 776–780 illus.
3 FOURIE J.F. & HOLZ G. (1987) Post-harvest fungicidal treatment of stone fruit for the control of *Monilinia laxa*, *Botrytis cinerea* and *Rhizopus stolonifer*. *Phytophylactica* **19**, 73–78.
4 HALL R. (1971) Pathogenicity of *Monilinia fructicola*. 2. Penetration of peach leaf and fruit. *Phytopathologische Zeitschrift* **72**, 281–290.
5 KABLE P.F. (1983) New fungicides for suppression of sporulation by *Monilinia fructicola* from overwintering sites in peach trees in spring. *Journal of Horticultural Science* **58**, 45–50.
6 MICHAELIDES T.J., OGAWA J.M. & OPGENORTH D.C. (1987) Shift of *Monilinia* spp. and distribution of isolates sensitive and resistant to benomyl in California prune and apricot orchards. *Plant Disease* **71**, 893–896.
7 PUSEY P.L., WILSON C.L., HOTCHKISS M.W. & FRANKLIN J.D. (1986) Compatibility of *Bacillus subtilis* for postharvest control of peach brown rot with commercial fruit waxes, dicloran, and cold-storage conditions. *Plant Disease* **70**, 587–590.
8 TATE K.G. & CORBIN J.B. (1978) Quiescent fruit infections of peach, apricot, and plum in New Zealand caused by the brown rot fungus *Sclerotinia fructicola*. *New Zealand Journal of Experimental Agriculture* **6**, 319–325 illus.
9 WILLETTS H.J. & HARADA Y. (1984) A review of apothecial production by *Monilinia* fungi in Japan. *Mycologia* **76**, 314–325.
10 ZEHR E.I. (1982) Control of brown rot in peach orchards. *Plant Disease* **66**, 1101–1105 illus.

CLADOSPORIUM ROT of peaches, nectarines, apricots, plums and cherries caused by
Cladosporium herbarum (Pers.) Link

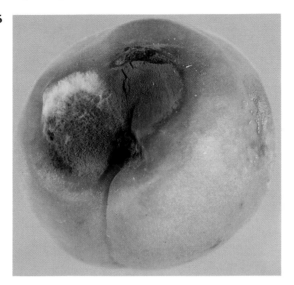

395

OCCURRENCE Cladosporium rot may affect stone fruits in all producer countries. *C. herbarum* (usually considered a weak parasite) is commonly found on plums and cherries, sometimes causing the predominant type of decay (2,4). It occurs less frequently on peaches, nectarines and apricots. Most references are from the USA.

SYMPTOMS Lesions are associated with injuries, and become covered with a white mould which later gives rise to a velvety mat of dark green spores. The rotted tissue remains limited in area at the surface but decay extends deeply towards the stone (2). Microscopic examination of the spores may be necessary to distinguish between cladosporium rot and alternaria rot (q.v.).

BIOLOGY The fungus is widespread and survives on dead plant material in the soil. Spores (conidia) are disseminated in the air and are able to cause infection of fruit which has been damaged by rain or by rough handling (3). High incidence of infection has been recorded in plums shaken from the tree and harvested from the ground (4). Decay can occur even at cold-storage temperatures (1,4).

CONTROL The main means of control is prevention of damage to the skin of the fruit, and every care should be taken during harvesting and handling. It may be advisable to modify the method of harvesting (4). Fungicide treatments are beneficial (4,5) and refrigerated storage serves to minimise the rate of mould development (1).

---

1 BERRY J.A. & MAGOON C.A. (1934) Growth of microorganisms at and below 0°C. *Phytopathology* **24**, 780–796.
2 BRATLEY C.O. (1931) Decay of sweet cherries from California. *Plant Disease Reporter* **15**, 73–74.
3 ENGLISH H. (1945) Fungi isolated from moldy sweet cherries in the Pacific Northwest. *Plant Disease Reporter* **29**, 559–566.
4 MICHAELIDES T.J., OGAWA J.M. & SHOLBERG P.L. (1987) Chemical control of fungi causing decay of fresh prunes during storage. *Plant Disease* **71**, 14–17.
5 PIERSON C.F. (1958) Fungicides for reduction of post-harvest decay of sweet cherries. *Proceedings of Washington State Horticultural Association* **54**, 115–116.

GREY MOULD ROT of peaches, nectarines, apricots, plums and cherries caused by
Botryotinia fuckeliana (de Bary) Whetzel
Conidial state: *Botrytis cinerea* Pers.

396

OCCURRENCE Grey mould rot is important on stone fruits in all producing areas, for example European countries (1,3), the USA (4), South America (5) and Japan. In South Africa it is the most prevalent post-harvest decay (2).

SYMPTOMS A pale brown circular spot may develop anywhere on the fruit. Under humid conditions there is copious growth of white mould, but in a drier atmosphere the fungus tends to produce a mass of grey-brown spores (conidia). Infected flesh is soft and moist.

BIOLOGY The fungus survives as resistant resting bodies (sclerotia) in the soil, or as mould strands in dead plant material. During cool wet weather spores are disseminated by wind and rain and may infect dying blossoms, later resulting in fruit rot. In addition, infection may occur via injuries sustained during harvesting and handling. Grey mould tends to be particularly prevalent in fruits which have been stored for some time, especially if they have been physiologically weakened by post-harvest hot water treatment (4). Decay spreads into adjacent sound fruits, causing 'nesting'.

CONTROL Orchard fungicide sprays prevent blossom infection and protect the fruits (1,3), while post-harvest fungicide dips help to reduce decay in store (2,5). Careful handling is essential if mechanical damage is to be minimised, and efficient cooling serves to retard the development of decay. Grey mould rot of cherries can be controlled to some extent by exposing them to a high concentration of carbon dioxide during storage and/or transit.

---

1 BOMPEIX G., COEFFIC M. & GREFFIER P. (1979) Lutte contre les pourritures des pêches à *Monilia* spp., *Botrytis* sp. et *Rhizopus* sp. *Fruits* **34**, 423–430.
2 FOURIE J.F. & HOLZ G. (1986) Post-harvest fungicidal treatments for the control of storage decay of stone fruit. *Deciduous Fruit Grower* **36**, 253–256.
3 SEKSE L. & GJAERUM H.B. (1987) Effects of preharvest fungicide sprays on grey mould (*Botrytis cinerea*) of sweet cherries. *Acta Agriculturae Scandinavica* **37**, 89–92.
4 SMITH W.L. & ANDERSON R.E. (1975) Decay control of peaches and nectarines during and after controlled atmosphere and air storage. *Journal of the American Society for Horticultural Science* **100**, 84–86.
5 SUAZO V., I., MOORE T., C. & MORALES M., A.R. (1983) Efectos de la aplicación de fungicidas en pre y postcosecha en el control de pudriciones en nectarinos. *Simiente* **53**, 43–47.

PUSTULAR SPOT of peaches, nectarines, apricots, plums and cherries caused by
Stigmina carpophila (Lév.) M.B. Ellis

OCCURRENCE Pustular spot (Coryneum blight) has been recorded on all the stone fruits but is especially important on apricots (1,10). The disease occurs in most producer countries, including the USSR (3), Romania (6), Iran (1), Pakistan (10) and the USA, particularly the western states (8,9). It has also been recorded in most European countries, North Africa, South Africa, South America, Australasia and Japan (CMI Map 188).

SYMPTOMS Small purplish-red circular spots develop and, whilst the edges darken, the centre of each lesion becomes cream-coloured as dry corky tissue forms. In apricots the lesions are raised and rough, whilst in peaches they become slightly sunken and there is usually an exudation of gum (7).

BIOLOGY The fungus survives dry periods in twig cankers or inside buds (8) and may also produce resting spores (chlamydospores) (4). In wet weather it produces numerous asexual spores (conidia) (2) which are disseminated in rain drops and, if moisture persists, can cause infection of flowers, leaves and fruits. Leaf symptoms include a 'shot-hole' effect, and there may be serious defoliation (8). Lesions appear on the fruit whilst they are still on the tree, the optimal temperature for disease development being approximately 20°C (6). Expansion of the lesions after harvest is very slow even at temperatures favourable to the fungus; the main effect is superficial disfigurement.

CONTROL The disease is controlled by orchard fungicide sprays and, with the aid of a forecasting system based on meteorological data, the applications of fungicide can be efficiently timed (1,5,6). After harvest the fruit should be rapidly cooled and kept under refrigeration.

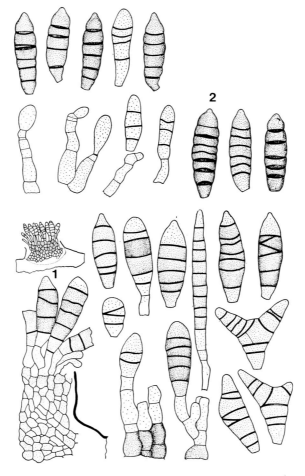

398 *Stigmina carpophila* (from Mycological Papers No. 72, Ellis 1959)
1 sporodochium x 100   2 conidiophores and conidia x 500

1 ASHKAN M. & ASSADI P. (1971) [Shot-hole of stone fruits in Iran. *Stigmina carpophila* (Lev.) Ellis = *Coryneum beijerinckii* Oud.] *Iranian Journal of Plant Pathology* **7**, 11–14, 39–63.
2 ELLIS M.B. (1959) *Clasterosporium* and some allied Dematiaceae–Phragmosporae. 2. *Mycological Papers* No. 72, 56–58 illus.
3 KALINICHENKO R.I. (1978) [The main fungus diseases of stone fruit leaves and fruits in the Ukraine.] *Mikologiya i Fitopatologiya* **12**(1), 27–32.
4 KOUL A.K. & NARAIN U. (1983) Existence of chlamydospores in the culture of *Stigmina carpophila*. *Indian Phytopathology* **36**, 769–770 illus.
5 OGAWA J.M., MacSWAN I.C., MANJI B.T. & SCHICK F.J. (1977) Evaluation of fungicides with and without adhesives for control of peach diseases under low and high rainfall. *Plant Disease Reporter* **61**, 672–674.
6 RAFAILA C & ZAHRIA A. (1979) [Biological and ecological features of *Stigmina carpophila* (Lev.) M.B. Ellis needed for the determination of forecasting and monitoring elements.] *Analele Institutului de Cercetări pentru Protecția Plantelor* **14**, 51–60.
7 SCHWABE W.F.S. & MATTHEE F.N. (1973) Gomvlek of Coryneum-versengingsiekte van steenvrugte. *NIVV Stellenbosch Inligtingsbulletin* No. 180.
8 WILSON E.E. (1937) The shot-hole disease of stone-fruit trees. *Bulletin of the University of California Agricultural Experiment Station* No. 608, 40 pp illus.
9 WRIGHT W.R., SMITH M.A., RAMSEY G.B. & BERAHA L. (1960) Observations on pustular spot on peaches. *Plant Disease Reporter* **44**, 424–425 illus.
10 ZAFAR S.I. & SUFI N.A. (1972) *Coryneum* blight and other diseases of apricot (*Prunus armeniaca*) in north-west Pakistan. *Pakistan Journal of Scientific and Industrial Research* **15**, 193–195.

RHIZOPUS ROT of peaches, nectarines, apricots, plums and cherries caused by
Rhizopus oryzae Went & Prinsen Geerligs
Rhizopus stolonifer (Ehrenb. ex Fr.) Lind

OCCURRENCE Rhizopus rot is one of the most important post-harvest diseases of stone fruits, and can be responsible for heavy losses, especially in peaches, nectarines and cherries. There are also records of the disease in apricots (3) and plums (1). The fungi are widespread (CMI Maps), and records on stone fruits include the USA (6,8,10), Australia (3,4), India (1), South Africa (5) and European countries (2,7).

SYMPTOMS Circular brown watersoaked areas appear on the fruit, and the skin is easily displaced from infected flesh. There is profuse development of coarse white mould strands, giving rise to globular white spore-heads (sporangia) which later turn black and are easily visible to the naked eye (6,7). They arise from dark, branched, root-like structures (rhizoids); an exception is a form of *R. oryzae* known as *R. arrhizus* Fischer, which does not possess rhizoids. Decayed tissue is soft and watery, releasing juices which have a fermented or acidic odour (1).

BIOLOGY The fungi exist on dead plant material and their spores (sporangiospores) are very common in the atmosphere. Immature fruits are generally resistant to attack (7); there are a few reports of infection of mature fruits in the orchard (2). Much more commonly, however, infection takes place via injuries (1), such as are sustained during harvesting and handling (8). Once established, these fungi can invade adjacent sound fruits, causing extensive 'nesting' (1). Infection and rotting are favoured by a warm, moist environment, the optimal growth temperature being approximately 35°C for *R. oryzae* and 25°C for *R. stolonifer* (1,9). Soft-rotting by these fungi is accomplished by means of enzymes, some of which are heat-tolerant and survive the sterilisation process in fruit-canning factories, leading to breakdown of the canned product (3).

CONTROL It may be necessary to spray the fruit with a fungicide shortly before maturity (2), in order to prevent infection in the orchard. A pre-harvest spray also helps to reduce post-harvest rotting (5,6). Strict hygiene in orchard and packhouse is important; fallen and rejected fruits should be removed and destroyed before they can become a source of spores. Careful handling of harvested fruit is essential if injuries are to be minimised (1). Mechanically harvested fruit suffers more damage than fruit picked by hand, so that chemical treatment becomes even more important (8). Various fungicide dips have proved successful (3,4) and heat treatment (for peaches and nectarines) serves to deactivate incipient infections (7,10). If washing or hydrocooling is used, the water should be regularly changed and disinfected. Recently biological control has been investigated; the fruit is treated with a suspension of antagonistic bacteria which suppress fungal growth (10). Post-harvest treatments should be carried out promptly so that fruit may be cooled as soon as possible after picking (6). Delay at high

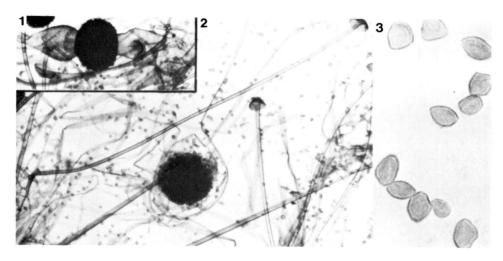

401 *Rhizopus stolonifer* (from CMI Descr. No. 524, Lunn 1977) 1 zygospore formed between two compatible isolates x 93   2 sporangium, sporangiophores and rhizoids x 60
3 sporangiospores x 933

temperature permits rapid development of any *Rhizopus* infections which are still viable (1,6). Once the fruit is cooled below about 5°C, however, these fungi can make no progress at all, and their growth is somewhat inhibited when the fruit is returned to higher temperatures for ripening (9). Severe rhizopus rot in a consignment of stone fruit is a clear indication that it has not been properly pre-cooled and refrigerated.

---

1 BHARGAVA K.S. & GUPTA S.C. (1957) Market diseases of fruits and vegetables in Kumaon. 2. Rhizopus rot of plums (*Prunus communis*). *Horticultural Advance* **1**, 65–67 illus.
2 FOSCHI S. (1950) Attacchi di *Rhizopus nigricans* su pesche pendenti. *Rivista della Ortoflorofrutticoltura Italiana* **34**, 184–186 illus.
3 HARPER K.A., BEATTIE B.B., PITT J.I. & BEST D.J. (1972) Texture changes in canned apricots following infection of the fresh fruit with *Rhizopus stolonifer*. *Journal of the Science of Food and Agriculture* **23**, 311–320.
4 HEATON J.B. (1980) Control of brown rot and transit rot of peaches with post-harvest fungicidal dips. *Queensland Journal of Agriculture and Animal Science* **37**, 155–159.
5 HEYNS A.J. (1968) Control of post-harvest decay of clingstone peaches by pre-harvest applications of dicloran. *Deciduous Fruit Grower* **18**, 21–22.
6 LUEPSCHEN N.S., ROHRBACH K.G., JONES A.C. & PETERS C.L. (1971) Methods of controlling Rhizopus decay and maintaining Colorado peach quality. *Bulletin of Colorado State University Experiment Station* No. 547 S, 24 pp illus.
7 NGUYEN-THE C. & SOUTY M. (1985) La pourriture des pêches par *Rhizopus stolonifer*. *Arboriculture Fruitière* **32** (376), 54–56 illus.
8 OGAWA J.M., SANDENO J.L. & MATHRE J.H. (1963) Comparisons in development and chemical control of decay-causing organisms on mechanical and hand-harvested stone fruits. *Plant Disease Reporter* **47**, 129–133.
9 PIERSON C.F. (1966) Effect of temperature on growth of *Rhizopus stolonifer* on peaches and agar. *Phytopathology* **56**, 276–278.
10 WILSON C.L., FRANKLIN J.D. & PUSEY P.L. (1987) Biological control of Rhizopus rot of peach with *Enterobacter cloacae*. *Phytopathology* **77**, 303–305.

RUST of peaches, nectarines, apricots and plums caused by
Tranzschelia pruni-spinosae (Pers.) Dietel var. *discolor* (Fuckel) Dunegan

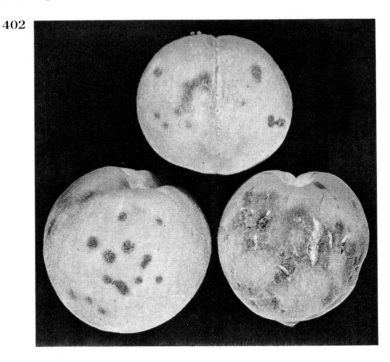

402

OCCURRENCE Rust of stone fruits has been recorded in the western USA (3,4,7), Australia (2,6), New Zealand (5), South Africa (9) and Israel (10). It also occurs in Japan and in several European countries (CMI Map 223), for example Germany (8) and Switzerland (1). The disease is of primary importance on the leaves (resulting in defoliation), but in peach, nectarine, apricot and plum, the fruits may also be attacked.

SYMPTOMS Rust lesions are roughly circular, a few mm in diameter and generally sunken, the majority having pustules of dark brown spores (urediniospores) at their centres (4,5). The disease does not develop further in the fruits after harvest, but the blemish causes a loss in market value. Furthermore, since infected tissue is tough and leathery below the skin, this may lead to rejection at the canning factory (4,9).

BIOLOGY This fungus may follow a complex life-cycle, involving two alternate hosts. On the stone fruit host, pustules are formed on the leaves and branches (and later on maturing fruit), giving rise to 'summer spores' (urediniospores) and then 'winter spores' (teliospores) which remain dormant until the spring (8). These then germinate to form basidiospores, which infect *Anemone coronaria* L., the alternate host (1). Pustules containing pycniospores generally occur on the upper surface of anemone leaves, while on the underside there are corresponding lesions producing aeciospores, which then cause infection of stone fruit foliage (1,5). In Israel it was observed that anemone rust is very common, and that development of symptoms on anemone slightly precedes the unfolding of new leaves on stone fruit trees (10). In the UK the disease occurs occasionally on plums, but appears to be of greater importance on the commercial anemone crop; in areas where corm production is threatened it may be advisable to eradicate nearby trees of plum and wild damson. In Australia and New Zealand the stage on anemone is said to be rare, and in South Africa it has not yet been recorded; in these regions it has been demonstrated that the fungus can survive the winter in twigs of stone fruit trees and cause leaf infection by urediniospores produced the following year (2,5,9). Cool humid rainy weather is conducive to the development of rust epidemics (4).

CONTROL Fungicides can be used during the growing season, and it is also important to spray the trees immediately after harvest, in order to prevent the fungus from surviving the winter (2,6,7,8). Infected leaf litter can be irrigated (which seems to kill spores) and ploughed in.

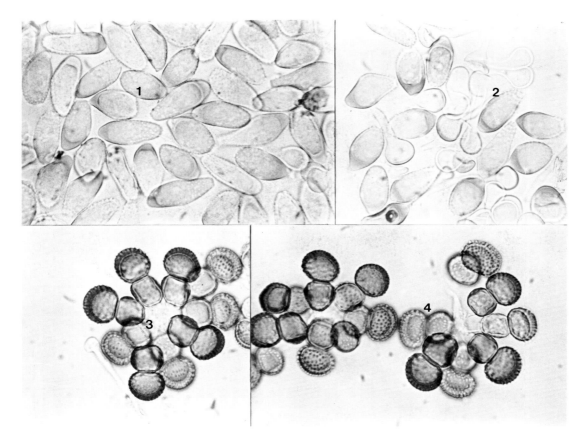

403 *Tranzschelia pruni-spinosae* var. *discolor* (from CMI Descr. No. 287, Laundon & Rainbow 1971)
1 urediniospores 2 paraphyses 3, 4 teliospores x 500

1 BLUMER S. (1960) Untersuchungen über die Morphologie und Biologie von *Tranzschelia pruni-spinosae* (Pers.) Dietel und *T. discolor* (Fuck.) Tranz. & Litv. *Phytopathologische Zeitschrift* **38**, 355–383 illus.
2 CARTER M.V., MOLLER W.J. & PADY S.M. (1970) Factors affecting uredospore production and dispersal in *Tranzschelia discolor. Australian Journal of Agricultural Research* **21**, 905–914.
3 DUNEGAN J.C. (1938) The rust of stone fruits. *Phytopathology* **28**, 411–427 illus.
4 GOLDSWORTHY M.C. & SMITH R.E. (1931) Studies on a rust of clingstone peaches in California. *Phytopathology* **21**, 133–168 illus.
5 JAFAR H. (1958) Studies on the biology of peach rust (*Tranzschelia pruni-spinosae* Pers.) in New Zealand. *New Zealand Journal of Agricultural Research* **1**, 642–651 illus.
6 KABLE P.F., KEEN B.J. & BAMBACH R.W. (1987) Evaluation of fungicides for curative activity against *Tranzschelia discolor*, cause of the rust disease of French prune (*Prunus domestica* L.). *Australian Journal of Agricultural Research* **38**, 577–585.
7 MICHAELIDES T.J. & OGAWA J.M. (1986) Chemical control of prune leaf rust (*Tranzschelia discolor* f.sp. *domesticae*). *Plant Disease* **70**, 307–309.
8 PAUL H.V. (1987) Studies on the symptomatology and morphology of *Tranzschelia pruni-spinosae* (Pers.) Diet. var. *discolor* (Fuckel) Dunegan, the causal organism of rust of stone fruits. *Pflanzenschutz-Nachrichten Bayer* **39**, 345–363 illus.
9 ROTH G. (1966) *Tranzschelia discolor* (Fuck.) Tranz. & Litv. on peaches in Transvaal Lowveld, South Africa. *Phytopathologische Zeitschrift* **56**, 141–150.
10 ZWIRN-HIRSCH H.E. (1945) Infection experiments with aeciospores of *Tranzschelia pruni-spinosae* (Pers.) Diet. in Palestine. *Palestine Journal of Botany, Jerusalem Series* **3**, 178–179.

OTHER DISEASES of peaches, nectarines, apricots, plums and cherries

BLACK MOULD ROT, caused by *Aspergillus niger* (**307**), is occasionally found on peaches, nectarines and cherries, but only on injured or weakened fruit stored at high temperatures (8). Control therefore involves careful handling and cold storage.

BOTRYODIPLODIA ROT of peaches is characterised by brown lesions in which the skin remains firmly attached to the flesh (1,2). Grey or white mould spreads over the fruit (which eventually shrivels), and gives rise to minute black spore-bearing bodies (pycnidia) typical of *Botryodiplodia theobromae* (**197**). Initial infection is via wounds but, once established, the fungus can spread into neighbouring sound fruits. Its optimal growth temperature is approximately 30°C and decay is arrested at temperatures below about 8°C.

BOTRYOSPHAERIA ROT of peaches may be caused by *Botryosphaeria dothidea* or by *B. obtusa* (page 176), both of which are reported from the southern USA and South Africa. Lesions are dark, sunken and elliptical; the skin becomes leathery (later wrinkled) and clings tightly to the underlying flesh, which remains firm. Early symptoms are indistinguishable from those of anthracnose (q.v.). The fungi cause bark cankers and gummosis (2), and pre-harvest infection of fruit can occur via wounds in the skin resulting from twig abrasion.

DIPLODINA ROT of peaches is caused by *Diplodina persicae* Horn, and the initial symptoms resemble those of brown rot (q.v.). As the lesion enlarges, however, shrivelling occurs in concentric rings, and minute black spore-bearing bodies (pycnidia) develop, containing two-celled spores (5). Decay is rapid at temperatures between 25° and 30°C.

MUCOR ROT may be caused by *Mucor piriformis* (page 190), which produces a coarse, erect, white mould growth giving rise to globular black spore-heads (sporangia). The disease has been recorded on nectarines from Chile and on peaches from California (7). In contrast to rhizopus rot (q.v.), mucor rot can develop even at 0°C. Control measures include improvements in orchard and packhouse hygiene, in order to minimise contamination of fruit.

PHOMOPSIS ROT is caused by *Diaporthe perniciosa* (conidial state: *Phomopsis mali*) (page 202). The disease has been reported on peaches in the USA (9) and in Japan; it is characterised by

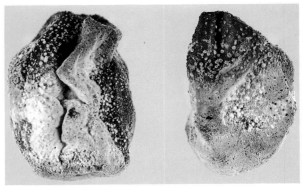

404 Phomopsis rot

sunken black lesions on the fruit, following direct infection in the orchard. The author has observed this fungus on plums in the UK, the fruits having previously been weakened by drought stress. Numerous minute bodies (pycnidia) formed beneath the skin, giving rise to characteristic spores (**144**).

PINK MOULD ROT of plums is caused by *Trichothecium roseum* (page 191) and has been reported from India (4). Rotting can be prevented by storing the fruit at temperatures below 10°C.

POWDERY MILDEW of peaches and apricots may be caused by *Sphaerotheca pannosa* (Wallr.: Fr.) Lév. (conidial state: *Oidium*) originating from mildewed roses in the vicinity of the orchard (6). Hard white patches appear on developing fruits, often resulting in cracks. Various species of *Podosphaera* (page 196) have been recorded on fruits of apricot, plum and cherry.

SCAB of peaches, nectarines, apricots and plums is caused by *Venturia carpophila* Fisher (conidial state: *Fusicladium carpophilum* (Thüm.) Oudem.). It has caused important losses in the eastern USA (3), Argentina, South Africa, Australia and the USSR (CMI Map 198). Cherry scab is caused by *Venturia cerasi* Aderh. In the UK the conidial state only has been recorded, *Fusicladium cerasi* (Rabenh.) Sacc. Symptoms on stone fruits consist of numerous circular spots, giving rise to dark spores. Even if the lesions remain limited, they constitute an unsightly blemish. Tissue beneath the lesions becomes corky and may crack, thus exposing the fruit to invasion by other fungi. Scab is essentially an orchard disease (3) and can be controlled by a fungicide spray applied shortly after petal fall.

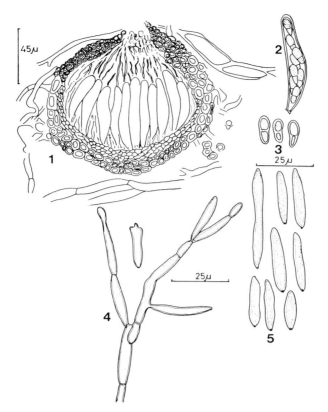

405 *Venturia carpophila* (from CMI Descr. No. 402, Sivanesan 1974)
1 pseudothecium 2 ascus 3 ascospores 4 conidiophore 5 conidia

SOUR ROT of plums, caused by *Geotrichum candidum* (page 78), has been recorded in India. Sour rot of peaches occurs in the USA and may be caused by *G. candidum* or by *Monilia implicata* Gilman & Abbott; mixed infections result in especially rapid decay (10). Lesions are soft and brown, later covered by quantities of moist white spores having an odour of vinegar. Infection by these fungi has been associated with poor sanitation in the packhouse, and losses can be reduced by careful handling and efficient chlorination of the water used to wash and cool the fruit.

---

1 ALLAM M.E. & YUSEF H.M. (1969) Studies on Diplodia peach fruit rot in U.A.R. *Journal of Phytopathology of the U.A.R.* **1**, 13–21 illus.
2 BRITTON K.O. & HENDRIX F.F. (1982) Three species of *Botryosphaeria* cause peach tree gummosis in Georgia. *Plant Disease* **66**, 1120–1121.
3 GOTTWALD T.R. (1983) Factors affecting spore liberation by *Cladosporium carpophilum*. *Phytopathology* **73**, 1500–1505 illus.
4 HASIJA S.K. & AGARWAL H.C. (1977) Pathological studies on *Trichothecium roseum* causing rot of apple and plum fruits. *Indian Phytopathology* **30**, 254–255.
5 HORN N.L. & HAWTHORNE P.L. (1954) A new fruit rot of peach. *Phytopathology* **44**, 134–136 illus.
6 KABLE P.F., FRIED P.M. & McKENZIE D.R. (1980) The spread of a powdery mildew of peach. *Phytopathology* **70**, 601–604.
7 MICHAELIDES T.J. & OGAWA J.M. (1987) Effect of soil temperature and moisture on the survival of *Mucor piriformis*. *Phytopathology* **77**, 251–256 illus.
8 MORALES M., A., CANCINO E., L., OEHRENS E., E. & PAIROA C., H. (1974) Identificación y patogenicidad de hongos aislados de pudriciones de cerezas Corazón de Paloma y Bing en almazenaje refrigerado. *Agricultura Técnica* **34**, 15–19.
9 ROSENBERGER D.A. & BURR T.J. (1982) Fruit decays of peach and apple caused by *Phomopsis mali*. *Plant Disease* **66**, 1073–1075 illus.
10 WELLS J.M. (1977) Sour rot of peaches caused by *Monilia implicata* and *Geotrichum candidum*. *Phytopathology* **67**, 404–408.

DISORDERS of peaches, nectarines, apricots, plums and cherries

BRUISING can occur when fruit is handled without sufficient care, such as when it is dropped into an unlined bucket during harvesting (7) or subjected to harsh impacts on the packing line. Bruising of peaches can lead to skin discoloration (q.v.), while in cherries the result may be surface pitting (q.v.) or internal browning (7). Bruised fruit is predisposed to decay. In contrast to vibration damage (page 207), impact damage is much greater in fruit handled cold (less than 5°C) than in fruit handled warm (about 20°C) (9). Thus, if fruit cannnot be packed on the day of harvest, it may be advisable to cool it for the holding period and then allow it to re-warm before being handled and packed (5). Cooling of packaged fruit is best achieved by the method of forced air or pressure cooling (page 21). Intrinsic susceptibility to impact damage can be reduced by ensuring an adequate calcium content through pre- and post-harvest treatments with a calcium compound (4).

CRACKING of cherries (and occasionally other stone fruits) is especially likely if they are subjected to rain or high humidity shortly before harvest. The exposed flesh may subsequently dry out but, if not, tends to become infected by fungi. Cultivars differ in susceptibility, and resistance may be enhanced by the application of certain orchard sprays (such as a calcium compound). There is no evidence to implicate hydro-cooling as a cause of cracking (3).

FREEZING INJURY is characterised by a translucent, water-soaked appearance of the flesh. Severely frozen fruit tends to break down on thawing, the injured tissue becoming soft, brown and watery. The temperature at which freezing occurs is determined, among other factors, by the sugar content of the fruit; thus sour cherries may freeze at −1°C whereas sweet black cherries may not freeze until −3°C. The pattern of damage in a consignment can indicate whether the freezing occurred before/after packing or before/after loading in a refrigerated container.

INSECT INJURY takes many forms, such as russeting, scarring or distortion of the fruit, depending on type of insect and time of attack. Infestation during early fruit development may lead to exudation of gum, while later insect injury can result in decay. The insects are usually no longer present in the harvested crop; an exception, however, is certain scale insects (of peaches and nectarines) which remain attached to the skin, especially in the stem-end region, and induce local reddening.

407 Insect injury

INTERNAL BREAKDOWN of peaches and nectarines is usually known as 'woolliness' (q.v.).

INTERNAL BREAKDOWN of plums was first investigated in the USA (10) and later in South Africa (2). A simliar disorder occurs in apricots (10) but most reports refer to plums. The first symptom is **internal browning** of the flesh near the stone. Subsequently the affected tissue breaks down into a gelatinous mass (2). Early symptoms are sometimes seen when the fruit is still on the tree, but more usually they develop only after harvest (10). Fruit is predisposed to the disorder by certain orchard and climatic factors, for example very hot weather; firm fruits are seldom affected (10) and it seems likely that calcium and other minerals are influential (2).

INTERNAL BROWNING of cherries can be a result of bruising (q.v.).

SHRIVELLING, resulting from moisture loss, detracts from appearance and, in fruits sold by

406 Freezing injury

weight, can be responsible for substantial monetary loss. Nectarines are especially prone to shrivelling, as are 'defuzzed' peaches. Most of the moisture loss occurs through evaporation, though some is the result of respiration. Losses can be minimised by prompt and rapid pre-cooling (hydro-cooling being especially appropriate) and by the maintenance of low temperature and high humidity. Film packaging helps to maintain high humidity in packs of cherries, which are at risk from moisture loss through stems as well as berries. Other stone fruits are sometimes waxed (page 218) to reduce evaporative losses.

SKIN DISCOLORATION of peaches has been reported from the USA (1). The blemish consists of streaks, speckles or spots, which may be purplish, brown or black. The disorder has been related to chemical injury, handling damage, the 'defuzzing' process, hydro-cooling and chlorination. There are probably multiple causes, but it has been suggested that an important factor is the breaking of the natural 'hairs' (trichomes), permitting subsequent local injury by chemical treatments. However, treatment with certain sodium or calcium salts has been used successfully to mask the discoloration resulting from rough handling.

SPLIT PITS can cause problems in peaches and plums. Adverse weather or the over-use of nitrogen fertiliser may result in slightly abnormal growth and development of the fruit and its stone (pit). Splitting of the pit is often followed by development of mould which, however, does not usually spread into the flesh.

SURFACE PITTING of cherries refers to the development of one to several irregular depressions, approximately 0.5 cm across, on the shoulders or sides of the fruit. The disorder is associated with low temperatures and bruising (q.v.), and may develop within a week of cold storage (8). Pre-harvest spraying with a growth-regulating chemical helps to maintain fruit firmness and reduce pitting; so also does a post-harvest calcium dip treatment.

'SWEATING' is the popular term for condensation which occurs when cold fruit is moved into a warm atmosphere. Contrary to general belief, these drops of moisture do not necessarily lead to increased decay (generally being present for too brief a period to permit germination of fungal spores). It is unwise to dispense with cooling in the mistaken belief that subsequent 'sweating' will be detrimental (3).

'WOOLLINESS' of peaches and nectarines is characterised by a mealy texture, a lack of flavour and a dearth of juice. Symptoms develop when susceptible fruits are brought out of cold storage for ripening. Susceptibility varies with cultivar, growing area and weather, but the main predisposing factor is immaturity of fruit at the time of harvest. Research has been carried out in New Zealand, the USA and South Africa. Development of woolliness can be reduced by intermittent warming during the storage period or by holding the fruit at 20°C for one or two days prior to cooling (6). A storage temperature of 0°C is optimal for control of woolliness. The most recent recommendation is to harvest only mature fruit and, after storage, to allow it to ripen at a temperature of at least 15°C.

---

1 DENNY E.G., COSTON D.C. & BALLARD R.E. (1986) Peach skin discoloration. *Journal of the American Society for Horticultural Science* **111**, 549–553 illus.
2 DODD M.C. (1984) Internal breakdown in plums. *Deciduous Fruit Grower* **34**, 255–256.
3 GERHARDT F., ENGLISH H. & SMITH E. (1945) Cracking and decay of Bing cherries as related to the presence of moisture on the surface of the fruit. *Proceedings of the American Society for Horticultural Science* **46**, 191–198.
4 LIDSTER P.D., TUNG M.A. & YADA R.G. (1979) Effects of preharvest and postharvest calcium treatments on fruit calcium content and the susceptibility of 'Van' cherry to impact damage. *Journal of the American Society for Horticultural Science* **104**, 790–793.
5 MITCHELL F.G. (1986) Influence of cooling and temperature maintenance on stone fruit quality. *Deciduous Fruit Grower* **36**, 205–211.
6 MOLLENDORFF L.J. von & VILLIERS O.T. de (1988) Physiological changes associated with the development of woolliness in 'Peregrine' peaches during low-temperature storage. *Journal of Horticultural Science* **63**, 47–51.
7 OGAWA J.M., BOSE E., MANJI B.T. & SCHREADER W.R. (1972) Bruising of sweet cherries resulting in internal browning and increased susceptibility to fungi. *Phytopathology* **62**, 579–580 illus.
8 PORRITT S.W., LOPATECKI L.E. & MEHERIUK M. (1971) Surface-pitting – a storage disorder of sweet cherries. *Canadian Journal of Plant Science* **51**, 409–414 illus.
9 SOMMER N.F., MITCHELL F.G., GUILLOU R. & LUVISI D.A. (1960) Fresh fruit temperatures and transit injury. *Proceedings of the American Society for Horticultural Science* **76**, 156–162 illus.
10 VERNER L., KOCHAN W.J., LONEY C.E., MOORE D.C. & KAMAL A.L. (1962) Internal browning of fresh Italian prunes. *Research Bulletin of the University of Idaho Agricultural Experiment Station* No. 56, 38 pp illus.

CHAPTER 6
SOFT FRUITS AND BERRY FRUITS

The fruits considered in this chapter are not related botanically, but all are native to the temperate zones (although some cultivars thrive in subtropical regions). Strawberries and raspberries are known as 'soft fruits'; they are not true berries. Blueberries, cranberries, currants and gooseberries are berries which grow on bushes, while grapes and kiwifruits are types of berry which grow on vines. With the exception of cranberries, the optimal storage temperature for all these fruits is 0° to 1°C, but they differ substantially in potential storage life.

Strawberries and raspberries

The **strawberry**, *Fragaria* x *ananassa* Duchesne, is much prized for its flavour and delicacy; as well as being consumed in the fresh state it is also important in the manufacture of conserves (4). Total world production is about 2 million tonnes, the main producer being the USA, followed by Poland, Japan, Spain, Italy, the USSR, France, Mexico, the UK and West Germany.

The **raspberry**, *Rubus idaeus* L., has an exquisite flavour, much of which is retained even if the fruit is frozen. A substantial proportion of the world's crop is processed (4). Total production is about 300,000 tonnes, of which over one third is grown in the USSR. Other producer countries include Yugoslavia, Hungary, West Germany, the UK, Canada, Poland, the USA, Bulgaria and France.

The strawberry is sometimes called a 'false fruit', consisting as it does chiefly of tissue derived from the flower receptacle; the true fruits (small one-seeded achenes) are to be found embedded in the surface of this fleshy structure. The raspberry is an aggregate fruit formed in a different way, being composed of many small fleshy fruitlets (drupelets, each with the remains of style and stigma). The differences in botanical structure have an important bearing on the mechanisms of infection by fungi such as grey mould (q.v.), which can become established in the flower parts (8). Early fungicide spraying is necessary for control (4), especially during wet seasons. The use of a mulch in strawberry culture is an effective means of reducing infection by soil fungi (8).

Specialised harvesting machines are sometimes used for soft fruits destined for processing (4). Generally, however, the crop is picked by hand. Strawberries for the fresh market are harvested with a short stem and the calyx in place (**408**); it should not be necessary to grasp the fruit itself. Raspberries, in contrast, are harvested by gentle pulling so that the calyx and 'plug' are left behind (**409**). For both fruits great care is required, the flesh being very delicate (2). In order to obviate the need for a second handling it is usual to place the fruits directly into the 'punnets' (or plastic baskets) in which they will be marketed. With untrained pickers, however, it may be necessary to reselect in a nearby packhouse. In any event, prompt and rapid cooling is essential for consignments going to distant markets (6,7,10).

Strawberries and raspberries have a very brief storage life, perhaps only one or two days under warm conditions. Respiration rate is very high (5), especially in ripe fruits; strawberries have been classified as 'late peak type' (page 15). Optimal maturity for harvest depends on requirements (9); colour development and ripening continue during storage (7). Rapid reduction of temperature to 1°C is best achieved by the method of forced air or pressure cooling (page 21) (10). Low temperature serves to minimise metabolic activity and inhibit fungal decay, which is the primary cause of loss in soft fruits (2,8). Unless processing is intended (page 246), post-harvest fungicides are little used because of the problems of residues and because of the preference for keeping the fruits dry. Storage life can be extended by means of modified atmospheres, and these fruits are able to withstand concentrations of carbon dioxide as high as 10 to 30% (10), although off-flavours sometimes develop (1,3). The main use of modified atmospheres is during transportation when optimal temperature cannot be guaranteed. The carbon dioxide can be supplied from gas cylinders or in the form of 'dry ice'. Alternatively, respiratory carbon dioxide may be allowed to accumulate within specially designed plastic film packaging. Soft fruit cultivars differ greatly in their post-harvest characteristics, and breeding work continues.

1. BROWNE K.M., GEESON J.D. & DENNIS C. (1984) The effects of harvest date and CO_2-enriched storage atmospheres on the storage and shelf-life of strawberries. *Journal of Horticultural Science* **59**, 197–204.
2. CEPONIS M.J., CAPPELLINI R.A. & LIGHTNER G.W. (1987) Disorders in sweet cherry and strawberry shipments to the New York market, 1972–1984. *Plant Disease* **71**, 472–475.
3. EL-KAZZAZ M.K., SOMMER N.F. & FORTLAGE R. J. (1983) Effect of different atmospheres on post-harvest decay and quality of fresh strawberries. *Phytopathology* **73**, 282–285.
4. GIVEN N.K. (1985) Effect of crop management and environment on berryfruit quality – a review. *New Zealand Journal of Experimental Agriculture* **13**, 163–168.
5. HALLER M.H., ROSE D.H. & HARDING P.L. (1941) Studies on the respiration of strawberry and raspberry fruits. *Circular of the United States Department of Agriculture* No. 613, 13 pp.
6. HARVEY J.M. & HARRIS C.M. (1976) Temperature maintenance in air shipments of strawberries to Far Eastern markets. *Bulletin de l'Institut International du Froid, Annexe* 1976–1, 559–565.
7. LUOTO L. (1984) Strawberry quality: effects of handling, packaging and storage on shelf-life. *Acta Horticulturae* No. 157, 79–81.
8. MAAS J.L. (Ed.) (1984) *Compendium of strawberry diseases.* St Paul, Minnesota: American Phytopathological Society, 138 pp illus.
9. SJULIN T.M. & ROBBINS J. (1987) Effects of maturity, harvest date, and storage time on post-harvest quality of red raspberry fruit. *Journal of the American Society for Horticultural Science* **112**, 481–487.
10. TONINI G. (1983) Nuovo tecniche di prerefrigeracione e trasporto per limitare le alterazioni della fragola. *Rivista di Frutticoltura e di Ortofloricoltura* **45**(11), 7–13.

See also:

CHILDERS N.F. (Ed.) (1981) *Strawberries – cultivars to marketing.* Gainesville, Florida: Horticultural Publications, 550 pp illus.

ANTHRACNOSE or BLACK SPOT of strawberries and raspberries
caused by
Colletotrichum spp.

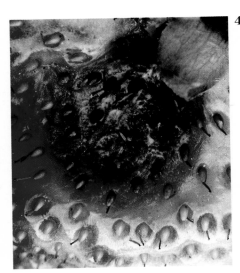

OCCURRENCE *Colletotrichum acutatum* Simmonds ex Simmonds causes serious losses of strawberry fruits in Australia (7,8,9) and South Africa (10). On several occasions in recent years this fungus was inadvertently introduced into the UK on strawberry plants imported from continental Europe and from California (6). Statutory action was immediately taken to contain the disease and prevent its establishment in Britain. Other species recorded on strawberries in the USA include *C. dematium* (Pers.: Fr.) Grove (1,6), *C. fragariae* Brooks (4,5,6) and *C. gloeosporioides* (page 130) (2,6). Some authors do not distinguish *C. acutatum* from *C. gloeosporioides*; outbreaks of black spot in Italy, for instance, have been attributed to the last named (3). *C. gloeosporioides* has been recorded on raspberries in the UK but is regarded as of minor importance on that host. The following account is based on references to strawberries.

SYMPTOMS Black spot is first visible as a circular, slightly sunken, brown watersoaked lesion, enlarging rapidly over a period of 2 or 3 days to involve most of the fruit (9). Mould growth may appear at the edges of the lesion, while at the centre there is copious production of spores. These are initially salmon-coloured (2,3) but, after a period of hot weather, turn dark brown to almost black (7). Infected fruits may dry up completely, but usually they are further rotted by other fungi (10).

BIOLOGY The disease is generally severe only on second year and older plantings, the fungus surviving unfavourable periods within the strawberry plant itself (5). Nearby weed species may also be colonised. During warm humid weather asexual spores (conidia) are produced in abundance, and are dispersed by water-splash (rain or sprinkle irrigation) and wind (3,9). Under moist conditions direct infection of sound young fruit can occur (2), though symptoms do not usually appear until the fruit ripens (8,9,10). Infected fruits may be in apparently good condition at harvest time and yet undergo rotting in the market; the disease does not, however, spread from fruit to fruit (7).

CONTROL As a means of reducing opportunities for infection, trickle irrigation (beneath a plastic mulch) is preferable to sprinkle irrigation. Since infection can occur at any time after petal fall, orchard fungicide spraying must be started at an early stage and repeated during the period of fruit development (3,10). Fruit should be harvested at optimal maturity and marketed without delay (10). Work is in progress to breed strawberry cultivars with resistance to this disease (4). Meanwhile, vigilance is necessary to prevent introduction of the disease into areas still free from it.

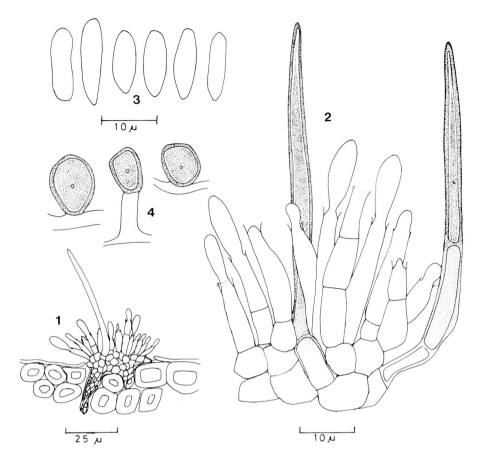

412 *Colletotrichum acutatum* (from CMI Descr. No. 630, Dyko & Mordue 1979)
1 acervulus 2 conidiogenous cells and setae 3 conidia 4 appressoria

---

1 BERAHA L. & WRIGHT W.R. (1973) A new anthracnose of strawberry caused by *Colletotrichum dematium*. *Plant Disease Reporter* **57**, 445–448 illus.
2 ELLIS M.A. & BULGER M.A. (1986) Anthracnose fruit rot (*Colletotrichum gloeosporioides*) of strawberry in Ohio. *Plant Disease* **70**, 475.
3 GULLINO M.L., ROMANO M.L. & GARIBALDI A. (1985) Identification and response to fungicides of *Colletotrichum gloeosporioides*, incitant of strawberry black rot in Italy. *Plant Disease* **69**, 608–609.
4 GUPTON C.L. & SMITH B.J. (1987) Inheritance of resistance to *Colletotrichum* spp. in strawberry. *HortScience* **22**, 724.
5 HORN N.L. & CARVER R.B. (1968) Overwintering of *Colletotrichum fragariae* in strawberry crowns. *Phytopathology* **58**, 540–541.
6 MAAS J.L. & HOWARD C.M. (1985) Variation of several anthracnose fungi in virulence to strawberry and apple. *Plant Disease* **69**, 164–166.
7 McGECHAN J.K. (1977) Black spot of strawberry. *Agricultural Gazette of New South Wales* **88**(3), 26–27 illus.
8 SIMMONDS J.H. (1965) A study of the species of *Colletotrichum* causing ripe fruit rots in Queensland. *Queensland Journal of Agricultural and Animal Sciences* **22**, 437–459 illus.
9 STURGESS O.W. (1957) A ripe fruit rot of strawberry caused by a species of *Gloeosporium*. *Queensland Journal of Agricultural Science* **14**, 241–251 illus.
10 ZYL H.E. van (1985) Anthracnose on strawberries. *Deciduous Fruit Grower* **35**, 404–406 illus.

CALYX BLOTCH and STEM-END ROT of strawberries caused by
Gnomonia fructicola (Arnaud) Fall
Conidial state: *Zythia fragariae* Laibach

OCCURRENCE This disease has been reported in the USA (1,3), Canada (5), and most European countries, being particularly important in Germany (2,4), the Netherlands (8), Spain (10) and the UK (9). It also occurs in New Zealand and Israel.

SYMPTOMS Small brown lesions appear at the tips of the sepals, spreading until the entire calyx is brown and shrivelled. The stem-end of the fruit may also be invaded, and minute flesh-coloured bodies (pycnidia) form in the fruit tissue beneath the calyx. Affected tissues then frequently support abundant growth of the grey mould fungus (q.v.), which obscures the primary cause.

BIOLOGY The leaf blotch fungus produces the sexual stage (perithecia giving rise to ascospores) (7) and the asexual stage (pycnidia giving rise to conidia) (6), and both types of spore-producing body have been found on strawberry plants (10). Spores are disseminated by water-splash, and infection occurs through stomata or via injury sites. The optimal growth temperature is 25°C and the fungus does not grow at all below 5°C (4).

CONTROL Since infection can take place early in the growing season, it is necessary to use a fungicide spray before the onset of flowering (2). Some strawberry cultivars are particularly susceptible to the disease, and none is immune (9).

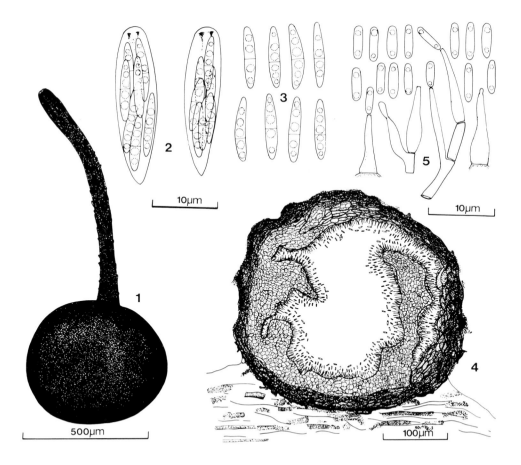

415 *Gnomonia comari* (from CMI Descr. No. 737, Punithalingam 1982) 1 perithecium 2 asci 3 ascospores 4 pycnidium 5 conidiophores, conidiogenous cells and conidia

---

1. ALEXOPOULOS C.T. & CATION D. (1952) *Gnomonia fragariae* in Michigan. *Mycologia* **44**, 221–223.
2. ANON. (1971) Gnomonia – eine neue Krankheit der Erdbeere. *Erwerbsgärtner* **25**, 677.
3. BARR M.E. (1978) The Diaporthales in North America with emphasis on *Gnomonia* and its segregates. *Mycologia Memoir of the New York Botanical Garden* No. 7, 232 pp illus.
4. BOLAY A. (1971) Contribution à la connaissance de *Gnomonia comari* Karsten (syn. *G. fructicola* (Arnaud) Fall). Etude taxonomique, phytopathologique et recherches sur sa croissance in vitro. *Berichte der Schweizerischen Botanischen Gesellschaft* **81**, 398–482 illus.
5. BOLTON A.T. (1954) *Gnomonia fructicola* on strawberry. *Canadian Journal of Botany* **32**, 172–181 illus.
6. FALL J. (1951) Studies on fungus parasites of strawberry leaves in Ontario. *Canadian Journal of Botany* **29**, 299–315 illus.
7. McONIE K.C. & SNYDER W.C. (1966) Production of perithecia by *Gnomonia fructicola* in culture. *Phytopathology* **56**, 197–202 illus.
8. SCHEER H.A.T. van der (1981) Gnomonia-vruchtrot. Een toenemend probleem bij de teelt van oardbeien in Nederland. *Fruitteelt* **71**, 1240–1241.
9. SHIPTON P.J. (1967) A fruit rot of strawberries caused by *Zythia fragariae*. *Plant Pathology* **16**, 123–125.
10. TUSET BARRACHINA J.J. (1972) El *Gnomonia fructicola* (Arnaud) Fall y el *Zythia fragariae* Laibach, productores de un marchitamiento del fresal en Andalucía Occidental. *Anales del Instituto Nacional de Investigaciones Agrarias, Protección Vegetal* **2**, 35–50 illus.

GREY MOULD ROT of strawberries and raspberries caused by
Botryotinia fuckeliana **(de Bary) Whetzel**
Conidial state: *Botrytis cinerea* Pers.

416

417

OCCURRENCE Grey mould rot is one of the most serious diseases of soft fruits and occurs wherever these crops are grown, for example the UK (5,6,7,10), Poland (2), Israel (1), the USA (4,8,9) and Canada (3).

SYMPTOMS The firm brown rot tends to remain dry, and there is no distinct line between healthy and diseased tissue. The surface of the lesion becomes covered with grey-brown spores, although if the atmosphere is humid there is more likely to be copious development of fluffy greyish-white mould bearing few spores.

BIOLOGY The fungus persists in the field as resting bodies (sclerotia) and also as mould strands in decaying plant material which, under favourable conditions, supports quantities of spores (conidia) (3,7). During periods of rapidly changing humidity, such as occur between showers of rain, spores are disseminated by air currents and impinge on all parts of the plant. Primary infection is characteristically via flower parts. In raspberry the fungus typically invades stigmas and styles (page 238), eventually resulting in colonisation of individual fleshy fruitlets (10). In strawberry it is often the stamens which become infected; the fungus remains quiescent in the base of the receptacle, and disease is later manifest as a stem-end rot of the ripe fruit (8). Another type of infection may occur when infected petals are shed and land on the surface of ripe fruit; under moist conditions they are likely to adhere to strawberries (though not to raspberries) and can cause immediate infection in this way (7). In wet seasons strawberry and raspberry crops may rot in the field. Alternatively, fruit may be harvested in apparently sound condition, only to decay during transit and marketing (1,9). Post-harvest decay involves further development of pre-harvest infections together with new infections arising from germination of spores on the fruit surface. Grey mould is capable of substantial development even at low storage temperatures (5) and spreads into neighbouring fruits causing 'nesting' (9).

CONTROL Field hygiene is very important, and the removal of crop debris helps to limit spore production (3). The use of a mulch not only discourages weed growth but also protects the plants from infection through soil-splash or contact with plant debris (9).

Since the flowers are susceptible even in the bud stage, fungicide spraying must be started early in the season (7). Several applications are necessary, especially during wet growing seasons (4). *B. cinerea* has developed strains tolerant of several fungicides (6), although many of these new strains show reduced ability to produce spores. Harvested fruit should be marketed immediately or else promptly cooled to 1°C. Carriage by refrigerated container does not ensure that such a temperature can be maintained, and readings of 5°C and above are more usual (9). The adverse effects of poor temperature control can be partially mitigated by the use of a carbon dioxide-enriched atmosphere (2,9) (page 23). A modified atmosphere can be generated by the fruit itself, if packed in suitable plastic film (1). Senescence and decay of the fruit is thereby delayed. Decay may also be postponed by the use of irradiation, but this is not widely practised. Some strawberry and raspberry cultivars (notably those with firm fruit) show resistance to grey mould rot.

---

1 AHARONI Y. & BARKAI-GOLAN R. (1987) Pre-harvest fungicide sprays and polyvinyl wraps to control *Botrytis* rot and prolong the post-harvest storage life of strawberries. *Journal of Horticultural Science* **62**, 177–181.
2 BORECKA H. & MILLIKAN D.F. (1981) Influence of storage conditions upon the development of post-harvest gray-mold rot of strawberries. *Phytoprotection* **62**, 11–16.
3 BRAUN P.G. & SUTTON J.C. (1987) Inoculum sources of *Botrytis cinerea* in fruit rot of strawberries in Ontario. *Canadian Journal of Plant Pathology* **9**, 1–5.
4 BULGER M.A., ELLIS M.A. & MADDEN L.V. (1987) Influence of temperature and wetness duration on infection of strawberry flowers by *Botrytis cinerea* and disease incidence of fruit originating from infected flowers. *Phytopathology* **77**, 1225–1230.
5 DENNIS C. & COHEN E. (1976) The effect of temperature on strains of soft fruit spoilage fungi. *Annals of Applied Biology* **82**, 51–56.
6 HUNTER T., BRENT K.J., CARTER G.A. & HUTCHEON J.A. (1987) Effects of fungicide spray regimes on resistance in grey mould (*Botrytis cinerea*) on strawberry plants. *Annals of Applied Biology* **110**, 515–525.
7 JARVIS W.R. (1962) The infection of strawberry and raspberry fruits by *Botrytis cinerea* Fr. *Annals of Applied Biology* **50**, 569–575.
8 POWELSON R.L. (1960). Initiation of strawberry fruit rot caused by *Botrytis cinerea*. *Phytopathology* **50**, 491–494.
9 SOMMER N.F., FORTLAGE R.J., MITCHELL F.G. & MAXIE E.C. (1973) Reduction of postharvest losses of strawberry fruits from gray mold. *Journal of the American Society for Horticultural Science* **98**, 285–288.
10 WILLIAMSON B. & McNICOL R.J. (1986) Pathways of infection of flowers and fruits of red raspberry by *Botrytis cinerea*. *Acta Horticulturae* No. 183, 137–141.

LEAK of strawberries and raspberries caused by
Mucor spp.
Rhizopus spp.

OCCURRENCE Species of *Mucor*, in particular *M. piriformis* (page 190) and occasionally *M. hiemalis* Wehmer, constitute a major cause of soft rot of strawberries and raspberries in the UK (2,5). In Canada, the USA and the USSR species of *Rhizopus* seem to be more prevalent than those of *Mucor* (1,6,7), and *R. stolonifer* (page 230) is frequently recorded. *R. stolonifer* is also common in the west of England, but in eastern and central areas of the UK *R. sexualis* (Smith) Callen is more important. Leak is generally associated with late season fruit, though disease incidence (and the relative importance of different species) is markedly influenced by the prevailing weather (3,4).

SYMPTOMS Each of the four fungi is capable of breaking down the fruit tissue so that juice leaks out, hence the common name for the disease. Strands of white mould develop on the surface of infected fruits, usually producing black spore-heads visible to the naked eye. Microscopic examination is required to identify the species (8,9).

BIOLOGY These fungi are soil-inhabitants, and persist either in the sexual state (zygospores) or as mould strands in crop debris. Direct infection can occur if ripe fruit is in contact with the ground; alternatively, during wet weather, fruit may become contaminated through splashing rain (4). In warm conditions infection and development take place rapidly, especially if the fruit has been injured. In *Mucor* species abundant spore-heads (sporangia) commonly form on the surface of infected fruits, and the asexual spores cause further infections either in the field or after harvest. If conditions become warm and dry then *Rhizopus* species are able to do likewise. (Under humid conditions *R. sexualis* tends to produce numerous zygospores, but these have never been observed to germinate and their relevance remains obscure.) All four species cause further post-harvest rotting as mould spreads from infected fruits to neighbouring healthy fruits, and still more infections arise through contamination with spore-laden juices. At temperatures above 20°C decay is very rapid (10). These same fungi are capable of causing serious disintegration of strawberries preserved in sulphite liquor and intended for the production of whole fruit jam (4).

CONTROL It is advantageous to use a mulch to minimise contact between fruits and soil. Pre-harvest sprays can be used on the growing crop, but many fungicides are totally ineffective against these particular fungi (2,5). The disease can be especially serious at the end of a season in which wet weather at the start of fruiting is followed by a warm dry period, allowing a heavy build-up of spores on ripe fruits (3). Under such circumstances it is particularly important to harvest promptly, picking early in the day whilst it is still cool, thus facilitating rapid cooling of harvested fruit (6). At storage temperatures below 10°C spore-production by *R. sexualis* and *R. stolonifer* is inhibited, and below 5°C neither fungus is able to grow (10). Furthermore, at 0°C the asexual spores of *R. sexualis* rapidly lose viability (4), and fruits stored at 0°C do not become infected by this species when moved to a higher temperature.

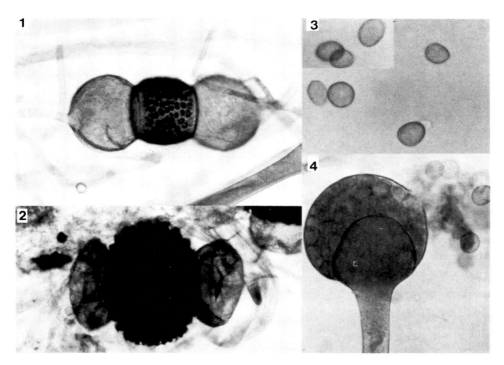

421 *Rhizopus sexualis* (from CMI Descr. No. 526, Lunn 1977) 1 developing zygospore 2 mature zygospore 3 sporangiospores 4 sporangium

1 DAUBENY H.A., PEPIN H.S. & BARRITT B.M. (1980) Post-harvest *Rhizopus* fruit rot resistance in red raspberry. *Horticultural Science* **15**, 35–37.
2 DENNIS C. (1975) Effect of pre-harvest fungicides on the spoilage of soft fruit after harvest. *Annals of Applied Biology* **81**, 227–234.
3 FREEMAN J.A. & PEPIN H.S. (1977) Control of post-harvest fruit rot of strawberries by field sprays. *Canadian Journal of Plant Science* **57**, 75–80.
4 HARRIS J.E. & DENNIS C. (1980) Distribution of *Mucor piriformis*, *Rhizopus sexualis* and *R. stolonifer* in relation to their spoilage of strawberries. *Transactions of the British Mycological Society* **75**, 445–450 illus.
5 MASON D.T. & DENNIS C. (1978) Post-harvest spoilage of Scottish raspberries in relation to pre-harvest fungicide sprays. *Horticultural Research* **18**, 41–53.
6 MITCHELL F.G., MAXIE E.C. & MAYER G. (1966) Low holding temperatures still vital with rapid marketing of strawberries. *California Agriculture* **20**(3), 13–14.
7 POPOVA I.V., KONSTANTINOVA A.F., ZEKALASHVILI A.U. & ZAITSEVA S.P. (1983) [Investigation of resistance of strawberry breeding material to fruit rot.] *Mikologiya i Fitopatologiya* **17**, 222–225.
8 SCHIPPER M.A.A. (1978) On certain species of *Mucor* with a key to all accepted species. *Studies in Mycology* No. 17, 1–52 illus.
9 SCHIPPER M.A.A. (1984) A revision of the genus *Rhizopus*. 1. The *Rhizopus stolonifer* group and *Rhizopus oryzae*. *Studies in Mycology* No. 25, 1–19 illus.
10 STEVENS N.E. & WILCOX R.B. (1917) *Rhizopus* rot of strawberries in transit. *Bulletin of the United States Department of Agriculture* No. 531, 22 pp illus.

LEATHER ROT of strawberries caused by *Phytophthora* spp.

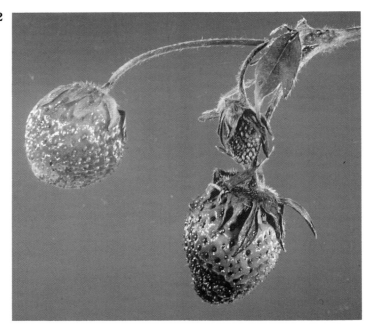

422

OCCURRENCE *Phytophthora cactorum* (Lebert & Cohn) Schröter can cause heavy losses of strawberries grown in the south-eastern USA, especially in wet growing seasons (2,8); the disease is of less importance on the west coast (10). In the UK *P. cactorum* causes crown rot in strawberry plants raised in the greenhouse (3), and there are occasional records of leather rot of the fruit (4). Fruit-rotting also occurs in Hungary (6) and Turkey (9). In Taiwan two species are responsible, viz. *P. cactorum* and *P. citrophthora* (page 65) (5), while in the highlands of West Malaysia an atypical species has been described, having some resemblance to *P. nicotianae* var. *parasitica* (page 64) (1).

SYMPTOMS Tissue affected by phytophthora rot is usually discoloured, tough and leathery, although a leaking soft rot has also been reported (1). Under humid conditions the fruit becomes covered with white mould. *P. citrophthora* produces a sparse aerial mould, while *P. cactorum* gives rise to a shorter denser growth in which there are frequently oospores as well as sporangia (5). Internally the vascular strands show a distinct browning (8). At no stage is there any clear line of demarcation between diseased and healthy tissue.

BIOLOGY These fungi produce both the sexual stage (oospores) and the asexual stage (sporangia giving rise to zoospores). The oospores, being thick-walled, are resistant to unfavourable conditions and can survive in the soil for many years. Germination takes place in moist conditions. Fruits become contaminated with sporangia, zoospores and mould fragments, splash-dispersed by rain (2) or irrigation

423

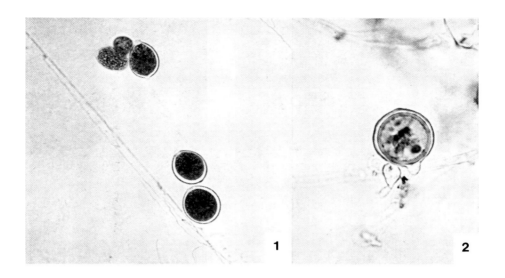

424 *Phytophthora cactorum* (from CMI Descr. No. 111, Waterhouse & Waterston 1966)
1 sporangia   2 antheridium and oogonium, oospore (stained with cotton blue in lactophenol)

water (10). If a film of water persists for 2 or 3 hours, infection can occur at any stage of fruit development (2,6). Fruit infected shortly before harvest may appear healthy when picked, only to decay during transport or marketing. The disease spreads from infected fruits into adjacent sound fruits (8).

CONTROL   It is important to use healthy planting material and to ensure that there is good field hygiene and adequate drainage (3). The use of a mulch serves to prevent soil from being splashed on to fruits, and this reduces the incidence of infection (7). Fungicide sprays can be used during the period of fruit development (4). Studies of the relationship between infection and weather are helpful in timing the sprays correctly (7). After harvest the disease can be held in check by a heat treatment followed by refrigeration and the use of a modified atmosphere (page 23); however, decay resumes when the fruit is returned to an ordinary environment.

---

1 CHEE K.H. & WASTIE R.L. (1976) Post-harvest fruit rot of strawberries. *FAO Plant Protection Bulletin* **24**, 58–59.
2 ELLIS M.A. & GROVE G.G. (1983) Leather rot in Ohio strawberries. *Plant Disease* **67**, 549 illus.
3 HARRIS D.C. & STICKELS J.E. (1981) Crown rot (*Phytophthora cactorum*) in glasshouse-grown strawberries at East Malling Research Station. *Plant Pathology* **30**, 205–212.
4 HUNTER T., JORDAN V.W.L. & PAPPAS A.C. (1979) Control of strawberry fruit rots caused by *Botrytis cinerea* and *Phytophthora cactorum*. *Proceedings of the British Crop Protection Conference (Pests and Diseases)* **1**, 177–183.
5 KAO C.W. & LEU L.S. (1979) Strawberry fruit rot caused by *Phytophthora cactorum* and *P. citrophthora*. *Plant Protection Bulletin, Taiwan* **21**, 239–243.
6 MÁRTON G. (1981) [Occurrence of *Phytophthora cactorum* (Leb. & Cohn) Schroet. on strawberry in Hungary.] *Kertészeti Egyetem Közlémenyei* **13**, 153–161.
7 REYNOLDS K.M., MADDEN L.V. & ELLIS M.A. (1988) Effect of weather variables on strawberry leather rot epidemics. *Phytopathology* **78**, 822–827.
8 ROSE D.H. (1924) Leather rot of strawberries. *Journal of Agricultural Research* **28**, 357–376 illus.
9 TURAN K. & DINÇ N. (1981) [Preliminary studies on causes of black rot of strawberry in the Aegean region.] *Bitki Koruma Bülteni* **21**, 211–217.
10 WRIGHT W.R., BERAHA L. & SMITH M.A. (1966) Leather rot on California strawberries. *Plant Disease Reporter* **50**, 283–287.

OTHER DISEASES of strawberries and raspberries

ALTERNARIA ROT of strawberries and raspberries has been recorded in the USA (3), the UK and West Germany. The disease, caused by *Alternaria alternata* (page 172), is generally of minor importance. Irregular, slightly sunken lesions develop on harvested fruits, and under bright moist conditions there is abundant production of dark green to black spores. Field spraying with an appropriate fungicide can give good control. Since the fungus can colonise straw, it is preferable that mulches be of plastic, for susceptible crops.

BLACK SEED DISEASE of strawberries is caused by *Mycosphaerella fragariae* (Tul.) Lindau (conidial state: *Ramularia tulasnei* Sacc.). It was first reported from the USA and also occurs in New Zealand (10). Infection occurs at flowering time and the fungus progresses into individual fruitlets, resulting in a discoloured seed surrounded by a hard brown lesion. Control of this fungus (which is the cause of the commonly occurring strawberry leaf spot) involves fungicide spraying and destruction of infected leaf debris.

426 Yeasty rot

BLUE MOULD ROTS of strawberries and raspberries are caused by species of *Penicillium*, which invade damaged fruit after harvest (4). Disease incidence varies greatly from season to season. *P. cyclopium* and *P. expansum* (page 178) have been identified on Israeli strawberries.

CLADOSPORIUM ROT of raspberries may be caused by *Cladosporium herbarum* (page 226) or by *C. cladosporioides* (Fres.) de Vries (4). The disease is characterised by a superficial olive-green velvety growth on the shoulder of the fruit. Initial infection usually occurs inside the cavity of the fruit, as a result of injury sustained when the 'plug' is removed during harvesting. Control measures which have proved effective include fumigation with sulphur dioxide gas followed by prompt cooling to 10°C or below (3).

COTTONY ROT of strawberries, recorded in the USA, is caused by *Sclerotinia sclerotiorum* (page 66). The fungus first appears on the calyx and subsequently invades the fruit (1). Affected flesh is firm but rather watery, and under warm humid conditions there is copious growth of white mould, spreading into neighbouring fruits and giving rise to black resting bodies (sclerotia).

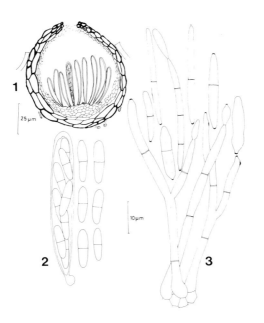

425 *Mycosphaerella fragariae* (from CMI Descr. No. 708, Sivanesan & Holliday 1981)
1 pseudothecium 2 ascus and ascospores
3 conidiophores and conidia

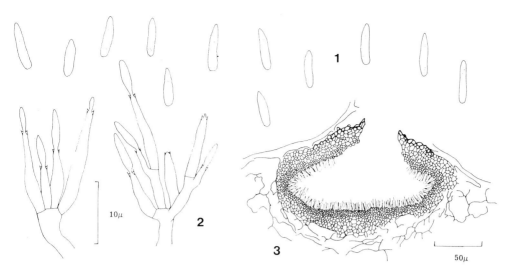

427 *Phomopsis obscurans* (from CMI Descr. No. 227, Sutton & Waterston 1970) 1 conidia 2 conidiophores 3 pycnidium

PESTALOTIA ROT of stored strawberry fruits, caused by *Pestalotia longisetula* Guba, was first noted in Israel, where it was considered insignificant as long as strawberry production was confined to the cool season. In the southern USA the disease was recorded on ripe fruits in the field, the lesions being initially light tan, irregular in shape, slightly sunken and dry, occasionally bordered by a more depressed soft band of tissue. White mould growth may develop, or there may be copious production of black spores (7).

PHOMA ROT, caused by the asexual stage of *Didymella applanata* (Niessl) Sacc., is occasionally found on raspberry fruits (4), but this fungus is chiefly important as the cause of spur blight of the canes.

PHOMOPSIS ROT can sometimes result in severe losses of strawberries in the southern USA. The causal fungus is *Phomopsis obscurans* (Ell. & Ev.) Sutton (syn. *Dendrophoma obscurans* (Ell. & Ev.) Anders.). Symptoms are similar to those caused by the calyx blotch fungus (q.v.) and the two diseases have previously been confused (6). They can be distinguished by microscopic examination of the spore-bearing bodies (pycnidia giving rise to conidia) which readily form on infected fruit tissue.

POWDERY MILDEW of strawberries and raspberries is caused by *Sphaerotheca macularis* (Wall.: Fr.) Lind, and has been reported in the USA and the UK. Infected fruits turn reddish brown with a dull dry appearance, and the flesh may crack (8). In humid conditions infected ripe fruits are soft and pulpy and may exhibit a white 'bloom' consisting of asexual spores (conidia). The sexual stage (cleistothecia giving rise to ascospores) has been observed on strawberry fruit stalks and leaves (8). Control is achieved by timely fungicide sprays and by destruction of infected crop debris. Work is in progress to breed resistant cultivars.

PYTHIUM ROT of strawberries has been reported from Japan, the causal fungus being *Pythium ultimum* (page 273). Infected fruits lose their colour and become soft (9).

RHIZOCTONIA BROWN ROT of strawberries occasionally causes severe losses in the southern USA. The disease is caused by *Rhizoctonia solani* (page 276), a soil fungus which thrives in warm wet conditions and penetrates fruits which happen to be in contact with the ground. Infected strawberries may appear perfect when seen from above, but show a hard brown rot on the underside, with a clear line of demarcation between healthy and affected tissue (5). Particles of soil adhere to the lesion, held together by fungal strands. The chief means of control is mulching to minimise contact between fruit and soil, and rapid cooling of harvested fruit in order to limit the spread of decay.

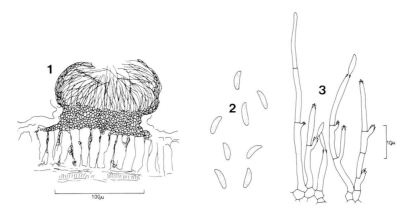

428 Pezizella oenotherae (conidial state: *Hainesia lythri*) (from CMI Descr. No. 535, Sutton & Gibson 1977) 1 acervulus 2 conidia 3 conidiophores

RUST of raspberries causes occasional losses in the USA, Canada and the UK (2). The fungus, *Phragmidium rubi-idaei* (DC.) Karsten, produces all five spore stages (page 45) on various parts of the raspberry plant. Yellow pustules (uredinia) may develop on mature fruits of susceptible cultivars, while other cultivars are completely resistant.

TAN BROWN ROT of strawberries is caused by *Pezizella oenotherae* (Cooke & Ell.) Sacc. (conidial state: *Hainesia lythri* (Desm.) Höhnel). The disease has been recorded on strawberry fruits in the USA, the UK and the temperate highlands of Malaysia. Lesions are tan-coloured, slightly sunken and water-soaked. The fungus penetrates deeply into the flesh, and there is a distinct boundary between healthy and infected tissue. The centre of the decayed area later becomes characteristically dry and spongy and can be removed intact. Mature lesions become covered with minute, pale brown, disc-shaped bodies (sporodochia) bearing quantities of asexual spores (conidia). Control is by mulching to prevent fruit from resting on the soil, and by prompt refrigeration of the harvested crop.

YEASTY ROT of strawberries causes infected tissue to turn pink and emit an odour of fermentation. The organisms responsible are species of *Saccharomyces* and other yeasts; they do not attack sound fruit but can colonise tissue which has been injured by modified atmosphere storage.

---

1 ALCORN J.W.S. (1966) *Sclerotinia sclerotiorum* on strawberry. *Horticultural Research* **6**, 128.
2 ANTHONY V.M., SHATTOCK R.C. & WILLIAMSON B. (1985) Life-history of *Phragmidium rubi-idaei* on red raspberry in the United Kingdom. *Plant Pathology* **34**, 510–520 illus.
3 CAPPELLINI R.A., STRETCH A.W. & WALTON G.S. (1961) Effects of sulphur dioxide on the reduction of post-harvest decay of Latham red raspberries. *Plant Disease Reporter* **45**, 301–303.
4 DENNIS C. & MOUNTFORD J. (1975) The fungal flora of soft fruits in relation to storage and spoilage. *Annals of Applied Biology* **79**, 141–147.
5 DODGE B.O. & STEVENS N.E. (1924) The Rhizoctonia brown rot and other fruit rots of strawberries. *Journal of Agricultural Research* **28**, 643–648 illus.
6 HOWARD C.M. & ALBREGTS E.E. (1973) A strawberry fruit rot caused by *Dendrophoma obscurans*. *Phytopathology* **63**, 419–421 illus.
7 HOWARD C.M. & ALBREGTS E.E. (1973) A strawberry fruit rot caused by *Pestalotia longisetula*. *Phytopathology* **63**, 862–863 illus.
8 HOWARD C.M. & ALBREGTS E.E. (1982). Cleistothecia of *Sphaerotheca macularis* on strawberry plants in Florida. *Plant Disease* **66**, 261–262.
9 KUSAKARI S. & TANAKA Y. (1985) [Soft rot of strawberry fruits caused by *Pythium ultimum* Trow var. *ultimum*.] *Annals of the Phytopathological Society of Japan* **51**, 39–42 illus.
10 NICHOLS M.A. (1960) 'Black seed' disease of strawberry fruit. *New Zealand Journal of Agriculture* **101**, 576.

Blueberries, cranberries, currants and gooseberries

Blueberries are mainly important in Canada and the USA (1,3,5,7), where the highbush blueberry, *Vaccinium corymbosum* L., is especially popular. Other producers include the countries of Scandinavia and eastern Europe (2), total world production being of the order of 150,000 tonnes. In recent years there has been increased planting in, for example, New Zealand, France and the UK.

One of the most important field diseases is 'mummy berry' caused by species of *Monilinia* (page 224). Recent research in the USA (1) has shown that bees are attracted to infected leaves (mistaking them for flowers) and that they and other insects serve to disseminate fungal spores during pollination. Flowers thereby become infected and give rise to diseased fruits. These shrivel and either fall to the ground or are harvested along with healthy berries. Blueberries are small, and hand-harvesting is tedious; in North America mechanical harvesting is widely practised. Much of the crop is processed, but there is also a market for fresh blueberries (3). If the fruit can be promptly cooled after harvest, the use of plastic film packaging permits a storage life of 2 to 4 weeks at 0°C (5). Concern about the effects of condensation is misplaced; this 'sweating' after removal from store is not deleterious (3). Storage at higher temperatures has the effect of curtailing storage life (5).

The most important causes of post-harvest decay are *Botrytis cinerea* (page 258), *Alternaria alternata* (page 172), *Colletotrichum* sp. (page 130) and *Phomopsis vaccinii* Shear (2,3,7). Several other fungal species have been recorded on blueberries, and it is the stem-scar which is the predominant site of infection (3). Control measures may include the use of fungicides as field sprays, a post-harvest fungicide dip and/or storage in a carbon-dioxide enriched atmosphere (3). The modified atmosphere technique can be used to protect blueberries in transit, and is especially valuable in situations where pre-cooling is not feasible (3). There have been several reports, however, of the development of off-flavours after modified atmosphere storage (2).

Cranberries are closely related to blueberries (q.v.), being also species of *Vaccinium*, for example *V. macrocarpon* Aiton. World production of cranberries is estimated at 130,000 tonnes. They are of particular importance in North America, and a traditional association with certain festivals means that there are brief periods of peak demand for the fresh berries. Much of the crop, however, is processed into juice and jelly.

429

Harvesting may be done by hand but labour costs are high (10). In some areas a more usual way is to flood the field (or 'bog', as it is usually called) and to harvest mechanically; berries are stripped from the plants by motorised rotating water reels, and the freed berries then float until removed from the water by conveyor belts. A disadvantage is the bruising caused by the reels (10). Furthermore, if the berries cannot be gathered within a few hours of stripping, then subsequent storage decay can reach unacceptable levels (10).

Numerous fungi attack cranberries, the prevalence of individual species varying greatly with location and season (9). Most of these fungi are unusual, unknown on other hosts (9). Two important examples are *Ceuthospora lunata* Shear and *Strasseria oxycocci* Shear, both of which cause black rot (10). The many post-harvest rots (**429**) can only be identified by culturing the fungi and examining their spores microscopically (8,9).

Cranberries stored at 0°C for longer than about 2 weeks are at risk from chilling injury, bruised fruit being especially susceptible. Low temperature breakdown is characterised by a dull external appearance, and watersoaked or rubbery flesh of a deep maroon colour (10). The optimal storage temperature for cranberries is between 2° and 4°C and they can be kept for 3 to 4 months. Modified atmosphere storage confers no extra benefit.

Currants, in the context of fresh fruit, are species of *Ribes* closely related to the gooseberry (q.v.). (The other meaning of currant is 'raisin of Corinth', the dried fruit of a seedless grape grown in the Levant.) The major producers of fresh currants are the USSR, eastern and north-west Europe, particularly Poland, Germany, Czechoslovakia and the UK. They are little grown in North America; indeed, cultivation of currants is prohibited in some states. This is because of their role as alternate host in an important fungal disease of pine trees, white pine blister rust;

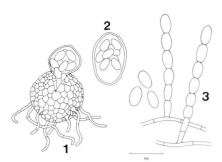

430 *Sphaerotheca mors-uvae* (from CMI Descr. No. 254, Purnell & Sivanesan 1970) 1 cleistothecium 2 ascus and ascospores 3 conidia

eradication of currants (and gooseberries) prevents the fungus from completing its life cycle (see also page 232). The European black currant is *R. nigrum* L., renowned for its high vitamin C content. Much of the crop is processed, particularly for juice. Also important is *R. rubrum* L., which includes red currant and white currant. World production of currants is nearly 600,000 tonnes.

Fresh currants should be marketed immediately or, alternatively, they may be held for up to 1 or 2 weeks at 0°C. Plastic film packaging is probably appropriate. Currants are remarkably tolerant of carbon dioxide; it is reported that, as a supplement to refrigeration, CO_2 concentrations as high as 30 to 40% are used to maintain quality in harvested black currants awaiting processing. The fungal diseases of currants are similar to those of gooseberry (q.v.).

Gooseberries are especially popular in the UK and in France, though they are not so widely grown as in earlier centuries. The European species is *Ribes uva-crispa* L. The berries of the common cultivars are green even when fully mature, but there are also varieties with reddish berries. Gooseberries are larger and more robust than the related currants (q.v.). Being quite tart, they are usually cooked with sugar.

The most important diseases are grey mould rot (page 258), alternaria rot (page 172) and, in particular, one of the powdery mildews, caused by *Sphaerotheca mors-uvae* (Schwein.) Berk. Mildewed berries initially display patches of white powdery mould growth, with abundant spores (conidia of the *Oidium* type). By the time of harvest, however, the mould may no longer be apparent; instead, the affected areas of skin are disfigured by dark brown russeting. Powdery mildew is the most serious disease of the growing crop. The same fungus attacks currant bushes, though in a different way and to a lesser extent. Control is by pruning and destruction of infected shoots, coupled with cultural methods to maintain plant vigour. Fungicide sprays need to be carefully timed in relation to fruit development and weather. *Ribes* crops affected by powdery mildew are predisposed to attack by the grey mould fungus. Infection by *Alternaria alternata* likewise takes place in the field, and can result in premature fruit drop. Infected gooseberries have dark centres, the seeds having been colonised by mould which entered via the calyx-end (4). Decay can progress further after harvest. A change of fungicide may be necessary to control this disease.

If harvested gooseberries cannot be sold or processed immediately, they may be held for 2 to 4 weeks at 0°C. Plastic film packaging is advantageous, and a slight build-up of carbon dioxide is beneficial (6).

---

1 BATRA S.W.T. (1987) Deceit and corruption in the blueberry patch. *Natural History* **96**(8), 56–59 illus.
2 BORECKA H.W., PLISZKA K. & NOWOSIELSKI S.B. (1985) Quality of blueberry fruit (*Vaccinium corymbosum* L.) stored under LPS, CA, and normal air storage. *Acta Horticulturae* No. 165, 241–249.
3 CEPONIS M.J. & CAPPELLINI R.A. (1983) Control of postharvest decays of blueberries by carbon-dioxide enriched atmospheres. *Plant Disease* **67**, 169–171.
4 DENNIS C., GROOM R.W. & DAVIS R.P. (1976) *Alternaria* fruit rot of gooseberry. *Plant Pathology* **25**, 57–58 illus.
5 HRUSCHKA H.W. & KUSHMAN L.J. (1963) Storage and shelf life of packaged blueberries. *Marketing Research Report of the United States Department of Agriculture* No. 612, 16 pp.
6 KENNY A. (1971) Storage of gooseberries in plastic bags. *Journal of Food Technology* **6**, 403–408.
7 MILHOLLAND R.D. & DAYKIN M.E. (1983) Blueberry fruit rot caused by *Phomopsis vaccinii*. *Plant Disease* **67**, 325–326.
8 SCHWARZ M.R. & BOONE D.M. (1983) Black rot of cranberry caused by *Strasseria oxycocci*. *Plant Disease* **67**, 31–32 illus.
9 SHEAR C.L., STEVENS N.E. & BAIN H.F. (1931) Fungous diseases of the cultivated cranberry. *Technical Bulletin of the United States Department of Agriculture* No. 258, 57 pp illus.
10 STRETCH A.W. & CEPONIS M.J. (1986) Fungal and physiological breakdown in six cranberry cultivars following water harvesting and cold storage. *HortScience* **21**, 265–267.

See also:

ECK P. (1988) *Blueberry science*. Gainesville, Florida: Horticultural Publications, 282 pp illus.

Grapes

The grapevine, *Vitis vinifera* L., had its origins near the Caspian Sea, and has been cultivated in the Near East for millennia. The Romans planted vines throughout their empire, appreciating both the fruit and the wine derived from it. Modern hybrids were obtained by crossing European varieties with native American species (9). The leading producers of grapes are Italy, France, Spain, the USSR, the USA, Turkey, Argentina, Romania, Greece and South Africa. Total world production is about 65 million tonnes (the highest figure for any fruit), but most of this is made into wine, while some is dried into raisins. The biggest exporters of fresh **grapes** are Italy and Chile. Research has been carried out in many countries, including the USA (1,3,5,7,9,10), South Africa (8), France (2), Cyprus (6) and Australia (4).

The clusters of berries vary in size, shape and colour, according to cultivar (9). Seedlessness may be inherent or else induced by spraying the crop with a growth-regulating chemical. Grapes are typical non-climacteric fruits (page 14); their rate of respiration is very low, and falls further after harvest (2). Sound healthy berries produce virtually no ethylene. The berries contain no starch and do not undergo further sweetening after removal from the vine. Hence they must be allowed to reach full maturity before being picked (6,9).

The major post-harvest hazards are bruising, desiccation and decay (1). If roughly handled, the berries lose their natural 'bloom', a superficial waxiness of the skin. Moisture loss is more rapid from stems than from berries, because of the distribution of natural openings (stomata and lenticels). Dehydrated stems shrink and turn brown (but see page 265) and eventually berries become flaccid (6). Berry drop (q.v.) is another common problem (1). By far the most important cause of decay is grey mould (q.v.), which can result in disastrous losses, especially after a wet growing season (3,5,7,8).

Harvested grapes must have their 'field heat' removed as rapidly as possible (2,6,8), and this is best achieved by pressure cooling (page 21). Plastic film packaging is effective in preventing undue moisture loss (2). The universal means of controlling decay is by the use of sulphur dioxide gas (SO_2). Treatment is applied by repeated fumigation of grapes in cold store, by fumigation in the transport vehicle, or by means of in-package SO_2 generators (2,4) (page 265). The type of modified atmosphere used for climacteric fruits is not appropriate for grapes, though trials have indicated that the addition of carbon monoxide can be beneficial. It is suggested (10) that this might prove superior to SO_2, which can cause injury (q.v.).

The optimal temperature for long-term storage is $-1°C$. Sweet grape berries may not freeze until $-3°C$ but their stems (containing much less sugar) can freeze at $-2°C$, leading to wilting and berry drop (2). Temperature control in refrigerated containers is often far from ideal, and it is usual to stipulate a carriage temperature well above $0°C$ in order to avoid the risk of freezing.

---

1 CAPPELLINI R.A., CEPONIS M.J. & LIGHTNER G.W. (1986) Disorders in table grape shipments to the New York market, 1972–1984. *Plant Disease* **70**, 1075–1079.
2 GÄRTEL W. & ULRICH R. (Eds) (1977) *Le raisin de table et le froid. Table grapes and refrigeration.* Paris: Institut International du Froid/Office International de la Vigne et du Vin, 248 pp.
3 HARVEY J.M. & PENTZER W.T. (1960) Market diseases of grapes and other small fruits. *Agriculture Handbook of the United States Department of Agriculture* No. 189, 37 pp illus.
4 HEDBERG P.R. (1977) Techniques for long-term storage of table grapes. *Australian Journal of Experimental Agriculture and Animal Husbandry* **17**, 866–870.
5 KOKKALOS T.I. (1986) Postharvest decay control of grapes by using sodium metabisulphite in cartons enclosed in plastic bags. *American Journal of Enology & Viticulture* **37**, 149–151.
6 NELSON K.E. (1979) Harvesting and handling California table grapes for market. *University of California Agricultural Sciences Publication* No. 4095, 67 pp illus.
7 PEARSON R.C. & GOHEEN A.C. (Eds) (1988) *Compendium of grape diseases.* St Paul, Minnesota: American Phytopathological Society, 93 pp illus.
8 PIENAAR P.J. (1972) Factors which influence the quality of table grapes between harvesting and sale. *Deciduous Fruit Grower* **22**, 39–41.
9 WINKLER A.J., COOK J.A., KLIEWER W.M. & LIDER L.A. (1974) *General Viticulture.* Berkeley: University of California Press, 410 pp illus.
10 YAHIA E.M., NELSON K.E. & KADER A.A. (1983) Post-harvest quality and storage life of grapes as influenced by adding carbon monoxide to air or controlled atmospheres. *Journal of the American Society for Horticultural Science* **108**, 1067–1071.

ASPERGILLUS ROT of grapes caused by
Aspergillus niger v. Tieghem

431

OCCURRENCE   This disease is important in the warmer grape-producing countries such as Egypt (1), India (5), South Africa (3) and Australia (4), as well as in temperate lands such as parts of Canada (2), Bulgaria and the USSR.

SYMPTOMS   Infected tissue is pale and water-soaked (2). The spores of *Aspergillus niger* are usually black, but there is also a brown-spored form which causes rotting in grapes. The spore-heads may be seen with the naked eye (**307**), and the spores are easily liberated when mature, resulting in soot-like deposits on adjacent berries (3). A sour odour accompanies the decay.

BIOLOGY   The fungus exists on plant debris in the soil, and is favoured by temperatures between 25° and 35°C (2). Spores (conidia) are disseminated throughout the vineyard in air currents, and characteristically cause infection via injuries such as insect punctures, splits or stem-end fractures (2). In some grape cultivars a substance in the natural wax coating stimulates growth of *A. niger*, but infection occurs only in mature berries, young berries being resistant to invasion even when wounded (1). Infection may also occur during and after harvest, through injuries (not necessarily visible) caused by careless handling (3). Under warm conditions the fungus spreads throughout the bunch.

CONTROL   Pre-harvest fungicide sprays give some control, and thinning of bunches helps to prevent overcrowding of berries and resultant cracking. It is important to prevent injury to the berries before, during and after harvest (3). Refrigerated storage is an effective measure, since the fungus makes no growth whatever at temperatures below 5°C (3). Protection is also possible with various fumigants such as sulphur dioxide (5).

1. ABDELAL H.R., EL-FAHL A.M., IBRAHIM A.N. & BADAWI M.F. (1980) The source and nature of resistance in vine fruit stages to grape rot disease. 1. Plant growth regulators and gibberellin. *Agricultural Research Review, Cairo* **58**(2), 163–172.
2. JARVIS W.R. & TRAQUAIR J.A. (1984) Bunch rot of grapes caused by *Aspergillus aculeatus*. *Plant Disease* **68**, 718–719.
3. MATTHEE F.N., THOMAS A.C. & CRAFFORD D.C. (1975) Black-mould rot in table grapes. *Deciduous Fruit Grower* **25**, 154–159 illus.
4. NAIR N.G. (1985) Fungi associated with bunch rot of grapes in the Hunter Valley. *Australian Journal of Agricultural Research* **36**, 435–442.
5. SINGH J.P., SHUSHIL SHARMA & YAMADAGNI R. (1985) Control of post harvest black mould disease of grapes. *Indian Phytopathology* **38**, 531–532.

BLUE MOULD ROT of grapes caused by *Penicillium* spp.

OCCURRENCE Blue mould rot probably occurs in all grape-producing countries, and several species have been recorded, for example *Penicillium canescens* Sopp (4) and *P. citrinum* Thom in India, *P. cyclopium* Westling in Israel (2) and *P. expansum* (page 178) in the USA, Chile (5) and West Germany (1).

SYMPTOMS The stems as well as the berries may support a white mould which gives rise to greenish-blue powdery spores. The decayed tissues are soft and watery and have a characteristic mouldy odour.

BIOLOGY Species of *Penicillium* live on decaying plant material, and their spores (conidia) are disseminated by wind, water and insects. Injuries predispose grape berries to infection either before, during or after harvest, and mould later spreads through the bunch, developing slowly in refrigerated storage. *P. expansum* produces a toxin (patulin), necessitating scrupulous quality control in grapes destined for juice or wine-making (1).

CONTROL It is important to minimise injury to the grapes at all stages of production and handling. Post-harvest measures include a fungicide dip (4), fumigation with sulphur dioxide gas (5) (page 265), the use of tissue wraps impregnated with a fungicide (3), and irradiation of fruit before storage (2). Refrigeration serves to slow down the rate of development of blue mould but cannot halt it completely.

432

---

1 ALTMAYOR B., EICHHORN K.W. & SCHWENK S. (1985) Die Bedeutung mykotoxinbildener Pilzarten für den Weinbau. *Nachrichtenblatt des Deutschen Pflanzenschutzdienstes* **37**(8), 117–122 illus.
2 BARKAI-GOLAN R. & KAHAN R.S. (1971) The effect of radiation on the pathogenicity of fungi and yeasts causing rot in stored grapes. *Israel Journal of Agricultural Research* **21**, 144.
3 SCOTT K.J. & ROBERTS E.A. (1965) An evaluation of fungistats for purple Cornichon grapes. *Australian Journal of Experimental Agriculture and Animal Husbandry* **5**, 296–298.
4 SHARMA R. C. & VIR D. (1984) Efficacy of fungicide. 22. Evaluation of benzimidazole, an antibiotic and other fungicides against post-harvest spoilage of grapes. *International Journal of Tropical Plant Diseases* **2**(1), 5–7.
5 SOTO A., E.M., PINTO de T., A. & CANCINO E., L. (1973) Control preventivo de pudriciones de postcosecha en uva Emperor y Almería e identificación y patogenicidad de los hongos aislados. *Agricultura Técnica* **33**, 176–182.

GREY MOULD ROT of grapes caused by
Botryotinia fuckeliana (de Bary) Whetzel
Conidial state: *Botrytis cinerea* Pers.

433

OCCURRENCE   Grey mould rot is the most widespread and important disease of grapes, occurring wherever the crop is grown, for example in France (2), Italy (4), Bulgaria (5), the USSR (6), the USA (3,9), Canada (10), Chile (1), South Africa (7) and Australia (8).

SYMPTOMS   The early stage of fruit-rotting is called 'slip-skin' since, when the flesh is first invaded, the overlying skin becomes loose and is easily detached (9). A brown discoloration occurs, and later there may be copious production of grey-brown spores (7). A humid atmosphere favours development of whitish mould growth without spores.

BIOLOGY   In some localities the grey mould fungus exists in both the sexual state (apothecia giving rise to ascospores) (6) and the asexual state (conidia) (7,9), the former being prominent in the spring and the latter during the autumn. Resting bodies (sclerotia) are to be found in the soil and in infected vines. Spore production and infection of vines (including leaves and shoots) are encouraged by moist conditions. However, even when no rain occurs at flowering time there may still be a high incidence of blossom infection in dense plantings (3). The fungus remains quiescent in the developing fruits (7), resuming growth when the berries are mature (8). New infections can occur if moribund flower parts remain in the cluster (10) or if berries undergo splitting (page 266); mature uninjured berries may also be attacked (9). In wet seasons fruit may rot in the vineyard (8), but it is also possible for apparently sound grapes to be packed and shipped, later undergoing severe post-harvest rotting (9). Grey mould is capable of significant growth even at low temperatures.

CONTROL   Strict hygiene in the vineyard is necessary to minimise the amount of crop debris on which the fungus can survive and form spores. Husbandry measures which reduce moisture retention in the crop have also proved beneficial in reducing levels of infection (3). The efficacy of fungicide applications depends on correct timing (starting pre-bloom) (8) and on the use of dissimilar chemicals as a strategy for delaying the appearance of insensitive fungal strains (4,10). If biological

434

435 *Sclerotinia fuckeliana* (conidial state: *Botrytis cinerea*) (from CMI Descr. No. 431, Ellis & Waller 1974) Conidia and conidiophores x 500

control (page 18) can be used early in the season, then the number of chemical treatments can be reduced (2). Important post-harvest measures include careful handling, prompt and rapid cooling, and the use of sulphur dioxide (page 265). The fungistatic effect of sulphur dioxide is such that the skin of the berries is protected from new infections, and established infections are contained (1). In wet seasons, however, when decay potential is high, control is often found to be inadequate. Grape cultivars with resistance to grey mould rot tend to have thick-skinned berries (5).

---

1 ALVAREZ A., M. & VARGAS B., V. (1983) Efecto de fungicidas aplicados en precosecha y SO$_2$ en postcosecha en el control de *Botrytis cinerea* Pers. en uva almacenada cv. Sultanina. *Agricultura Técnica* 43, 61–66.
2 DUBOS B., JAILLOUX F. & BULIT J. (1982) L'antagonisme microbien dans la lutte contre la pourriture grise de la vigne. *EPPO Bulletin* **12**, 171–175.
3 GUBLER W.D., MAROIS J.J., BLEDSOE A.M. & BETTIGA L.J. (1987) Control of *Botrytis* bunch rot of grape with canopy management. *Plant Disease* **71**, 599–601.
4 GULLINO M.L. & GARIBALDI A. (1982) Use of mixture or alternation of fungicides with the aim of reducing the risk of appearance of strains of *Botrytis cinerea* resistant to dicarboximides. *EPPO Bulletin* **12**, 151–156.
5 KARADIMCHEVA B. (1981) [Characteristics of the anatomical structure of grape skin in relation to resistance to grey mould.] *Gradinarska i Lozarska Nauka* **18**, 94–99.
6 KUBLITSKAYA M.A. & RYABTSEVA N.A. (1970) [The biology of the winter state of *Botrytis cinerea* Fr.] *Mikologiya y Fitopatologiya* **4**, 291–293 illus.
7 MARAIS P.G. (1985) Infection of table grapes by *Botrytis cinerea*. *Deciduous Fruit Grower* **35**, 166–170 illus.
8 NAIR N.G., EMMETT R.W. & PARKER F.E. (1987) Programming applications of dicarboximides to control bunch rot of grapes caused by *Botrytis cinerea*. *Plant Pathology* **36**, 175–179.
9 NELSON K.E. (1956) The effect of Botrytis infection on the tissue of Tokay grapes. *Phytopathology* **46**, 223–229 illus.
10 NORTHOVER J. (1987) Infection sites and fungicidal prevention of *Botrytis cinerea* bunch rot of grapes in Ontario. *Canadian Journal of Plant Pathology* **9**, 129–136.

RHIZOPUS ROT of grapes caused by
Rhizopus oryzae Went & Prinsen Geerligs
Rhizopus stolonifer (Ehrenb. ex Fr.) Lind

436

OCCURRENCE   Rhizopus rot can probably occur in grapes from any source. Both species have been reported on grapes in the USA and in India (4,5). Other records include Australia (1) and Egypt (3).

SYMPTOMS   Coarse white mould strands can be distinguished with the naked eye. Smooth spherical spore-heads (sporangia), which are at first white and later black, are borne on stalks (sporangiophores) arising in clusters (page 230).

BIOLOGY   These fungi exist in the soil and on plant debris, and their spores (sporangiospores) are disseminated in air currents. Infection can occur before, during or after harvest. Primary infection usually occurs via injuries (1) although, in the presence of exuded grape juice, *R. oryzae* is capable of penetrating the intact skin of mature berries. Once established, these moulds are able to spread into adjacent sound berries and, under warm conditions, decay is very rapid. The optimal growth temperatures lie between 30° and 35°C for *R. oryzae* and between 20° and 25°C for *R. stolonifer*.

CONTROL   Pre-harvest fungicide sprays can be used, although only a few chemicals are effective against *Rhizopus* species (4). Stem-end fractures and splitting of berries should be minimised by judicious thinning of bunches (1). Harvested fruit must be handled gently in order to prevent injury. Sulphur dioxide treatment serves to kill spores on the exterior of the berries. Refrigeration effectively suppresses development of these fungi (3), but they resume growth on return to higher temperatures (2).

---

1   BARBETTI M.J. (1980) Bunch rot of Rhine Riesling grapes in the lower south-west of Western Australia. *Australian Journal of Experimental Agriculture and Animal Husbandry* **20**, 247–251.
2   CEPONIS M.J. & BUTTERFIELD J.E. (1974) Retail losses in California grapes marketed in metropolitan New York. *United States Department of Agriculture Publication* ARS-NE-53, 7 pp.
3   EL-HELALY A.F., ASSAWAH M.W. & WASFY E.-E.H. (1965) Studies on certain fungal diseases of grape berries. *Phytopathologia Mediterranea* **4**, 154–162.
4   MUTHUSWAMY S., PALANISWAMY A., SUNDARARAJ J.S. & KRISHNAMURTHY C.S. (1971) Pre-harvest sprays of fungicides for the control of storage decay in grape (*Vitis vinifera* L.). *Indian Journal of Agricultural Sciences* **41**, 711–715.
5   SINGH J.P. & KAINSA R.L. (1983) Microbial flora of grapes in relation to storage and spoilage. *Indian Phytopathology* **36**, 72–76.

OTHER DISEASES of grapes

437 Alternaria rot

ALTERNARIA ROT is caused by *Alternaria alternata* (**339**), and may involve the stem-end or any part of the berry. A dry brown rot has been described in Venezuela (4). In the Western USA it was observed that dying flowers were infected and became a source of airborne spores. The fungus is capable of direct penetration of sound mature grapes even in the absence of rain, but wet weather results in a higher incidence of disease (5). Post-harvest decay can be held in check by fumigation with sulphur dioxide in a humid atmosphere (2).

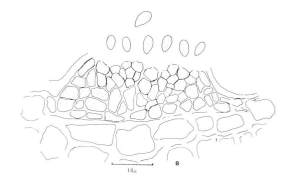

439 *Elsinoë ampelina* (from CMI Descr. No. 439, Sivanesan & Critchett 1974) Acervulus and conidia

438 Anthracnose

ANTHRACNOSE may be caused by *Elsinoë ampelina* Shear (conidial state: *Sphaceloma ampelinum* de Bary) (10). Only immature berries are susceptible to infection. Lesions are raised with a dark margin, hence the alternative name of bird's eye rot. Early spraying is necessary for control. Anthracnose is also caused by *Glomerella cingulata* (page 130) which produces salmon-pink spore masses on mature berries, hence the alternative name of ripe rot. Since infection by this fungus can occur at all stages of fruit development it is necessary to apply fungicide sprays from the time of flowering until just before harvest (3).

BITTER ROT is caused by the fungus *Greeneria uvicola* (Berk. & Curt.) Punith. (formerly known as *Melanconium fuligineum*), which usually spreads from the stem-end throughout the berry. Rotted fruits are soft and taste bitter, and the skin of purple cultivars is bleached to a pink colour. Eventually minute black bodies (pycnidia) form in the skin of the grape. The disease is controlled by fungicide sprays (6).

440 Bitter rot

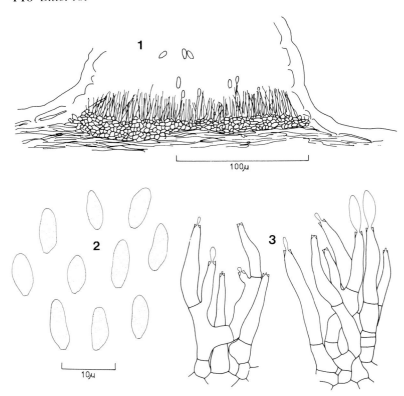

441 *Greeneria uvicola* (from CMI Descr. No. 538, Sutton & Gibson 1977)
1 acervulus 2 conidia 3 conidiophores

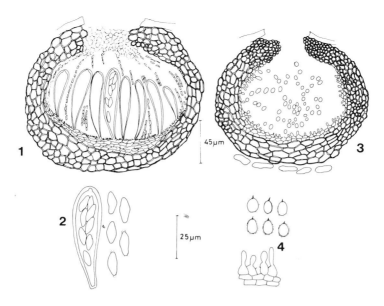

442 *Guignardia bidwellii* (from CMI Descr. No. 710, Sivanesan & Holliday 1981) 1 pseudothecium 2 ascus and ascospores 3 pycnidium 4 conidiogenous cells and conidia

BLACK ROT is caused by *Guignardia bidwellii* (Ell.) Viala & Ravaz (conidial state: *Phyllosticta ampelicida* (Engelman) van der Aa). It is essentially a field disease, and is of great importance on grapevines in Europe and the eastern USA (8). Lesions on berries are at first pale with a dark border, similar to those of bird's eye rot (q.v.). Later the rotting berries shrivel into mummies bearing minute black bodies (pycnidia). Affected bunches are usually culled during the packing operation and, in any event, the disease does not spread in store. Control measures include the pruning of infected branches and the timely application of fungicide sprays.

BOTRYODIPLODIA ROT has caused important losses in the USA, where it is known as summmer bunch rot (5). The causal organism is *Botryodiplodia theobromae* (page 81). Infection typically occurs at blossoming, and the fungus remains quiescent in the developing fruit, becoming active as the berry matures. Less commonly, sound mature grapes are infected directly. As rotting ensues the flesh turns pinkish, the skin cracks and juice leaks out, attracting fruit flies and beetles. Secondary organisms proliferate, and a complex sour rot (q.v.) develops, spreading into neighbouring bunches. Control measures include pruning to remove blighted wood, destruction of prunings instead of incorporation into the soil, and trellising of vines so that clusters hang separately (5).

CLADOSPORIUM ROT is common in stored grapes and is important because the causal fungus, *Cladosporium herbarum* (page 226), is able to grow (albeit slowly) even at 0°C. Circular black spots appear beneath the skin of the berries and, under humid conditions, a velvety olive-green mould develops on the lesions. Primary infection occurs before harvest, and the fungus is capable of direct penetration through the intact skin; high incidence of disease is associated with wet seasons. Post-harvest decay is minimised by the use of sulphur dioxide treatment and refrigeration (5).

443 Cladosporium rot

CONIELLA ROT, caused by *Coniella diplodiella* (Speg.) Petrak & Sydow, has been recorded in the USA, India, South Africa, and several European countries including Spain (10). The fungus attacks injured berries, and serious losses can occur following hailstorm damage (9). The first symptom is a yellow spot surrounded by brownish haloes; infected berries then take on a bleached and shrivelled appearance, and minute black bodies (pycnidia) develop. Such berries drop from the bunch, but by this time the fungus may have progressed into the stem, causing browning and shrivelling. Diseased bunches are occasionally seen in the market. Control measures include destruction of infected prunings and the use of a protective fungicide immediately after a hailstorm (9).

DOWNY MILDEW is one of the most important diseases of the vine in moist growing areas. The causal fungus is *Plasmopara viticola* (Berk. & Curt. ex de Bary) Berl. & de Toni. It can infect both leaves and fruit. If mature grapes are infected there is no obvious external mould; brownish-purple patches develop and eventually the entire berry becomes shrunken and brown. Downy mildew is essentially a field disease (10), and infected fruit is rarely seen in the market.

PHOMOPSIS ROT of grapes is caused by *Phomopsis viticola* Sacc., which is also responsible for 'dead-arm rot' of the vine. Spore-bearing bodies (pycnidia) form in infected branches and, during wet weather, spores are disseminated and can cause direct infection of intact mature grapes (5,7). Decay of the berries resembles that caused by the black rot fungus (q.v.) but the pycnidia are larger, less numerous and sometimes arranged in concentric circles. The disease is controlled by destruction of infected prunings and the application of fungicide sprays.

POWDERY MILDEW is caused by the fungus *Uncinula necator* (Schwein.) Burr., which produces the sexual stage (cleistothecia giving rise to ascospores) on shoots and leaves. Vines can suffer serious damage even in dry weather, following infection by asexual spores (conidia). Although powdery mildew is essentially a field disease (10), late season infection of berries can result in a light brown lace-like russeting of the skin which may be encountered in stored grapes; the berries may be mis-shapen owing to uneven growth. The bunch-stems may also be affected, the lesions turning from grey to brown, with web-like lines radiating outwards. Control of powdery mildew usually entails repeated fungicide spraying, according to the weather during the growing season (10).

RIPE ROT may be caused by any of several fungi, including *Botryosphaeria ribis* (page 96) reported from the USA and Chile. Small brown sunken spots appear on mature berries and, later, in the centre of each spot are minute black bodies (pycnidia). Eventually the entire berry undergoes a soft brown rot which may later dry out leaving a hollow shell covered with pycnidia.

SOUR ROT is the result of fermentation by acetic acid bacteria and yeasts (1,5). The disease tends to affect dense clusters shortly before harvest, cultivars with thin-skinned berries being especially susceptible (1). Control is by means of fungicide sprays and prevention of injury. Sour rot is also the final stage of botryodiplodia rot (q.v.).

---

1 BISIACH M., MINERVINI G. & ZERBETTO F. (1986) Possible integrated control of grapevine sour rot. *Vitis* **25**, 118–128.
2 COUEY H.M. (1965) Inhibition of germination of *Alternaria* spores by sulphur dioxide under various moisture conditions. *Phytopathology* **55**, 525–527.
3 DAYKIN M.E. & MILHOLLAND R.D. (1984) Ripe rot of muscadine grape caused by *Colletotrichum gloeosporioides* and its control. *Phytopathology* **74**, 710–714 illus.
4 DÍAZ POLANCO C. & BASTIDA P., R. (1971) Una especie de *Alternaria* infecciosa en granos de uva. *Agronomía Tropical* **21**, 129–134 illus.
5 HEWITT W.B. (1974) Rots and bunch rots of grapes. *Bulletin of California Agricultural Experiment Station* No. 868, 52 pp illus.
6 KATO K., MIYAGAWA T., NAKAGAMI K., HIROTA K. & TOMITA I. (1978) [Chemical control of grape bitter rot.] *Research Bulletin of Aichi-Ken Agricultural Research Centre B* No. 10, 69–75.
7 LAL B. & ARYA A. (1982) A soft rot of grapes caused by *Phomopsis viticola*. *Indian Phytopathology* **35**, 261–264.
8 LUTTRELL E.S. (1946) Black rot of Muscadine grapes. *Phytopathology* **36**, 905–924 illus.
9 MATTHEE F.N. & THOMAS A.C. (1981) Rot blanc in vines: a new disease in South Africa. *Deciduous Fruit Grower* **31**, 268–273 illus.
10 RUIZ CASTRO A. (1950) La lucha contra las enfermedades del viñedo (su estado actual en España). *Boletín de Patología Vegetal y Entomología Agrícola* **18**, 243–280 illus.

SULPHUR DIOXIDE INJURY of grapes

OCCURRENCE Sulphur dioxide gas (SO_2) is an effective fungistat; although it cannot arrest established infections, it kills spores and inhibits the growth of mould on the grape surface, thereby preventing new infections (3,4). The treatment is applied either by fumigation (1,3) or by means of in-package SO_2 generators (2,4,5). These rely on sodium or potassium metabisulphite, and can be sheets impregnated with the chemical, plastic sachets containing a solution, or dry sachets containing a powder formulation. The higher the temperature and humidity the more rapidly is the gas evolved (2,4). Grapes are almost the only fruit able to withstand exposure to this chemical, and there is a fine margin between the concentration required for effective suppression of mould and the concentration which leads to fruit injury (1,3).

SYMPTOMS There may be numerous small, sunken, bleached spots on each berry, with a clear line of demarcation between bleached and normal skin. More usually, however, it is the stem-end which becomes bleached. Whereas most berries in an affected bunch show partial discoloration extending a few mm from the stem-end, some berries may be bleached throughout. Injured berries are predisposed to moisture loss and tend to shrivel. The stems of treated bunches retain their fresh green colour for much longer than do those of untreated bunches.

BIOLOGY The severity of injury is determined by the amount of SO_2 which enters the berry. The weakest point is the junction between berry and capstem, and preferential diffusion occurs here. If the skin of a berry happens to be punctured, the gas enters without hindrance and the entire berry is affected. As well as bleaching the skin, SO_2 imparts a disagreeable off-taste to the flesh.

CONTROL It is not easy to provide an appropriate concentration of SO_2. Cultivars differ in their tolerance, and batches in their decay potential; furthermore, post-harvest environmental conditions are critical (1). To minimise the risk of injury in store it may be advisable to use reduced concentrations at frequent intervals (for example, 200 ppm 3 times a week) (3). With regard to in-package treatment, this originally consisted of single-stage generators which tended to be depleted within a day or two of packing; high initial concentrations sometimes led to SO_2 injury, while the subsequent low levels were insufficient to check the growth of mould. It was therefore recommended that grapes be given some cooling *before* being packed, so as to reduce the rate of SO_2 emission at the outset and prolong its protective effect (2). An alternative strategy is to have 2-stage generators, made from materials of differing permeabilities so that one releases SO_2 immediately and the other emits it during the later stages of transport or storage (1,4).

---

1 HARVEY J.M. & UOTA M. (1978) Table grapes and refrigeration. Fumigation with sulphur dioxide. *International Journal of Refrigeration* **1**, 167–171.
2 LASZLO J.C., COMBRINK J.C., EKSTEEN G.J. & TRUTER A.B. (1981) Effect of temperature on the emission of sulphur dioxide from gas generators for grapes. *Deciduous Fruit Grower* **31**, 112–119.
3 MAROIS J.J., BLEDSOE A.M., GUBLER W.D. & LUVISI D.A. (1986) Control of *Botrytis cinerea* on grape berries during postharvest storage with reduced levels of sulphur dioxide. *Plant Disease* **70**, 1050–1052.
4 NELSON K.E. (1983) Effects of in-package sulphur dioxide generators, package liners, and temperature on decay and desiccation of table grapes. *American Journal of Enology and Viticulture* **34**, 10–16.
5 PAULIN A. (1966) Conservation frigorifique du raisin de table en emballage de polyéthylène et en présence d'une émission d'anhydride sulfureux continuée et controlée. *Fruits* **21**, 127–137 illus.

OTHER DISORDERS of grapes

445 Berry drop

BERRY DROP can cause serious losses. Individual grapes become detached from the bunch during handling, transport and marketing, and the problem is sometimes known as 'shatter'. Investigations have been carried out in several countries, for example South Africa (5), India (2) and Japan (3). Susceptibility is genetically determined, but predisposing factors are moisture stress during the growing season, high temperature at harvest time, and delay in cooling the harvested crop (5). Seedless berries which have been produced by treating the vines with gibberellic acid (a growth-regulating chemical) are especially prone to berry drop, since a side effect of the treatment is a hardening of the stem-tissue (3). Certain other growth-regulating chemicals can, however, reduce the incidence of berry drop (2).

In the disorder described above, short lengths of vascular strands may be torn from the interior of the berry and project from the capstem in the form of a 'brush' (3). In some cultivars, however, there is a clean break at the point of attachment (abscission layer). This second type of berry drop may also be induced by extraneous ethylene, which promotes senescence of the connective tissue (3). A third possible cause of berry drop is alternaria rot (q.v.) of the stem-end, in which the 'brush' becomes infected. Yet another possibility is mild freezing (page 255), causing injury to the stem tissue.

CRACKING and **SPLITTING** of berries may occur before or after harvest. Some authors define cracking as fine concentric breaks in the skin around the stem-end or stylar-end, and splitting as a deeper and more severe rupture which frequently leads to fungal infection (4). Some cultivars are genetically prone to cracking (1), while splitting can be induced by wet weather shortly before harvest.

PHYSIOLOGICAL BROWNING of berries may be a sign of senescence (page 14) or it can be the result of bruising or freezing. Some cultivars are liable to develop internal browning after several weeks in store. Susceptible cultivars should be harvested at optimal maturity and samples removed from store at intervals for observation.

---

1 COMBRINK J.C., EKSTEEN G.J., SCHLIEMANN G.K.G.E., LASZLO J.C., KOCK I.S. de & CARREIRA J. (1982) Influence of pre- and post-harvest treatments on berry splitting in Alphonse Lavallée grapes. *Deciduous Fruit Grower* **32**, 109–111, 113–114.
2 KUSHAL SINGH, MANN S.S. & BAJWA M.S. (1985) Effect of auxins, sodium benzoate and calcium chloride on post-harvest berry drop in Himrod grapes. *Acta Horticulturae* No. 158, 413–418.
3 NAKAMURA M. & HORI Y. (1981) Postharvest berry drop of seedless berries produced by GA treatment in grape cultivar 'Kyoho'. 1. Relationship between postharvest berry drop and rachis hardness. *Tohoku Journal of Agricultural Research* **32**, 1–13.
4 SWIFT J.G., MAY P & LAWTON E.A. (1974) Concentric cracking of grape berries. *Vitis* **13**, 30–35 illus.
5 WAGENER G.N. (1985) Progress with the problem of loose berries in table grapes. *Deciduous Fruit Grower* **35**, 329–333.

Kiwifruits

Kiwifruits, formerly known as Chinese gooseberries (9), are berry fruits of the perennial vine *Actinidia deliciosa* (Chevalier) Liang & Ferguson. They have thin brown hairy skins, bright green flesh and small dark seeds. Their vitamin C content is higher than that of oranges. Commercial production was first started in New Zealand (6), and there are now plantings in other warm temperate countries such as Australia (10), Japan (7), South Africa, the western USA (2,5), Chile, Israel (3), France (8) and Italy (4). In cooler climates, for example in the Channel Islands, the crop is grown under glass.

Like pears, kiwifruits are picked when still hard and are ripened after harvest. Maturity criteria include flesh firmness (measured with a penetrometer) (2), soluble solids content (measured with a refractometer) (6) and total solids content (10). However, these are not infallible indicators of potential storage life, which shows considerable variation according to season and growing area (5).

Kiwifruits for export (which sometimes follows a period of cold storage) must be sufficiently hard to withstand the journey (2). The fruits are usually packed in a single-layer wooden tray which incorporates a polyethylene film wrap to minimise dehydration. Storage life is maximised by prompt and rapid cooling of the fruit to $0°C$. Healthy kiwifruits harvested at the correct stage of maturity and cooled without delay can be stored for several months (3). A post-harvest calcium chloride dip helps to delay the onset of softening (8).

The earliest storage studies suggested that kiwifruits had few disease problems. With expanding cultivation of the crop, however, diseases have assumed increasing importance and several post-harvest rots have been described (1). Furthermore, any fungal infection contributes to the main post-harvest problem which is premature ripening (2,8).

It has been established, after much experimentation and discussion (9), that kiwifruits show a climacteric rise in respiration (page 14). The ripening process is triggered by ethylene; therefore, if softening is to be prevented during storage, the environment must be kept free of any trace of this gas (4,7). Hence care must be taken to exclude damaged, diseased or overmature fruits (which produce ethylene) and, for the same reason, to avoid storage with commodities such as apples and pears (1). An ethylene absorbent can be beneficial, whether in-store (4) or in-package (3). Storage life can be further extended by the use of a modified atmosphere (3) or controlled atmosphere (2,8), which should be established as soon as possible after harvest (2).

Kiwifruits are often transported in refrigerated containers (5), the recommended carriage temperature being $0°$ to $1°C$ (kiwifruits freeze at approximately $-1.5°C$) (4). The exporter must see to it that the fruit is thoroughly pre-cooled and carefully stowed in the container, and ships' personnel must ensure that no container in the same 'bay' is carrying a commodity which produces ethylene (1).

---

1 ABOLHODA A. & KÜHNE H. (1986) Pilzbefall an importierten Kiwifrüchten, *Actinidia chinensis* Planch. *Nachrichtenblatt des Deutschen Pflanzenschutzdienstes* **38**, 145–150 illus.
2 ARPAIA M.L., LABAVITCH J.M., GREVE C. & KADER A.A. (1987) Changes in the cell wall components of kiwifruit during storage in air or controlled atmosphere. *Journal of the American Society for Horticultural Science* **112**, 474–481.
3 BEN-ARIE R. & SONEGO L. (1985) Modified-atmosphere storage of kiwifruit (*Actinidia chinensis* Planch.) with ethylene removal. *Scientia Horticulturae* **27**, 263–273.
4 GORINI F. (1983) Conservazione dell'actinidia. *Atti dell'Istituto Sperimentale per la Valorizzazione Tecnologica dei Prodotti Agricoli, Milano* **6**, 125–133.
5 HARVEY J.M., HARRIS C.M. & RASMUSSEN G.K. (1986) Protective storage and transit environments for kiwifruit (*Actinidia chinensis* Planch.) and their effect on fruit quality. *Crop Protection* **5**, 277–282.
6 HOPKIRK G., BEEVER D.J. & TRIGGS C.M. (1986) Variation in soluble solids concentration in kiwifruit at harvest. *New Zealand Journal of Agricultural Research* **29**, 475–484.
7 HYODO H. & FUKASAWA R. (1985) Ethylene production in kiwifruit. *Journal of the Japanese Society for Horticultural Science* **54**, 209–215.
8 NICOLAS J., BURET M., DUPRAT F., NICOLAS M., ROTHAN C. & MORAS P. (1986) Effects of different conditions of cold storage upon physicochemical changes of kiwifruit. *Acta Horticulturae* No. 194, 261–272.
9 PRATT H.K. & REID M.S. (1974) Chinese gooseberry: seasonal patterns in fruit growth and maturation, ripening, respiration and the role of ethylene. *Journal of the Science of Food and Agriculture* **25**, 747–757.
10 SCOTT K.J., SPRAGGON S.A. & McBRIDE R.L. (1986) Two new maturity tests for kiwifruit. *CSIRO Food Research Quarterly* **46**(2), 25–31.

GREY MOULD ROT of kiwifruits caused by
Botryotinia fuckeliana **(de Bary) Whetzel**
Conidial state: *Botrytis cinerea* Pers.

OCCURRENCE This disease occurs in kiwifruits from all growing areas, for example New Zealand (1,4), the USA (3) and Italy (2,5).

SYMPTOMS The most usual symptom is a soft rot starting at the stem-end or at wound sites. Affected tissue becomes dark and water-soaked. Even in the absence of decay, there may be superficial white mould growth or grey-brown spores on the remains of the calyx (3).

BIOLOGY Initial infection can occur via senescent flower parts, at any time from the end of blossoming until harvest (2,3). Moist conditions are necessary for infection, after which the fungus may remain quiescent for several months, appearing only after a period of storage (2). Alternatively (4), or in addition (2), infection can occur via the cut stem at harvest time and through wounds in the skin. Post-harvest dipping in a bath of citric acid (to remove water-staining and enhance appearance) carries with it the disadvantage of increasing opportunities for infection (4). Grey mould is capable of slow growth even at 0°C and, during long-term storage, can spread into healthy fruits, causing 'nesting'.

CONTROL Recommendations include pre-harvest fungicide sprays (starting at blossom time) (1) and, if legislation permits, a fungicide treatment after harvest (4).

---

1 BEEVER D.J., McGRATH H.J.W., CLARKE D.L. & TODD M. (1984) Field application and residues of fungicides for the control of *Botrytis* storage rot of kiwifruit. *New Zealand Journal of Experimental Agriculture* **12**, 339–346.

2 BISIACH M., MINERVINI G. & VERCESI A. (1984) Biological and epidemiological aspects of the kiwifruit (*Actinidia chinensis* Planchon) rot, caused by *Botrytis cinerea* Pers. *Rivista di Patologia Vegetale* **20**(2), 38–55 illus.

3 OPGENORTH D.C. (1983) Storage rot of California-grown kiwifruit. *Plant Disease* **67**, 382–383.

4 PENNYCOOK S.R. (1986) Citric acid dipping of kiwifruits promotes *Botrytis* storage rot. *New Zealand Journal of Experimental Agriculture* **14**, 205–207.

5 SCAPIN I., MANCINI G. & GALLIANO A. (1983) Il marciume da *Botrytis cinerea* dei frutti di *Actinidia*. *Informatore Fitopatologico* **33**(4), 63–65 illus.

OTHER DISEASES of kiwifruits

447 Blue mould rot

BLUE MOULD ROT is caused by species of *Penicillium* (page 178), and decay can occur even at 0°C (3). The fungus is associated with wounds in the skin, and a mass of blue-green spores forms at the injury site. Mature fruits are particularly susceptible (3). Control measures include close clipping of stems, careful design of protective packaging, gentle handling and avoidance of prolonged storage.

PHOMOPSIS ROT may be caused by *Diaporthe perniciosa* (page 202) or by *Diaporthe actinidiae* Sommer & Beraha (conidial state: *Phomopsis actinidiae*). The disease was first recorded on New Zealand kiwifruits imported into the USA, and typically takes the form of a stem-end rot which develops as the fruit ripens (1). There may be a growth of white mould, and exudation of liquid can cause staining of the skin. A sour or fermented odour accompanies the decay (1). In storage trials stem-end rot was held in check for 2 months at 1°C, but fungal growth resumed at ambient temperature (2). Spore-formation has been observed only in laboratory cultures (5). Control measures are as for ripe rot (q.v.).

RIPE ROT may be caused by any of several fungi, including species of *Botryosphaeria* (page 176), *Colletotrichum* (page 130), *Phoma* (page 52) and *Phomopsis* (q.v.). These fungi are capable of colonising dying flower parts, and remain quiescent in the fruit until it ripens after harvest (2). Control is therefore dependent on early fungicide sprays (2) and improved orchard hygiene (4). Other recommendations include scrupulous quality control during grading and packing (to achieve good general storage quality), and careful handling throughout the distribution chain to ensure that ripening is not accelerated at too early a stage (4).

---

1 BERAHA L. (1970) Stem-end rot of Chinese gooseberry (*Actinidia chinensis*) on the market. *Plant Disease Reporter* **54**, 422–423 illus.
2 HAWTHORNE B.T., REES-GEORGE J. & SAMUELS G.J. (1982) Fungi associated with leaf spots and post-harvest fruit rots of kiwifruit (*Actinidia chinensis*) in New Zealand. *New Zealand Journal of Botany* **20**, 143–150 illus.
3 PADFIELD C.A.S. & BAILEY F.L. (1952) Chinese gooseberries (*Actinidia chinensis*): a survey of their behaviour in cool-storage at all stages of harvest maturity from May to July. *New Zealand Journal of Science and Technology* **33**, 113–116.
4 PENNYCOOK S.R. (1985) Fungal fruit rots of *Actinidia deliciosa* (kiwifruit). *New Zealand Journal of Experimental Agriculture* **13**, 289–299 illus.
5 SOMMER N.F. & SUADI J.E. (1985) Postharvest disease and storage life of kiwifruits. *Acta Horticulturae* No. 157, 295–302.

CHAPTER 7
MELONS AND WATERMELONS

Melons and watermelons are members of the family Cucurbitaceae, and are thus closely related to cucumbers, marrows, squashes and pumpkins (9). The diseases and disorders of this group are considered in detail in Volume 2, and so are dealt with here only briefly.

The **melon**, *Cucumis melo* L., comprises a range of types, the common names of which have various connotations. The three varietal names are perhaps unjustified, since all the forms hybridise readily and there are many intermediate forms.

'Cantaloupes', *C. melo* var. *cantalupensis*, are characterised by a thick rough rind with regularly spaced ribs but without surface 'netting'; Charentais is a typical cultivar, having a yellowish-green, slightly ribbed skin and fragrant orange flesh. True cantaloupes are popular in continental Europe; in the USA the term 'cantaloup' is applied to muskmelons (in either of the senses of that word) (q.v.).

'Muskmelons', *C. melo* var. *reticulatus*, are green or yellow with finely netted rinds and very shallow ribs; they have a musky fragrance. In American terminology 'muskmelon' encompasses all types of *C. melo* as distinct from watermelon (q.v.). In order to avoid ambiguity it might be helpful to use the term 'netted melons' for var. *reticulatus*.

'Winter melons', *C. melo* var. *inodorus*, mature late and can be stored for several weeks. They are oval or elongated, and do not possess a strong aroma. Important cultivars include Honey Dew with smooth ivory skin and green flesh, and Tendral which has a dark green, irregularly grooved rind and yellowish-green flesh.

The place of origin of the melon is not known with certainty but the evidence points to Africa. It subsequently spread into Asia, where many new forms developed (9). Melons were not cultivated in Europe until the late Middle Ages, except perhaps in Moorish Spain. Present world production is of the order of 9 million tonnes, one quarter of which is grown in China. Other important producers are Spain, the USA, Egypt, Iran, Romania, Japan, Italy, Mexico and Iraq.

Melons are climacteric fruits but they have no starch reserves and cannot become significantly sweeter after harvest (9). Post-harvest ripening is restricted to softening of flesh and development of flavour volatiles (7). Picking is done by hand and if suitable illumination is available the work can be carried out at night rather than in the heat of the day; the benefits include enhanced retention of fruit quality and significant reduction in the cost of pre-cooling for onward shipment. Cantaloupes and netted melons are best harvested at 'full-slip', when the fruit stalk is ready to separate naturally from the vine. At this stage they are approaching the climacteric peak, with the potential to develop full aroma (7). Harvesting at 'half-slip', when the stem-crack is still forming, allows for shipment to more distant markets yet still permits ripening to an acceptable quality. Winter melons behave in a different way and it is more difficult to judge their maturity. One indication is the changing colour of the ground-spot, the area of the fruit which rests on the ground. Honey Dew melons, even though harvested mature, do not necessarily have the capacity for self-ripening; in order to ensure uniform ripening in a batch of fruits it is necessary to apply ethylene treatment before shipment (7). Other post-harvest treatments may include a fungicide dip, some formulations being more effective if heated (6).

Prompt cooling of melons is highly desirable since it retards the breakdown of sugars (4). For netted melons the optimal temperature for storage and carriage is approximately $5°C$, with high humidity (or a plastic film wrap) to minimise moisture loss (4). Under such conditions they will keep for up to 2 weeks. Winter melons may need to be stored at $7°$ to $10°C$. An exception is Honey Dew melons which have had the benefit of ethylene treatment; they show reduced susceptibility to chilling injury and can be held at $5°C$. Some melon cultivars benefit from storage in controlled atmosphere, a 20% concentration of carbon dioxide being inhibitory to mould growth (8). This technique offers an alternative to the use of post-harvest fungicides. Common causes of deterioration in melons are fungal and bacterial blemishes and rots (1,5,6,8,10). For a detailed classical study of storage diseases of melons, see reference 10.

The **horned melon** or **kiwano**, *Cucumis metuliferus* E. Mey. ex Naud, is usually classed as an 'exotic fruit', and began to receive attention only recently, first in New Zealand and subsequently in California. The fruits are oval, 100 to 150 cm long, and are characterised by numerous protuberances. On ripening, the skin turns yellow or orange but the jelly-like flesh remains green.

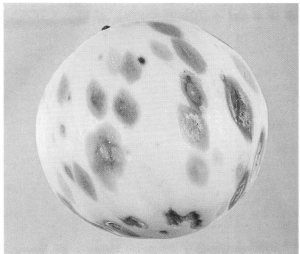

The **watermelon**, *Citrullus lanatus* (Thunb.) Matsum. & Nakai var. *lanatus*, is native to Africa. It has been cultivated in the Middle East for thousands of years. World production is currently nearly 30 million tonnes, grown mainly in China, Turkey, the USSR, Egypt, the USA, Iran, Japan, Italy, Spain and Syria. It is appreciated for its thirst-quenching properties and attractive appearance; some cultivars have yellow or orange flesh but most are deep pink with black seeds.

Until recently (3) it was believed that watermelons were climacteric fruits. Harvest maturity is judged on several criteria, including the colour of the groundspot (2,9). The fruits should be cut from the vine with a sharp knife. If stem-end rot (**448**) is likely to be a problem, it can be advantageous to disinfect the cut surface or to leave a long stalk so as to delay fungal advance into the fruit (9). Watermelons can be very large and heavy, up to about 20 kg, and even moderately sized fruits are difficult to handle without bruising. Small sized watermelons (between 2 and 5 kg) have recently been bred in the USA.

The optimal temperature for shipment and storage lies between 10° and 15°C, but shipment at ambient temperature is often feasible. Relative humidity is not critical, since watermelons do not readily lose moisture. They are sensitive to ethylene, however, and carriage with certain other commodities can lead to premature senescence (q.v.) of the watermelons. In the absence of ethylene they remain edible for as long as 2 or 3 months, but optimal quality is retained for 2 weeks at most.

---

1 CEPONIS M.J., CAPPELLINI R.A. & LIGHTNER G.W. (1986) Disorders in muskmelon shipments to the New York market, 1972-1984. *Plant Disease* **70**, 605–607.
2 COREY K.A. & SCHLIMME D.V. (1988) Relationship of rind gloss and groundspot color to flesh quality of watermelon fruits during maturation. *Scientia Horticulturae* **34**, 211–218.
3 ELKASHIF M.E., HUBER D.J. & BRECHT J.K. (1989) Respiration and ethylene production in harvested watermelon fruit: evidence for nonclimacteric respiratory behavior. *Journal of the American Society for Horticultural Science* **114**, 81–85.
4 LINGLE S.E., LESTER G.E. & DUNLAP J.R. (1987) Effect of postharvest heat treatment and storage on sugar metabolism in polyethylene-wrapped muskmelon fruit. *HortScience* **22**, 917–919.
5 MIMBELA-LEYVA L., PASSAM H.C., REILLY P.J.A. & WALLBRIDGE A. (1975) Quality problems of South American honeydew melons imported into Britain. *Tropical Science* **17**, 61–74 illus.
6 MORRIS S.C. & WADE N.L. (1983) Control of postharvest disease in cantaloups by treatment with guazatine and benomyl. *Plant Disease* **67**, 792–794.
7 PRATT H.K. (1971) Melons. In *The biochemistry of fruits and their products* (Ed. by A.C. Hulme), pp 207–232. London: Academic Press.
8 STEWART J.K. (1979) Decay of muskmelons stored in controlled atmospheres. *Scientia Horticulturae* **11**, 69–74.
9 WHITAKER T.W. & DAVIS G.N. (1962) *Cucurbits: botany, cultivation and utilization*. London: Leonard Hill, 249 pp illus.
10 WIANT J.S. (1937) Investigations of the market diseases of cantaloups and Honey Dew and Honey Ball melons. *Technical Bulletin of the United States Department of Agriculture* No. 573, 47 pp illus.

DISEASES of melons and watermelons

450 Alternaria rot

ALTERNARIA ROT may be caused by *Alternaria alternata* (page 172) or *Alternaria cucumerina* (Ell. & Ev.) Elliott. In a humid atmosphere the lesions become covered with dark spores. Closely related fungi, *Stemphylium* spp. and *Ulocladium chartarum* (Preuss) Simmons, cause similar rots. Control measures include seed treatment, crop spraying, careful handling, and storage at an appropriate temperature. Fruits which have suffered chilling injury (q.v.) are especially prone to infection.

ANTHRACNOSE is usually caused by *Colletotrichum gloeosporioides* f. sp. *cucurbitae*, often known as *C. lagenarium* (Pass.) Ell. & Halst or *C. orbiculare* (Berk. & Mont.) v. Arx. Periods of rain or sprinkle irrigation can lead to epidemics in the growing crop, and symptoms may also develop in harvested fruits. Lesions tend to be numerous and sunken, and are eventually covered with pinkish spore-masses. Control measures include seed treatment, strict hygiene in field or greenhouse, the application of fungicide sprays, and careful attention to irrigation regime (4).

BACTERIAL ROTS may be caused by species of *Erwinia*, which tend to gain entry via insect punctures or other injuries. *E. carotovora* subsp. *carotovora* (page 136) incites a soft rot, often with release of foul-smelling juices. *E. ananas* (page 158) may also cause an invasive soft rot but is more usually confined to the rind, on which develop smooth, firm, brownish spots (9). Infection by *E. herbicola* (page 136) characteristically results in hard brown areas within the rind and, in melons, also scattered in the flesh. *Pseudomonas syringae* pv. *lachrymans* (Smith & Bryan) Young, Dye & Wilkie is the cause of **bacterial spot**, in which the fruit surface becomes covered with circular, sunken, watersoaked spots. The interior of the fruit may undergo a rapid soft rot, with browning of the vascular tissue. *Xanthomonas campestris* pv. *melonis* Neto, Sugimori & Oliveira causes small, dark, depressed spots which may coalesce to form large areas of soft odourless watery tissue. Control methods for bacterial rots include treatment of seed with a bactericide, and attention to irrigation regime (since an uneven water supply can result in the cracking of fruit and bacterial invasion).

BLUE MOULD ROT is caused by *Penicillium* spp. Species isolated from melons include *P. crustosum* Thom, *P. cyclopium* (page 136), *P. expansum* (page 178), *P. pallidum* Smith and *P. viridicatum* Westling. Lesions are at first circular to oval, often watersoaked and with a tendency to rupture; blue-green spores are produced in the cracks so formed, and there is a typical musty odour. Some strains of these fungi form restricted lesions, while others spread over much of the fruit and cause a wet collapse. Infection is typically via wounds, or directly through skin which has been weakened by chilling injury or prolonged storage.

BOTRYODIPLODIA ROT can cause serious losses. The causal fungus is *Botryodiplodia theobromae* (**197**), which gains entry via the stem-end or through wounds and incites a dry rot. Lesions are initially buff-coloured, and the affected area is spongy with a watersoaked margin. Dark grey mould may develop on the surface. The fungus also produces a tough layer beneath the skin of the fruit, causing it to become severely wrinkled. Minute black bodies (pycnidia) are usually produced, and a sour odour accompanies the decay. It may be necessary to treat the seed, and to apply fungicide to the growing crop and the harvested fruits. Refrigerated storage is effective in preventing post-harvest decay.

CHARCOAL ROT is caused by *Macrophomina phaseolina* (Tassi) Goid. Lesions appear water-soaked but initially remain firm. A bright purple or pink coloration may be associated with the affected area, both externally and internally (cf. fusarium rot, q.v.). As the decay advances, the skin becomes wrinkled and there is profuse development of very minute spherical black bodies (microsclerotia). This superficial crust tends to rupture, exposing watery flesh which emits a pungent odour and eventually turns black. Sometimes infected fruits become hollow shells, of which the interior is also blackened. The fungus, which is seed- and soil-borne, can cause epidemics during hot dry weather (5). Seed treatment and fungicide sprays can give partial control. If rain is infrequent, irrigation should be carried out regularly, firstly so that the plants are not weakened by drought stress, and secondly because the charcoal rot fungus survives poorly in wet soil. Fruits can be protected by means of a post-harvest fungicide dip and by the use of refrigerated storage.

CLADOSPORIUM ROT (or sometimes a superficial blemish) may be caused by several species. *Cladosporium tenuissimum* Cooke has been recorded on watermelons, while those species attacking melons include *C. cladosporioides* (Fresen.) de Vries (8), *C. cucumerinum* Ell. & Arth. (10) and *C. herbarum* (page 226). Symptoms vary with fungus and with host cultivar. Netted melons tend to be particularly vulnerable to superficial colonisation of the stem-scar, which may become thickly covered with dark mould; sparse mould growth may also develop between the netted areas on the fruit surface (8). In Honey Dews the usual symptom is discrete dark circular spots, each of which may be surrounded by a pale brown halo. Cladosporium rot is often associated with storage at too low a temperature (10).

COTTONY LEAK and **PYTHIUM ROT** may be caused by any of several species, including *Pythium aphanidermatum* (Edson) Fitzp., *P. butleri* Subram., *P. myriotylum* Drechsler and *P. ultimum* Trow. These fungi are common soil inhabitants and thrive in wet conditions. Infection of melons can occur by direct penetration of the rind, while invasion of watermelons is usually via the stem-end or through wounds. Decay spreads rapidly during transit and storage, with the formation of 'nests' of mouldy fruits exuding watery juices. Control measures include crop rotation, seed treatment, improvements in drainage, the use of a mulch, fungicide spraying, post-harvest dipping, and prompt pre-cooling.

DIDYMELLA BLACK ROT is caused by *Didymella bryoniae* (Auersw.) Rehm (conidial state: *Phoma*

451 Didymella black rot

cucurbitacearum (Fr.) Sacc.). Lesions are at first watersoaked and roughly circular, sometimes zonate, later becoming dark, sunken and cracked. A gummy exudate may be produced (although this is not usual in watermelons) and, under humid conditions, a whitish mould may develop. Infected tissue shrivels and gives rise to minute black bodies (perithecia and pycnidia). Control measures include destruction of infective plant debris, the use of clean or treated seed, crop spraying, greenhouse ventilation, and hygienic techniques of harvesting and handling (6).

452 Cottony leak

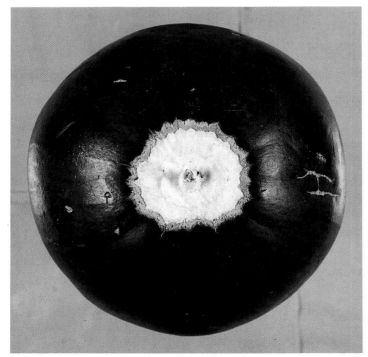

453 Fusarium rot

FUSARIUM ROT may be caused by any of several species, including *F. equiseti* (**296**), *F. oxysporum* Schlecht., *F. pallidoroseum* (7) (page 110) and *F. solani* (Mart.) Sacc. (3,8). Lesions can occur anywhere on the fruit but are frequently found at the stem-end. Affected tissue tends to be spongy or corky and, under humid conditions, becomes covered with a whitish or pinkish mould. Decay may be shallow or extend to the seed cavity, and there is usually a distinct line of demarcation between healthy and rotted tissue. A reddish-purple pigment is sometimes produced (cf. charcoal rot, q.v.). It may be necessary to disinfect the soil, to use a seed dressing, and to protect the crop from insect damage, which would otherwise predispose the fruit to infection (3). Fungicide sprays (7) and post-harvest fungicide dips (preferably heated) (8) are recommended.

454 Grey mould rot

GREY MOULD ROT of melons is caused by *Botrytis cinerea* (**435**). In common with the causal agent of watery soft rot (q.v.), the grey mould fungus typically colonises dying blossoms and is subsequently able to invade the fruit. Infection may also occur via wounds or through the cut stem at harvest time. Crop protection involves control of irrigation and ventilation to prevent the build-up of excessive moisture; fungicide sprays should be carefully timed and rationally used (page 20) so as to delay development of insensitive fungal strains. If incipient infections are suspected, fruits can be treated with a fungicide immediately after harvest. Refrigeration can retard rotting but does not stop it.

MYROTHECIUM ROT is caused by *Myrothecium roridum* Tode ex Fr. The dark sunken lesions support discrete greenish-black spore-masses (sporodochia) which are 1 to 1.5 mm in diameter and fringed with white (2). These bodies may coalesce into a sticky charcoal-black sheath, or become dry, hard and flaky. There is a gradual rotting of the flesh, which becomes watery but does not have a pronounced odour. Under humid conditions there is a luxuriant growth of white mould, interspersed with spore-masses. The disease may be seed-borne, making seed treatment advisable. Field hygiene is important; diseased plant debris should be removed and destroyed before the new crop is planted. A fungicide spray can be effective if applied at the onset of flowering. If infected fruit is inadvertently packed, deterioration can be delayed by the use of refrigeration.

PHOMOPSIS ROT of melons is caused by *Diaporthe melonis* Beraha & O'Brien. The fungus infects developing fruits but remains quiescent until the melons mature. Post-harvest rotting can be controlled by means of a heated fungicide dip (1). Phomopsis rot of watermelons may be caused by *Phomopsis cucurbitae* McKeen which incites a black rot.

PHYTOPHTHORA ROT may be caused by any of several species, including *Phytophthora cactorum* (**424**), *P. capsici* Leonian, *P. drechsleri* Tucker and *P. nicotianae* var. *parasitica* (page 64). Lesions may form on any part of the fruit, and are usually brown with an irregular margin, later becoming zoned or mottled. Under humid conditions a whitish mould develops, on which are produced characteristic spore-heads (sporangia). The rind may rupture, and eventually the entire fruit decays. Infected flesh is soft, water-soaked, rarely discoloured, and almost odourless unless invaded by secondary organisms. *Phytophthora* spp. are common soil inhabitants, which thrive in warm wet conditions. Fields should therefore be well-drained and properly irrigated so that water does not accumulate. An appropriate fungicide may be applied to the soil or to the growing crop, and decay in harvested fruits can be minimised by the use of refrigeration.

PINK MOULD ROT is caused by *Trichothecium roseum* (page 191). Lesions form at the blossom-end, stem-end or side of the fruit, and may extend over much of the surface. Affected rind becomes somewhat tough and shrivelled, and a viscous liquid may ooze from the lesion. Diseased flesh is spongy, slightly brown and extremely bitter in taste. Under humid conditions there is profuse development of pale salmon-pink spores. The fungus is a common soil inhabitant, and control involves good field hygiene, for example ensuring that the mould is not inadvertently introduced in mulching materials. Care should be taken to maintain the plants in vigorous condition, and to prevent insect injury or mechanical damage. Fungicide sprays applied to the growing crop can provide partial control of post-harvest decay. Deterioration can be further delayed by the use of post-harvest chemical treatments and refrigerated storage. Pink mould is extremely sensitive to low dose γ-irradiation.

455 Pink mould rot

456 Rhizopus rot

RHIZOPUS ROT is of particular importance on ripe melons, and may be caused by *Rhizopus oryzae* (8) or *Rhizopus stolonifer* (**401**). Lesions are soft and watersoaked, with a fairly distinct boundary. The skin often cracks open, and a coarse white mould develops on the exposed surface, later giving rise to a mass of globular spore-heads (sporangia) which turn from white to black. Diseased flesh is soft, very wet, and usually sharply delimited from healthy tissue, but it is held together by strands of mould and so does not disintegrate completely. A faintly sour odour accompanies the decay. Control measures may include spraying the crop with fungicide at flowering time, to prevent infection via dying blossoms. Harvested fruits should be handled with care and subjected to a heated fungicide dip (8). Since these fungi grow very slowly at temperatures below $10°C$, decay can also be held in check by refrigeration.

SCLEROTIUM ROT occasionally causes losses in fruits grown under warm wet conditions. The causal fungus is *Corticium rolfsii* Curzi (sclerotial state: *Sclerotium rolfsii* Sacc.). It can gain entry through insect injuries (3). Lesions are soft, watery and discoloured, and may exude amber droplets. The skin of the fruit often cracks open, and a fine white silky mould spreads over the surface, characteristically flat and fan-like in appearance. Seeds can become infected (3). Spherical bodies (sclerotia) are produced, approximately 1 mm in diameter, white at first and later brown. It is not easy to rid the soil of sclerotia, but partial control may be achieved with chemicals or by means of solarization. In the latter method a plastic tarpaulin is spread on the field for several weeks before planting; during hot dry weather the resulting soil temperatures can be lethal to the fungus. An alternative strategy is to prevent fruits from coming into contact with infective soil by using a mulch or by raising them off the ground with an arrangement of sticks. Insect control may also be important (3). Fungicide treatment of harvested fruits has been recommended, and cool storage holds decay in check.

SOIL ROT, caused by *Thanatephorus cucumeris* (Frank) Donk (sclerotial state: *Rhizoctonia solani* Kühn) occurs in warm wet growing areas. Lesions are at first firm, brownish and watersoaked, enlarging into irregular-shaped sunken cankers. In a warm humid atmosphere rotting is rapid, and infected fruits become covered with a dense greyish-brown mould. Decay in harvested fruits may be held in check by storage at $10°C$ or below. The fungus may be seed-borne, necessitating treatment of seed before planting. Cultural practices are important in control. The land can be deep-ploughed so as to bury infected material, or covered with plastic sheeting to prevent contact between fruits and soil. Since infection is facilitated by humid conditions, it is sometimes advisable to prune dense foliage which would otherwise tend to create a moist atmosphere near the soil surface. Partial control of the disease can be achieved by the addition of certain fungi to the soil; for example species of *Trichoderma* are antagonistic to the causal fungus but otherwise harmless. A combination of various control measures is usually necessary.

SOUR ROT is caused by *Geotrichum candidum* (**194**). Lesions are soft and watersoaked, and later become covered with a flat, dull, white, powdery or waxy mould growth. Infected flesh is not discoloured but undergoes a soft watery breakdown. A sour odour may accompany the decay. Since infection can take place via insect punctures, it may be advisable to use an insecticide on the growing crop. Care should be taken to minimise injury to the fruits during harvesting and handling. Most fungicides are ineffective against *G. candidum*, although some of the more recent chemicals do give control if applied within 4 to 8 hours of harvest (8). Refrigeration serves to inhibit decay, as the fungus makes little progress at temperatures below 10°C.

WATERY SOFT ROT is caused by *Sclerotinia sclerotiorum* (page 66), which is common in temperate countries and in subtropical regions possessing a cool wet season. Infected tissue is soft and watery, becoming covered with a dense bright white mould. Irregular-shaped bodies (sclerotia) develop; they are several mm across, white at first, exuding drops of liquid and finally turning black. Sclerotia can persist in the soil for many years, and integrated control measures are necessary. If the soil is drenched with a systemic fungicide before the crop is sown, the developing plants can then absorb the chemical and thereby be protected against infection. Alternatively, or in addition, aerial sprays can be applied during the growing season, usually in conjunction with control measures against grey mould (q.v.). For crops grown in plastic houses, the use of special film can be effective; if certain

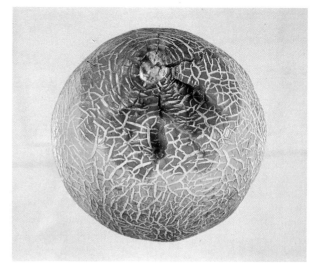

457 Sour rot

wavelengths can be cut out, 'germination' of sclerotia is inhibited and spore production is thereby prevented. An alternative practice involves covering the ground with plastic sheeting so that ascospores cannot be effectively released (**68, 70**). If primary control measures fail (owing, for example, to the development of fungal strains tolerant of the fungicide used) and infection is noticed in the growing crop, all infected plant parts should be removed and destroyed, before sclerotia can proliferate and fall on to the soil. Fruits which may have incipient infection can be treated with a fungicide immediately after harvest.

---

1 BERAHA L. & O'BRIEN M.J. (1979) *Diaporthe melonis* sp. nov., a new soft rot of market cantaloupes. *Phytopathologische Zeitschrift* **94**, 199–207 illus.

2 CARTER W.W. (1980) Incidence and control of *Myrothecium roridum* on cantaloup in relation to time of fungicide application. *Plant Disease* **64**, 872–874.

3 IKEDIUGWU F.E.O. (1980) *Corticium rolfsii* and fruit rot of *Citrullus lanatus* in the field in Nigeria. *Transactions of the British Mycological Society* **75**, 316–319 illus.

4 PEREGRINE W.T.H. & AHMAD K. bin (1983) Chemical and cultural control of anthracnose (*Colletotrichum lagenarium*) in watermelon. *Tropical Pest Management* **29**, 42–46.

5 REUVENI R., NACHMIAS A. & KRIKUN J. (1983) The role of seedborne inoculum on the development of *Macrophomina phaseolina* on melon. *Plant Disease* **67**, 280–281.

6 SANZ B.-M., H. (1971) *Mycosphaerella melonis* (Pass.) Chiu & Walker, una nueva enfermedad de las cucurbitáceas en el país. *Agricultura Técnica* **31**, 142–144.

7 TROUILLON P. & TISSUT M. (1972) Contribution à l'étude d'une fusariose de fruits de cucurbitacées et de ses possibilités de traitement. *Phytiatrie-Phytopharmacie* **21**, 61–67.

8 WADE N.L. & MORRIS S.C. (1982) Causes and control of cantaloupe postharvest wastage in Australia. *Plant Disease* **66**, 549–552.

9 WELLS J.M., SHENG W.-S., CEPONIS M.J. & CHEN T.A. (1987) Isolation and characterization of strains of *Erwinia ananas* from Honeydew melons. *Phytopathology* **77**, 511–514.

10 WIANT J.S. (1938) Market storage studies of Honey Dew melons and cantaloups. *Technical Bulletin of the United States Department of Agriculture* No. 613, 18 pp illus.

DISORDERS of melons and watermelons

BLOSSOM-END ROT of watermelons has recently been studied in Iraq and in Italy (1). Cylindrical-fruited cultivars are especially susceptible, the main symptom being a dark sunken area at the distal end of the fruit. The disorder develops as the fruits mature during late summer, a period characterised by high temperature, low humidity and warm dry winds. By harvest time 75% of the crop may be affected and although the disfigurement does not progress after harvest it constitutes a serious market blemish. Blossom-end rot is a nutritional disorder related to the concentration of calcium in fruit tissues; lime-deficient soils may require amendment with gypsum (calcium sulphate), and the frequency of irrigation may need to be increased to permit efficient uptake of minerals. While appropriate crop husbandry can give partial control, the simplest measure is to grow spherical-fruited cultivars which are not generally susceptible to the disorder (1).

CHILLING INJURY may take the form of reddish-brown patches on the skin (3,4); ripening may be prevented and taste and texture impaired. Chill-injured tissue shows an increased tendency to decay, manifest as multiple lesions affecting much of the surface area (**449**). Typical post-chilling infections are alternaria rot (q.v.) and cladosporium rot (q.v.). The critical temperature, below which chilling injury can occur, is between 10° and 14°C for watermelons; melons are less susceptible, and their cultivars differ substantially. Sensitivity to chilling injury can be reduced by plastic film wraps (netted melons), pre-shipment treatment with ethylene ('winter melons'), or pre-shipment 'conditioning' for a few days at moderate temperature (watermelons). Such treatments permit storage of netted melons at 4°C, Honey Dew melons at 5°C, and watermelons at 7°C (4).

PREMATURE SENESCENCE of watermelons occurs in the presence of ethylene. Watermelons are usually harvested when already ripe (or very nearly so), and do not have much capacity for improvement in flesh quality. On the contrary, once the fruits are harvested it is not long before there is a decline in colour, texture and flavour of the ripe flesh. Such changes (fading, softening and the development of off-flavours) are accelerated by exposure to ethylene (5). It is therefore important, during shipment and storage, to segregate watermelons from fruits such as melons, tomatoes, apples and bananas which (in comparison with watermelons) produce substantial quantities of ethylene (5).

SOLAR INJURY of melons may result from ultraviolet radiation or heat, or a combination of both (2). Symptoms on netted melons consist of a patchy ground colour or 'bronzing', and discoloration of the net. In 'winter melons', such as Honey Dew, the affected area may be brown and blotched or else greyish as if bleached. Severely injured tissue becomes sunken or wrinkled, and flesh quality may be impaired. In hot arid growing areas solar injury can be a problem in crops where there is inadequate foliage cover to shade the maturing fruits. One effective measure is to apply whitewash to the melons so that sunlight and heat are reflected (2). The whitewash is easily removed by washing the melons after harvest. Harmful exposure to the sun may also occur at harvest time if fruits are not brought under cover promptly; flesh temperatures can rise several degrees above ambient air temperature, sometimes resulting in irreversible damage. It is therefore important to provide shade and to transport melons to the packhouse without delay.

VEIN-TRACT BROWNING of netted melons is a post-harvest disorder in which there is a darkening of the longitudinal grooves between the netted areas. Completely netted cultivars may suffer a similar disorder but the discoloration is less obvious. The condition is a consequence of ageing, accelerated by exposure to the sun during maturation and to high temperatures after harvest (2). Control measures are as for solar injury (q.v.). Studies in Australia have demonstrated that high humidity storage is beneficial in minimising development of this disorder.

---

1 CIRULLI M. & CICCARESE F. (1981) Effect of mineral fertilizers on the incidence of blossom-end rot of watermelon. *Phytopathology* **71**, 50–53 illus.
2 LIPTON W.J. (1977) Ultraviolet radiation as a factor in solar injury and vein tract browning of cantaloupes. *Journal of the American Society for Horticultural Science* **102**, 32–36.
3 LIPTON W.J. & WANG C.Y. (1987) Chilling exposures and ethylene treatment change the level of ACC in 'Honey Dew' melons. *Journal of the American Society for Horticultural Science* **112**, 109–112.
4 PICHA D.H. (1986) Postharvest fruit conditioning reduces chilling injury in watermelons. *HortScience* **21**, 1407–1409.
5 RISSE L.A. & HATTON T.T. (1982) Sensitivity of watermelon to ethylene during storage. *HortScience* **17**, 946–948.

Appendix 1 Glossary of terms
(Origins of the terms are indicated in the Lexicon)

acaricide, a substance which kills mites.

acervulus, an erumpent, cushion-like mass of hyphae producing conidia on short conidiophores (Class Melanconiales).

aecium, a cup-shaped structure which produces aeciospores (Stage 1 in the rust fungi, Order Uredinales).

agar (from Malay **agar-agar**), a gelatine-like material obtained from seaweed and used to prepare culture media on which micro-organisms are grown and studied.

agronomy, the science of crop production.

alternate host, either of the two hosts of a heteroecious rust fungus.

alternative host, one of several plant species hosts of a given pathogen.

anamorph, asexual or imperfect state of a fungus.

antagonism, a general term for counteraction between organisms.

antheridium, the male sexual organ in some fungi (Class Oomycetes).

anthracnose, a plant disease having characteristic limited black lesions, usually sunken, with asexual spores in acervuli (Order Melanconiales).

antibiotic, a substance produced by a micro-organism and able to inhibit or kill other micro-organisms.

apothecium, a cup- or saucer-like structure containing asci and ascospores (Class Discomycetes).

appressorium, a thick-walled fungal cell formed prior to penetration.

Ascomycetes or, more strictly, Subdivision Ascomycotina, fungi which produce their spores in asci.

ascus, a sac-like structure containing ascospores (usually 8).

autoecious, (of rust fungus) able to complete its entire life cycle on one host species. cf. heteroecious.

bactericide, a substance which kills bacteria.

bacterium, a unicellular microscopic organism which lacks chlorophyll and which multiplies by splitting in two.

Basidiomycetes or, more strictly, Subdivision Basidiomycotina, fungi which produce their spores on basidia.

basidium, a club-shaped structure on which basidiospores are borne (Stage 4 in the rust fungi, Order Uredinales).

calyx, floral parts, consisting of sepals, remaining on some fruits such as apples, pomegranates, aubergines.

chilling injury, injury caused by low but non-freezing temperatures. cf. freezing injury.

chlamydospore, a hyphal cell, enveloped by a thick cell wall, which becomes separated from the parent hypha and behaves as a resting spore.

cleistothecium, a completely closed structure containing asci and ascospores (Class Plectomycetes).

climacteric, a period in the development of certain fruits, during which there is a pronounced increase in respiration, associated with the production of ethylene and leading to ripening.

clone, the aggregate of individual organisms produced asexually from one sexually produced individual.

conidiophore, a specialised hypha on which one or more conidia are borne.

conidium, an asexual spore which forms at the tip or side of a hypha.

controlled atmosphere (CA), an atmosphere in which oxygen, carbon dioxide and nitrogen concentrations are regulated, as well as temperature and humidity. cf. modified atmosphere.

cotyledon, the seed leaf of higher plants, one in the Monocotyledons and two in the Dicotyledons.

cultivar, a variety of a cultivated plant.

culture, growth of a micro-organism on artificial medium.

Deuteromycetes or, more strictly, Deuteromycotina, forms of fungi which produce asexual spores or no spores.

disease, harmful deviation from normal functioning of physiological processes, caused by one or more pathogenic organisms or viruses.

disorder, harmful deviation from normal functioning of physiological processes, arising from causes other than pathogenic organisms or viruses.

dormancy, state of minimal physiological activity.
downy mildews, diseases caused by fungi in the Family Peronosporaceae.
endoconidium, an asexual spore formed inside a hypha and extruded.
ethylene, a gas (C_2H_4) produced in small quantities by plant tissues.
exudate, liquid discharge from diseased or healthy plant tissue.
flagellum, a whip-like structure projecting from a bacterium or zoospore, serving to propel it through water.
freezing injury, injury caused by temperatures below the freezing point of the produce. cf. chilling injury.
fungicide, a substance which kills fungal spores or mycelium.
fungistat, a substance which prevents the growth of a fungus without killing it.
fungus, a type of organism which lacks chlorophyll and conductive tissues.
growth regulator, a natural substance which affects the enlargement, division, or activation of plant cells.
herbicide, weed-killer.
hermaphrodite, an individual bearing both functional male and female reproductive organs.
heteroecious, (of rust fungus) requiring two different host species for the completion of its life cycle. cf. autoecious.
horticultural maturity, the stage of development when a plant or plant part possesses the prerequisites for utilisation by consumers for a particular purpose.
host, an organism harbouring a parasite.
host range, the various kinds of host plants that may be attacked by a parasite.
hydathodes, structures with one or more openings that discharge water from the interior of the leaf to its surface.
hydro-cooling, cooling with cold water.
hypha, one of the threads of a mycelium.
immune, exempt from infection by a given pathogen.
imperfect fungus, a fungus which is not known to produce sexual spores.
imperfect state (anamorph), the state in which asexual spores (e.g. conidia) or no spores are produced.
infect, to enter and establish a permanent or temporary parasitic relationship with an organism.
infest, to over-run the surface of a plant, or to be dispersed through soil or other substrate.
inoculate, to introduce a micro-organism into an organism or into a culture medium.
isolate, to separate a micro-organism from its host or substrate and to establish it in pure culture.
lenticels, small pores in some plant organs (e.g. apple fruit, potato tuber), allowing the passage of water vapour and other gases.
lesion, a localised area of diseased or disordered tissue.
maturation, the stage of development leading to the attainment of physiological or horticultural maturity.
metabolism, the process by which cells or organisms utilise nutritive material to build living matter or to break down cellular material into simple substances to perform special functions.
modified atmosphere (MA), an atmosphere in which oxygen, carbon dioxide and nitrogen concentrations are different from those in air but are not precisely regulated. cf. controlled atmosphere.
mould (mold), superficial mycelium of a fungus.
mycelium, mass of fungal threads (hyphae).
mycology, the study of fungi.
mycotoxins, poisonous substances produced by certain fungi.
natural openings, stomata, lenticels, hydathodes.
nematicide, a substance which kills nematodes.
nematodes (eelworms), minute thread-like organisms living in soil, water or as parasites of plants and animals.
oogonium, the female sexual organ in some fungi (Class Oomycetes).
oospore, a sexual spore produced by the fusion of two morphologically dissimilar cells (oogonium and antheridium).
parasite, an organism living on or in, and getting its food from, its host, another living organism.
pathogen, a parasite able to cause disease in a particular host or range of hosts.
perfect state (teleomorph), the state of the life cycle in which spores (such as ascospores and basidiospores) are formed after nuclear fusion or by parthenogenesis.
perithecium, globose or flask-like structure (with an opening) containing asci and ascospores (Class Pyrenomycetes).
photosynthesis, the process by which carbon dioxide and water are combined in the presence of light and chlorophyll to form carbohydrate.
physiological maturity, the stage of development when a plant or plant part will continue to develop even if detached.
plant pathology (phytopathology), the study of plant disease.

plasmodium, a naked, slimy mass of protoplasm.

powdery mildews, diseases caused by fungi in the Family Erysiphaceae.

pre-cooling, removal of field heat prior to shipment.

pycnidium, globose or flask-like structure (with an opening) containing conidiophores and conidia (Order Sphaeropsidales).

pycnium (spermagonium), a flask-shaped structure containing pycniospores (spermatia) (Stage 0 in the rust fungi, Order Uredinales).

quarantine, (1) control of import and export of plants to prevent spread of diseases and pests; (2) holding of imported plants in isolation for a period to ensure their freedom from diseases and pests.

relative humidity (%), the ratio of the actual amount of water vapour in the air to the maximum it can hold at a given temperature, multiplied by 100.

resistant, (host plant) able to suppress or retard the activity of a pathogen; (micro-organism) able to withstand, completely or in some degree, the presence of a given chemical or other harmful factor.

respiration, a series of chemical oxidations within the cell, controlled and catalysed by enzymes, in which carbohydrate and fats are broken down, releasing energy to be used by the cell or organism in its various functions.

resting spore, a sexual or other thick-walled spore which is resistant to extremes of temperature and moisture and which tends to germinate only after a lapse of time.

ripening, the final stages of maturation when a fruit develops optimal appearance and eating quality.

rotting, the disintegration of tissues by the action of one or more micro-organisms.

rusts, diseases caused by the rust fungi (Order Uredinales).

sclerotium, a resting body of variable size, composed of a hard mass of hyphae with or without host tissue, usually with a darkened rind, from which may develop reproductive bodies, stromata, conidiophores or mycelium.

senescent, (tissue) growing old and dying.

sensitive, (host plant) reacting with severe symptoms to infection by a given pathogen; (micro-organism) succumbing to the effect of a given chemical or other harmful factor.

septate, possessing septa or cross-walls.

septum, a cross-wall in a hypha or spore.

seta, a bristle or stiff hair, generally thick-walled and dark coloured.

smuts, diseases caused by the smut fungi (Order Ustilaginales).

sooty moulds, sooty coatings on foliage and fruit, formed by the dark hyphae of fungi which live on the honeydew secreted by insects such as aphids, mealybugs and scales.

sorus, compact mass of spores, found especially in the rust and smut fungi.

spermagonium (pycnium), flask-shaped structure containing spermatia (pycniospores) (Stage 0 in rust fungi, Order Uredinales).

sporangiophore, a specialised hypha bearing one or more sporangia.

sporangium, a structure borne on a sporangiophore and containing sporangiospores (Class Oomycetes and Class Zygomycetes).

spore, the reproductive unit in fungi, consisting of one or more cells. It is analogous to the seed of green plants.

sporodochium, a cushion-like structure consisting of a mass of hyphae supporting conidiophores and conidia.

stoma, stomata, minute pores on the surfaces of leaves, stems and some fruits, allowing the passage of water vapour and other gases.

stroma, stromata, compact mycelial structures, with or without host tissue, in or on which spores are produced.

susceptible, subject to infection.

symptom, a detectable abnormality arising from disease.

syndrome, the totality of effects produced in a plant by a disease.

synergism, the concurrent parasitism of a host by two pathogens, in which the symptoms produced are of greater magnitude than the sum of the effects of each pathogen acting alone.

systemic, (of plant pathogen or chemical) occurring throughout the plant, having been translocated via the conductive system.

teleomorph, sexual or perfect state of a fungus.

teliospore, the sexual thick-walled resting spore of the rust and smut fungi.

telium, structure in which teliospores are produced (Stage 3 in the rust fungi, Order Uredinales).

tolerant, (host plant) able to endure infection by a particular pathogen, without showing severe symptoms; (micro-organism) able to withstand, completely or in some degree, the presence of a given chemical or other harmful factor.

transpiration, loss of water vapour from the surface of leaves and other plant organs.

uredinium, structure in which urediniospores are produced (Stage 2 in the rust fungi, Order Uredinales).

vacuum cooling, cooling by evaporation of water in a partial vacuum.
variety, (1) a subdivision below the level of subspecies; (2) a cultivar.
vascular, pertaining to the plant's conductive system.
vegetative, asexual.
vesicle, a bubble-like structure produced by a zoosporangium and in which the zoospores are released or are differentiated.
virulent, strongly pathogenic.
virus, a nucleoprotein entity able to pass through bacterium-retaining filters and able to replicate within the living cells of a host.
wilt, loss of rigidity and drooping of plant parts, generally caused by insufficient water in the plant or plant organ.
wound parasite, a parasitic organism which can invade a host only if it can first become established in damaged tissues.
yeasts, fungi which reproduce by budding.
zoosporangium, sporangium which contains or produces zoospores.
zoospore, spore possessing one or more flagella and capable of moving in water; a 'swimming' spore.
zygospore, sexual spore produced by the fusion of two morphologically similar cells (Class Zygomycetes).

AGRIOS G.N. (1988) *Plant Pathology*. 3rd ed. New York: Academic Press, 845 pp illus.

AINSWORTH G.C. (1976) *Introduction to the history of mycology*. Cambridge University Press, 359 pp illus.

AINSWORTH G.C. (1981) *Introduction to the history of plant pathology*. Cambridge University Press, 315 pp illus.

FBPP (1973) A guide to the use of terms in plant pathology. *Phytopathological Papers* No. 17, 55 pp.

HAWKSWORTH D.L., SUTTON B.C. & AINSWORTH G.C. (1983) *Ainsworth and Bisby's Dictionary of the fungi*. 7th ed. Farnham Royal, Slough, UK: Commonwealth Agricultural Bureaux, 445 pp illus.

HOLLIDAY P. (1989) *A dictionary of plant pathology*. Cambridge University Press, 369 pp.

JOHNSTON A. & BOOTH C. (Eds) (1983) *Plant pathologist's pocketbook*. 2nd ed. Farnham Royal, Slough, UK: Commonwealth Agricultural Bureaux, 439 pp.

LIDDELL H.G. (1889) *An intermediate Greek–English lexicon, founded upon the 7th edition of Liddell and Scott's Greek–English lexicon*. Oxford: Clarendon Press, 910 pp.

ONIONS A.H.S., ALLSOPP D. & EGGINS H.O.W. (1981) *Smith's Introduction to industrial mycology*. 7th ed. London: Edward Arnold, 398 pp illus.

SIMPSON D.P. (1968) *Cassell's New Latin–English English–Latin dictionary*. London: Cassell, 883 pp.

Appendix 2 Lexicon of Greek and Latin words
(from which are derived scientific names and technical terms)

If an organism has been well-named it is often possible to deduce some of its attributes. In the fungi, for example, *Albugo candida* is clearly *whiter* than *white*, *Phyllo/sticta* spp. cause *leaf spots*, while *Thanate/phorus cucumeris brings death* to *cucumbers* (and much else besides). *Cerato/cystis fimbriata* is characterised by the formation of a type of structure resembling a tough *horny bag* with a *fringe* (**50**). *Colleto/trichum circinans* possesses *sticky* spores (hence splash-dispersed rather than wind-borne), often interspersed with *hair-like* structures (setae); furthermore, the spore-masses tend to be arranged in concentric *circles* (**129**). Less usefully, *Botryo/diplodia theo/bromae* is supposed to resemble in some aspects a *bunch of grapes*, its spores are (eventually) *2-celled*, and it lives on (amongst many other things) the *food* of the (Amerindian) *gods* i.e. cocoa, *Theobroma cacao* L. (**154**).

Technical terms (which are defined in the Glossary) tend to have Latin endings even though the original word may be Greek, for example **ascos** becomes **ascus**, **hyphe** becomes **hypha** and **basidion** becomes **basidium**. Latin plurals of words ending in **-us**, **-a** and **-um** are, respectively, **-i**, **-ae** and **-a**. A few technical terms remain pure Greek, for example **calyx** (plural **calyces**), **stoma** (plural **stomata**) and **stroma** (plural **stromata**). In the lists which follow, verbs are given in several forms and some nouns are quoted additionally in the genitive case where this is closer to the derived word. Greek transcription is largely based on Bodmer F. (1944) *The Loom of Language* (Ed. by L. Hogben), London: George Allen & Unwin.

GREEK

a-, an-	not, without	**chalaros**	slack, loose
acares	too short to be cut	**chlamys, chlamydos**	cloak
aecia	injury	**chloros**	green, yellowish green
agros	field		
ampelos	vine	**choane**	funnel
ana	up	**chrysos**	gold
angion	box, chest	**cinnabari**	vermilion
antheros	flowering	**circinos**	circle
anthos	flower	**clados**	branch, sprout
anthrax, anthracos	coal	**cleistos**	closed
anti	against, opposed to	**climacter**	rung of a ladder, critical stage
apothece	storehouse		
ascos	bag (wineskin)	**climax, climacos**	ladder, ascent
aster (astron)	star	**clon**	shoot
autos	self	**coccos**	berry, grain
bacter (bactron)	rod	**cochlias**	snail
basidion	small pedestal	**coelos**	hollow
bios, biotos	life	**colla**	glue
botane	herb, pasture	**conis**	dust
botrys	bunch of grapes	**coryne**	club
broma	food	**cotyle**	cup
bryone	moss	**cryptos**	hidden
calyx, calycos	covering	**cyclos**	circle
carpos	fruit	**cystis**	bladder, bag
cata	down	**derma, dermatos**	skin
caulos	stalk	**deuteros**	second, next
cephale	head	**di**	two
ceras, ceratos	horn	**dia**	among, through
cercos	tail	**didymos**	twin
chaete	long hair, mane	**diploos**	double

discos	dish, quoit, disc	**phlyctaena**	blister
dochion	receptacle	**phorein**	to bear, bring, carry
endon	inside	**phos, photos**	light
ergon	work	**phragma, phragmatos**	fence
erythros	red	**phthora**	destruction
ge	earth	**phycos**	seaweed
genete	birth	**phylaxis, phylactica**	guarding
genos	kind, race, offspring	**phyllon**	leaf
gloea	glue	**phyma, phymatos**	wart, pustule
gone	birth	**physa**	bellows
gonos	offspring	**physis**	nature
helios	sun	**phyton**	plant
helix, helicos	spiral, coil, snail	**plasma, plasmatos**	form
helmis, helminthos	worm	**plectos**	plaited, twisted
hermaphroditos	hermaphrodite (see Glossary) (the god Hermes and the goddess Aphrodite)	**pleon**	more
		pleos	full, infected with
		polys	many
		porthein	to destroy
heteros	other, different	**pous, podos**	foot
homos	same, similar	**pseudos**	false
hydor, hydros	water	**pycnos**	compact
hyper	above, over, beyond	**pyren, pyrenos**	fruit stone, pip
hyphe	web, tissue	**rhiza**	root
hypo	beneath, under (diminutive)	**rhodon**	rose
idion		**scleros**	hard
idios	distinct	**sitos**	bread, food
leptos	slender, delicate	**soros**	pile
leucos	white	**sperma**	seed
logos	discourse, reasoning, word	**sphacelos**	gangrene
		sphaera	globe, sphere
lysis, lyticos	loosening	**sphongos**	fungus, mushroom
macros	long, large	**spongia**	sponge
melas, melanos	black	**sporos**	seed
meli, melitos	honey	**stasis, staticos**	standstill
metabole	change	**stemphylon**	mass of pressed olives
micros	small		
monas	unit	**stenos**	narrow
monos	alone, single	**stictos**	spotted
morphe	form, shape	**stoma, stomatos**	mouth
myces, mycetos	mushroom	**streptos**	twisted, pliant
myron	unguent, ointment	**stroma, stromatos**	mattress
nema, nematos	thread	**syn**	together, with
nomos	law, custom	**teleos**	complete, perfect
oecos	house	**telos**	end
oidion	little egg	**thanatos**	death
oon	egg	**thece**	box, case, sheath
opsis	appearance	**theos**	god
oxys	sharp, acid	**thrips, thripos**	wood-worm
para	beside	**thrix, trichos**	hair
pathos	suffering	**thyreos**	shield
peri	around	**toxicon**	poison
pezis	puff-ball	**trachys**	rough
phaeos	dusky, grey	**ule**	scar
phagein	to eat, devour	**xanthos**	yellow
pherein	to bear, bring, carry	**zoon**	animal
phiale	bowl, saucer	**zygon**	yoke
philos	loving	**zyme**	yeast

LATIN

acarus	mite	humidus	wet, moist, damp
acervulus	little heap	imperfectus	incomplete
albus	white	inficere, infectum	to poison, taint
alternus	one after the other	infuscere	to make dark, to blacken
ante	before		
apprimere, appressum	to press against	latens, latentis	concealed, hidden
armilla	bracelet	latericius	of brick
artus	tight, close	latex, laticis	fluid, liquid
asper, asperis	rough	laxus	loose, wide
aspergere	to sprinkle	lunatus	like the crescent moon
aspergillum	holy water sprinkler		
atramentum	black ink	macula	spot
bacillus (bacillum)	little rod	monile, monilis	necklace of beads
bi	two	moribundus	dying
caedere, cecidi	to kill	mors, mortis	death
caeruleus	blue	necare	to kill
candidus	shining white	niger, nigra	black, dark
canescens	becoming white	pallidus	pale
caput, capitis	head	pannosus	ragged, tattered
charta	paper	penicillus	little brush
cinis, cineris	ashes	perfectus	complete, perfect
cingula	girdle	pirus	pear tree
clava	club	planta	green twig, cutting
clavus	nail, spike	pomus	fruit tree
colo, colere	to inhabit	post	after
comosus	hairy	purpureus	purple, dark red
cortex, corticis	bark, rind, shell	pustula	blister
creberrimus	crowded together	quiescere	to rest, repose
digitus	finger	racemus	cluster (especially of grapes)
exiguus	small, little, scanty		
falcatus	sickle-shaped	ramulus	small branch
fero, ferre	to bear, bring, carry	reticulum	little net
fimbriae (pl.)	fringe	roseus	rose-coloured
flagellum	whip	scopae, scoparum (pl.)	broom made of twigs
flavus	golden yellow	senescere	to grow old
forensis	relating to the law	septum (saeptum)	wall, barrier, partition
fovea	pit	seta (saeta)	bristle, stiff hair
fuligo, fuliginis	soot	squamus	scale
fulvus	tawny, yellowish brown	stria	furrow, groove
		torulus	small swelling
fumigatus	smoky	trans	across, over, through
fungus	mushroom	uncinula	little hook
funiculus	thin rope, cord, string	uredo, uredinis	blight upon plants
furfur, furfuris	scurf	uro, urere	to burn
fuscus	dark	ustulare	to burn, scorch
fusus	spindle	uva	bunch of grapes
galla	gall	vasculum	small vessel
gemma	bud	vesicula	little bladder
glomerella	little ball	viridis	green
herba	green plant, weed	virus	poison, venom
hiemalis	of winter	voro, vorare	to devour

Appendix 3 Recommended Storage Conditions

The conditions given are those which permit the longest possible duration of storage for the particular commodity, without undue loss of quality and allowing sufficient 'shelf-life' after the product has entered the marketing channels. If, however, a short period of storage is intended, it is often feasible to store produce at higher or lower temperatures and relative humidities than those recommended in the table.

Commodity Variety/Type/Origin	Temperature °C	°F	Relative Humidity %	Storage Period
Apples				
Bramley's Seedling	3 to 4	37 to 39	90–95	4–5 months
Cox's Orange Pippin	3 to 3.5	37 to 38	90–95	3–4 months
Crispin (Mutsu)	1.5 to 2	35 to 36	90–95	4 months
Discovery	3.5	38	90–95	1 month
Egremont Russet	3 to 3.5	37 to 38	90–95	3–4 months
Golden Delicious	1.5 to 2	35 to 36	90–95	4–6 months
Granny Smith	−1 to 0	30 to 32	90–95	3–8 months
Idared	3.5 to 4.5	38 to 40	90–95	5 months
James Grieve	3 to 4	37 to 39	90–95	1 month
Jonathan	3 to 3.5	37 to 38	90–95	3–5 months
Laxton's Superb	1.5 to 3.5	35 to 38	90–95	5 months
McIntosh	3.5 to 4	38 to 39	90–95	4–6 months
Mutsu (Crispin)	1.5 to 2	35 to 36	90–95	4 months
Red Delicious	0 to 1	32 to 34	90–95	4–6 months
Rome Beauty	1 to 2	34 to 36	90–95	6 months
Spartan	0 to 0.5	32 to 33	90–95	3–5 months
Tydeman's Late Orange	3 to 4	37 to 39	90–95	5 months
Worcester Pearmain	0 to 1	32 to 34	90–95	2–3 months
Apricots	−1 to 0	30 to 32	90–95	1–4 weeks
Avocados				
unripe				
Booth 1 and 8, Taylor	4.5	40	85–90	2–4 weeks
Ettinger	5.5	42	85–90	3–4 weeks
Fuchs, Pollock, Waldin	10 to 13	50 to 55	85–90	2 weeks
Fuerte, Hass	5.5 to 8	42 to 46	85–90	3–4 weeks
Lula	4.5	40	85–90	4–5 weeks
ripe				
Fuerte, Hass	2 to 5	36 to 41	85–90	1–2 weeks
Babacos				
green	10	50	90	5–6 weeks
turning	7	45	90	4 weeks
Bananas				
green				
Cavendish	13	55	85–90	10–20 days
Gros Michel	13	55	85–90	10–20 days
Poyo Robusta	13	55	85–90	10–20 days
Valery	13	55	85–90	10–20 days
Lacatan	13 to 15	55 to 59	85–90	1 month
coloured	13 to 16	55 to 61	85–90	5–10 days

Commodity Variety/Type/Origin	Temperature °C	°F	Relative Humidity %	Storage Period
Blueberries	0	32	90–95	2–4 weeks
Breadfruit	13	55	95	1–3 weeks
Cape Gooseberries	14	58	80	
Carambolas	5 to 10	41 to 50	90	3–4 weeks
Cherimoyas	8 to 9	46 to 48	90	1–2 weeks
Cherries				
Sour	0 to 1	32 to 34	90–95	1–2 weeks
Sweet	–1 to 0	30 to 32	90–95	2–3 weeks
Clementines	4 to 5	39 to 41	90	4–6 weeks
Cranberries	2 to 4	36 to 40	90–95	2–4 months
Currants	0	32	90–95	1–2 weeks
Custard apples				
Philippino	5 to 7	41 to 44	85–90	4–6 weeks
West Indian	12	54	85–90	
Dates				
fresh	0	32	85–90	1–2 months
dried	0 or lower	32 or lower	70–75	12 months
Durians	4 to 6	39 to 43	85–90	1–2 months
Feijoas	4	39	90	3–4 weeks
Figs	0	32	90–95	1–2 weeks
Gooseberries	–0.5 to 0	31 to 32	90–95	2–4 weeks
Granadillas	7 to 10	45 to 50	85–90	3–5 weeks
Grapefruits				
California and Arizona	14 to 15	58 to 60	90	4–6 weeks
Florida and Texas	10	50	90	4–6 weeks
Israeli	10 to 12	50 to 54	90	10–16 weeks
South African	11	52	90	12–14 weeks
Grapes				
Almeria	–1 to –0.5	30 to 31	90–95	3–6 months
Alphonse Lavallée	–1 to –0.5	30 to 31	90–95	3 months
Barlinka	–1 to –0.5	30 to 31	90–95	4–6 months
Red Emperor	–1	30	90–95	4–6 months
Thompson Seedless	–1	30	90–95	2–3 months
Waltham Cross	–1 to –0.5	30 to 31	90–95	3 months
Guavas	5 to 10	41 to 50	90	2–3 weeks

Commodity Variety/Type/Origin	Temperature °C	°F	Relative Humidity %	Storage Period
Horned melons, Kiwanos	10	50	90	
Jackfruits	13	55	95	1–3 weeks
Kiwifruits	−0.5 to 0	31 to 32	90–95	2–3 months
Lemons				
green	10 to 14	50 to 57	90	2–5 months
yellow	11	52	90	3–6 months
Limes	9 to 10	48 to 50	90	1–2 months
Litchis	5 to 10	41 to 50	90 to 95	4–6 weeks
Longans	4	40	90	
Loquats	0	32	90–95	2–3 weeks
Mandarins	4 to 8	39 to 46	90	3–8 weeks
Mangoes				
Alphonso, Indian	7 to 9	45 to 48	90	7 weeks
USA	13	55	90	2–3 weeks
Bangalore, Safeda	5.5 to 7	42 to 45	90	7 weeks
Haden, Keitt	12 to 14	54 to 57	90	2 weeks
Julie	11 to 12	52 to 54	90	2 weeks
Neelum, Raspuri	5.5 to 9	42 to 48	90	5–6 weeks
Zill	10	50	90	3 weeks
Mangosteens	10	50	90	3–4 weeks
Melons				
Cantaloupes	4 to 5	39 to 41	90	1–3 weeks
Galia, Ogen	6	43		
Honey Dew, South African	10 to 15	50 to 59	90	3 weeks
USA	5	41	90	2–4 weeks
Minneolas	4 to 5	39 to 41	90	3–5 weeks
Naranjillas	7 to 10	45 to 50	90	4–6 weeks
Nectarines	−1 to 0	30 to 32	90–95	2–7 weeks
Oranges				
Camargo				
Brazilian	2	36	90	3 months
South African	4	39	90	2–3 months
Castellana, Spanish	1 to 2	34 to 36	90	2–3 months
Jaffa, Israeli	8	46	90	3 months
Navel				
Australian	4.5 to 5.5	40 to 42	90	1–2 months
Moroccan, Spanish	2 to 3	36 to 37	90	2–3 months

Commodity Variety/Type/Origin	Temperature °C	°F	Relative Humidity %	Storage Period
Salustiana, Spanish	2	36	90	3–4 months
Valencia Late				
California	2 to 7	36 to 45	90	1–2 months
Cyprus	2 to 3	36 to 37	90	2–3 months
Florida, Texas	0 to 1	32 to 34	90	2–3 months
Israeli	2	36	90	2–3 months
Moroccan	2 to 3	36 to 37	90	2 months
South African	4.5	40	90	2–4 months
Spanish	2	36	90	3–4 months
Verna, Spanish	1 to 2	34 to 36	90	3–4 months
Washington Navel, USA	2 to 7	36 to 45	90	1–2 months
Ortaniques	4 to 5	39 to 41	90	3–5 weeks
Papayas				
green	10	50	90	3–4 weeks
turning	7	45	90	2–3 weeks
Passion fruits	7 to 10	45 to 50	85–90	3–5 weeks
Peaches	−1 to 0	30 to 32	90–95	2–6 weeks
Pears				
d'Anjou	−1	30	90–95	4–6 months
Bartlett (William's Bon Chrétien)	−1 to −0.5	30 to 31	90–95	1–3 months
Beurre Bosc	−1	30	90–95	3 months
Beurre Hardy	−1 to −0.5	30 to 31	90–95	2–3 months
Clapps Favourite	−1 to 0	30 to 32	90–95	1–2 months
Comice	−1 to −0.5	30 to 31	90–95	3–4 months
Conference	−1 to −0.5	30 to 31	90–95	4–6 months
Josephine	0	32	90–95	5–6 months
Jules Guyot	−1 to −0.5	30 to 31	90–95	1–3 months
Packham's Triumph	−1	30	90–95	3–6 months
Passecrassane	−1 to 0	30 to 32	90–95	5–6 months
William's Bon Chrétien (Bartlett)	−1 to −0.5	30 to 31	90–95	1–3 months
Winter Cole	−1	30	90–95	4 months
Winter Nelis	−1 to −0.5	30 to 31	90–95	4–6 months
Pepinos	7 to 10	45 to 50	90	4–6 weeks
Persimmons	−1 to 0	30 to 32	90–95	2–4 months
Pineapples				
mature green	10 to 13	50 to 55	90	3–4 weeks
turning	7 to 10	45 to 50	90	3–4 weeks
ripe	7	45	90	2–4 weeks
Plantains				
green	10	50	85–90	5 weeks
coloured	11 to 15.5	52 to 60	85–90	1–3 weeks

Commodity Variety/Type/Origin	Temperature °C	°F	Relative Humidity %	Storage Period
Plums				
Californian,				
Casselman				
Late Santa Rosa	0	32	90–95	3–4 weeks
South African				
Gaviota, Santa Rosa	0	32	90–95	10 days followed by
	7	45	90–95	10 days
Giant Prune, Kelsey	0	32	90–95	10 days followed by
	10	50	90–95	10 days
UK				
Victoria	0 to 1	32 to 34	90–95	3 weeks
Pomegranates	5 to 6	41 to 43	90	3–4 months
Prickly pears	8 to 10	46 to 50	90	2 weeks
Pummelos	10 to 15	50 to 60	90	
Rambutans	10 to 12	50 to 54	95	1–2 weeks
Raspberries	0	32	90–95	1–7 days
Sapodillas				
turning	20	68	90	2–3 weeks
ripe	0	32	90	2 weeks
Satsumas	4 to 5	39 to 41	90	6–8 weeks
Sour sops	10 to 15	50 to 60	90	1–2 weeks
Strawberries	0	32	90–95	2–7 days
Sweet sops	10 to 15	50 to 60	90	1–2 weeks
Tamarillos	3 to 4	37 to 39	90	6–10 weeks
Tangelos	4 to 5	39 to 41	90	3–5 weeks
Tangerines				
California and Arizona	4.5 to 7	40 to 45	90	3–5 weeks
Florida and Texas	0 to 4.5	32 to 40	90	3–5 weeks
Tangors	4 to 5	39 to 41	90	3–5 weeks
Ugli fruits	4 to 5	39 to 41	90	3–5 weeks
Watermelons	10 to 15	50 to 60	90	2–3 weeks

Index

(Page numbers are in roman type, illustration numbers in **bold type**)

Acervulus, 50-51, 279, **135**
Acetobacter sp.
   on pineapples, 160-161
Acremonium sp., **92**
   on bananas, 110
Aflatoxin
   see Mycotoxins
Ageing
   in citrus, 91, **213**
Agonomycetes, 44, 45
Akaropeltopsis sp.
   on avocados, 101
Albedo browning
   of citrus, 87
Albugo candida, 283
Albugo sp., 41, **28**
Alcohol formation
   in apples, 216, **381**
   in pears, 216
Alternaria alternata, **339, 340**
   on apples, 172-173, 182-183, 213, **321, 338**
   on apricots, 220, **386**
   on blueberries, 253
   on Cape gooseberries, 166
   on carambolas, 166
   on cherries, 220
   on citrus, 56-57, **159, 160, 161**
   on figs, 168, **315**
   on gooseberries, 254
   on grapes, 261, **437**
   on mangoes, 136, **275**
   on melons, 272, **450**
   on nectarines, 220
   on papayas, 148, **291, 297**
   on passion fruits, 152
   on peaches, 220
   on pears, 172-173, 213
   on persimmons, 154, **301**
   on plums, 220
   on raspberries, 250
   on strawberries, 250
   on watermelons, 272
Alternaria citri, **162**
   on citrus, 56-57
Alternaria cucumerina
   on melons, 272
Alternaria kikuchiana, **322**
   on pears, 172-173, **320**
Alternaria mali
   on apples, 172-173
Alternaria passiflorae, **298**
   on passion fruits, 152, **299**
Alternaria sp., **122, 126**
   on avocados, 100
Anamorph (conidial state), 47, 279
Angiosorus sp., 45, **74**
Annonas
   see Sour sops, Sweet sops
Anthracnose, 34, 51, **135**
   of avocados, 94, **216, 217**
   of bananas, 106-107, **10, 232, 233**

   of cherries, 221
   of citrus, 58, **163**
   of feijoas, 167
   of grapes, 261, **438**
   of guavas, 123
   of mangoes, 130-131, **266, 267**
   of melons, 272
   of papayas, 141, **279, 280**
   of peaches, 221
   of plums, 221, **387**
   of pomegranates, 163
   of raspberries, 240
   of strawberries, 240-241, **410, 411**
   of sweet sops, 164
   of tamarillos, 169
   of watermelons, 272
Apothecium, 42, 43, 279, **68, 70**
Apples, 170-217, 286
   alcohol formation, 216, **381**
   alternaria rot, 172-173, 182-183, 213, **321**
   aspergillus rot, 201
   bacterial blister spot, 201
   bitter pit, 203, **1, 369**
   bitter rot, 174-175, **323, 324**
   black rot, 176-177, **326**
   blue mould rot, 178-179, **329, 332**
   botryodiplodia rot, 201
   Brooks fruit spot, 201
   brown core, 209, **374**
   brown heart, 28, 206, **318**
   brown rot, 180-181, **333, 334**
   brown spot, 207
   bruising, 16, 207
   calyx-end rot, 201
   carbon dioxide injury, 28, 206, **318**
   chemical injury, 216
   chilling injury, 208, **373**
   cladosporium rot, 201, 213, **378**
   core flush, 209, **374**
   core rot, 182-183, **337**
   cylindrocarpon rot, 184-185, **342**
   deep scald, 213, **378**
   external carbon dioxide injury, 216, **382**
   flyspeck, 201
   freezing injury, 210
   friction discoloration, 207
   fusarium rot, 201, **368**
   gloeosporium rot, 186-187, **345, 346**
   grey mould rot, 188-189, **349, 350**
   Jonathan breakdown, 216
   Jonathan spot, 211, **376**
   lenticel spot, 211, **376**
   low temperature breakdown, 208, **373**
   McIntosh breakdown, 217
   mouldy core, 182-183, **338**
   mucor rot, 190, **352**
   nigrospora rot, 201
   phacidiopycnis rot, 192-193
   phomopsis rot, 202
   phytophthora rot, 194-195, **357**
   pink mould rot, 191, **353**
   powdery mildew, 196, **360**
   pyrenochaeta rot, 202
   rhizopus rot, 197, **362**

   ribbon scald, 213, **378**
   ripe spot, 186-187, **345, 346**
   russeting, 217
   scab, 198-199, **363, 364**
   senescent breakdown, 212, **377**
   side rot, 200
   snow-mould rot, 202
   soft scald, 213, **378**
   sooty blotch, 202
   Spartan breakdown, 217
   stemphylium rot, 202
   sunburn, 217
   sunscald, 217, **383**
   superficial scald, 214, **379**
   trichoderma rot, 202
   vibration damage, 27, 207, **372**
   water core, 215, **380**
   white rot, 176-177
Apricots, 218-237, 286
   see also Stone fruits
Ascochyta pisi
   on citrus, 84
Ascochyta sp., 51, 84, **151, 155**
Ascomycetes, 42-43, 47, 51, 279
Aspergillus flavus
   on citrus, 84
   on guavas, 123
   on litchis, 127
   on papayas, 148
   on pineapples, 160
   on pomegranates, 163
Aspergillus niger, 160, **124, 307**
   on apples, 182, 201
   on cherries, 234
   on citrus, 59, **164**
   on figs, 168
   on grapes, 256, **431**
   on guavas, 123
   on litchis, 127
   on mangoes, 133, **271**
   on nectarines, 234
   on papayas, 148
   on passion fruits, 153
   on peaches, 234
   on pears, 201
   on pineapples, 160
   on pomegranates, 163, **310**
Aspergillus niveus
   on citrus, 84
Aspergillus sp., 47, **120, 124**
Aspergillus variecolor
   on citrus, 84
Asperisporium caricae, **292**
   on papayas, 148
Atemoyas, 164-165
Auxins, 20
Avocados, 93-103, 286
   alternaria rot, 100
   anthracnose, 94, **216, 217**
   bacterial soft rot, 100
   blotch, 95, **218, 219**
   blue mould rot, 100
   cercospora spot, 95, **218, 219**
   chilling injury, 93, 102, **228**
   'cukes', 103, **229**

dothiorella rot, 96-97, **220, 221**
fusarium rot, 100, **226**
internal darkening, 103, **230**
pestalotiopsis rot, 101
phytophthora rot, 101
pink mould rot, 101
rhizopus rot, 101, **227**
rind discoloration, 103
scab, 98, **223**
sooty blotch, 101
stem-end rots, 99, **224**

Babacos, 166, 286
Bacteria, 12, 16-20, 29, 39, **25, 26**
  see also under individual names
Bacterial black spot
  of mangoes, 132, **269, 270**
Bacterial blister spot
  of apples, 201
Bacterial canker
  of citrus, 60, **165**
Bacterial rot
  of guavas, 123
  of mangoes, 132, 136
  of melons, 272
  of pineapples, 160
  of watermelons, 272
Bacterial spot
  of apricots, 222
  of cherries, 222
  of melons, 272
  of nectarines, 222
  of passion fruits, 153
  of peaches, 222, **388, 389**
  of plums, 222
  of watermelons, 272
Bactericides, 20
Bananas, 104-121, 286
  anthracnose, 106-107, **10, 232, 233**
  black heart, 116
  black sigatoka, 114-115
  black tip, 116
  brown spot, 117
  chilling injury, 105, 120, **255**
  cigar-end rot, 108-109, **235, 236**
  corky scab, 121, **256**
  crown rot, 110-111, **239, 240**
  diamond spot, 117
  dothiorella rot, 117
  finger drop, 121
  finger rot, 112, **242**
  finger-stalk rot, 110-111, **239, 240**
  freckle, 117
  high temperature injury, 121
  Johnston spot, 113, **243**
  phytophthora finger rot, 117
  pink mould rot, 117
  pitting, 113, **243**
  premature ripening, 27, 104-105, **231**
  rhizopus rot, 117, **250**
  sigatoka, 114-115, **244, 245**
  'sinkers', 121
  sooty mould, 117, **251, 252**
  speckle, 118
  splitting, 121
  squirter, 119
  swamp spot, 118
  watery soft rot, 119
Basidiomycetes, 44-45, 279
Benomyl (Benlate), 20

Berry drop
  of grapes, 266, **445**
Biological control, 20
  of *Botrytis cinerea*, 189, 258-259
  of *Monilinia* spp., 225
  of *Penicillium expansum*, 179
  of *Rhizoctonia solani*, 276
  of *Rhizopus stolonifer*, 230
  of *Venturia inaequalis*, 199
  of *Xanthomonas campestris* pv.
    mangiferaeindicae, 132
Biphenyl (diphenyl), 20
Bitter pit
  of apples, 203, **1, 369**
  of pears, 204, **370**
Bitter rot
  of apples, 174-175, **323, 324**
  of cherries, 221
  of grapes, 262, **440**
  of peaches, 221
  of pears, 174-175
  of plums, 221, **387**
Black canker
  of sweet sops, 164, **311**
Black currants, 254
Black-end
  of pears, 216
Black heart
  of bananas, 116
  of pineapples, 162
Black mould rot
  of apples, 201
  of cherries, 234
  of citrus, 59, **164**
  of grapes, 256, **431**
  of mangoes, 133, **271**
  of nectarines, 234
  of peaches, 234
  see also Aspergillus niger
Black pit
  of citrus, 61, **166**
Black rot
  of apples, 176-177, **326**
  of cranberries, 253
  of grapes, 263
  of papayas, 142-143, **281, 282**
  of pears, 176-177
  of pineapples, 156-157, **302**
Black seed
  of strawberries, 250
Black sigatoka
  of bananas, 114-115
Black spot
  of citrus, 62-63, **167, 168**
  of papayas, 148
  of raspberries, 240
  of strawberries, 240-241, **410, 411**
Black tip
  of bananas, 116
  of mangoes, 139
Blossom-end rot
  of watermelons, 278
Blotch, 35
  of avocados, 95, **218, 219**
  see also Sooty blotch
Blueberries, 253, 287
Blue mould rot
  of apples, 178-179, **329, 332**
  of apricots, 223
  of avocados, 100

  of cherries, 223
  of citrus, 70-72, **178, 179**
  of grapes, 257, **432**
  of kiwifruits, 269, **447**
  of mangoes, 136
  of melons, 272
  of nectarines, 223
  of papayas, 149, **294**
  of peaches, 223, **391**
  of pears, 178-179, **330**
  of plums, 223, **390**
  of raspberries, 250
  of strawberries, 250
  of watermelons, 272
Botryodiplodia sp., 51, **154, 158**
Botryodiplodia theobromae, 283, **158, 197**
  on apples, 201
  on avocados, 99
  on bananas, 110-111, 112, **242**
  on carambolas, 166, **313**
  on citrus, 80-81, **195**
  on grapes, 263
  on guavas, 123
  on litchis, 127
  on loquats, 168
  on mangoes, 134, 135, **272**
  on mangosteens, 168, **317**
  on melons, 272
  on papayas, 147
  on peaches, 234
  on pears, 201
  on pineapples, 160
  on rambutans, 169
  on sour sops, 165
  on sweet sops, 165, **312**
  on watermelons, 272
Botryosphaeria berengeriana
  syn. *Botryosphaeria ribis*
Botryosphaeria dothidea
  on apples, 176-177
  on peaches, 234
Botryosphaeria obtusa, **328**
  on apples, 176-177, **326**
  on peaches, 234
  on pears, 176-177
Botryosphaeria ribis, **222**
  on apples, 176-177
  on avocados, 96-97, **220, 221**
  on bananas, 117
  on citrus, 80-81
  on grapes, 264
  on pears, 176-177, **327**
Botryosphaeria sp., **53**
  on kiwifruits, 269
Botryotinia fuckeliana, 258-259
  see also Botrytis cinerea
Botrytis cinerea, **435**
  on apples, 188-189, **349, 350**
  on apricots, 227
  on blueberries, 253
  on cherries, 227, **396**
  on citrus, 73, **185**
  on feijoas, 167
  on figs, 168
  on grapes, 258-259, **433, 434**
  on kiwifruits, 268, **446**
  on melons, 274, **454**
  on nectarines, 227
  on peaches, 227
  on pears, 188-189, **351**

on plums, 227
on raspberries, 244-245, **417, 418**
on strawberries, 244-245, **416**
see also Grey mould rot
Botrytis sp., **96, 104**
on persimmons, 154
on pomegranates, 163
Breadfruits, 166, 287
Breakdown
of pears, 205, **371**
Bremia sp., 41, **31**
Bronzing
of citrus, 90
of melons, 278
Brooks fruit spot
of apples, 201
Brown core
of apples, 209, **374**
of pears, **319**
Brown heart
of apples, 28, 206, **318**
of pears, 206
Brown pitting
of citrus, 87, **208**
Brown rot
of apples, 180-181, **333, 334**
of apricots, 224-225
of cherries, 224-225
of citrus, 64-65, **170, 171**
of nectarines, 224-225
of peaches, 224-225, **392**
of pears, 180-181, **335**
of plums, 224-225, **393**
Brown spot
of apples, 207
of bananas, 117
of passion fruits, 152, **299**
of pears, 207
Brown staining
of citrus, 87, 90
Bruising
of apples, 16, 207
of cherries, 236
of peaches, 236
of pears, 207

Calcium treatment, 20, 30, 33
Calyx blotch
of strawberries, 242-243, **413, 414**
Calyx-end rot
of apples, 201
of pears, 201
Calyx separation
in persimmons, 154
Candida krusei
on citrus, 86
Canker, 35
of citrus, 60, **165**
of guavas, 123, **258**
Cantaloupes, 270
Cape gooseberries, 166, 287
Capnodium sp.
on citrus, 86
on mangoes, 137
Carambolas, 166, 287
Carbon dioxide injury
of apples, 28, 206, 216, **318, 382**
of pears, 206
Cargo claims, 7, 26, 28
Cargo surveys, 26-28

Cave storage, 22
Ceratocystis fimbriata, 283
on citrus, 84
Ceratocystis paradoxa, **303**
on bananas, 110-111
on guavas, 125
on pineapples, 156-157, **302**
Ceratocystis sp., 43, **50**
Cercospora angolensis
on citrus, 84
Cercospora hayi
on bananas, 117
Cercospora pappaea, **293**
on papayas, 149
Cercospora purpurea,
syn. *Pseudocercospora purpurea*
Cercospora sp., **117**
Ceuthospora lunata
on cranberries, 253
Chalaropsis sp., **108**
Charcoal rot
of citrus, 84
of mangoes, 136
of melons, 273
of watermelons, 273
Chemical injury
of apples, 216
of citrus, 90, **211**
of pears, 216
Chemical treatments, 20
Cherimoyas, 164-165, 287
Cherries, 218-237, 287, **385**
see also Stone fruits
Chilling injury, 16, 32, 279
of apples, 208, **373**
of avocados, 93, 102, **228**
of babacos, 166
of bananas, 105, 120, **255**
of breadfruits, 166
of carambolas, 166
of citrus, 27, 87, **7, 208**
of guavas, 122
of litchis, 126
of mangoes, 128, 138, **277**
of melons, 278, **449**
of papayas, 140, 151, **297**
of passion fruits, 152
of pepinos, 169
of persimmons, 154
of pineapples, 155, 162, **309**
of plums, 218-219
of pomegranates, 163
of rambutans, 169
of sour sops, 164
of sweet sops, 164
of watermelons, 278
Chinese gooseberries
see Kiwifruits
Choanephora cucurbitacearum
on guavas, 123
Choanephora sp., 41, **34**
Cigar-end rot
of bananas, 108-109, **235, 236**
Citrus, 54-91
ageing, 91, **213**
albedo browning, 87
alternaria rot, 56-57, **159, 160**
anthracnose, 58, **163**
aspergillus rot, 59, 84, **164**
bacterial canker, 60, **165**

black mould rot, 59, **164**
black pit, 61, **166**
black spot, 62-63, **167, 168**
blue mould rot, 70-72, **178, 179**
bronzing, 90
brown pitting, 87, **208**
brown rot, 64-65, **170, 171**
brown staining, 87, 90
ceratocystis rot, 84
cercospora spot, 84
charcoal rot, 84
chemical injury, 90, **211**
chilling injury, 27, 87, **7, 208**
cottony rot, 66, **174**
creasing, 90
deformation, 90
degreening, 20, 55
freckle, 84
freezing injury, 88, **209**
fusarium rot, 67, **175**
granulation, 90
greasy spot rind blotch, 68-69, **176**
green mould rot, 70-72, **178, 181, 184**
grey mould rot, 73, **185**
melanose, 74-75, **186, 187**
membranous stain, 87
oil gland darkening, 87
oleocellosis, 89, **210**
peteca, 90, **212**
pink mould rot, 84, **203**
pitting, 87, **7, 208**
pleospora rot, 85, **204**
puffiness, 90
rhizopus rot, 85, **205**
rind-oil spotting, 89, **210**
rind-staining, 87, 90
russeting, 90
scab, 76, **19, 189**
scald, 87
scale insects, 91
septoria spot, 77, **191**
soiling, 20, 55, 71, **181**
sooty blotch, 86
sooty mould, 86, **207**
sour rot, 78-79, **192, 193**
stem-end rind breakdown, 91, **213**
stem-end rots, 80-81, **195, 196**
stylar-end breakdown, 91
thrips damage, 91, **214**
trichoderma rot, 82-83, **198, 199**
watery breakdown, 87
yeasty rot, 86
Cladosporium cladosporioides
on bananas, 118, **251, 252**
on Cape gooseberries, 166
on carambolas, 166
on melons, 273
on raspberries, 250
Cladosporium cucumerinum
on melons, 273
Cladosporium herbarum
on apples, 201, 213, **378**
on apricots, 226, **395**
on cherries, 226
on grapes, 263, 443
on melons, 273
on nectarines, 226
on papayas, 149, **295**
on peaches, 226
on pears, 201, 213

on plums, 226
on raspberries, 250
Cladosporium sp., 47, **121, 125**
   on citrus, 86
   on dates, **314**
   on figs, 168
   on passion fruits, 153, **300**
   on persimmons, 154
   on pomegranates, 163
Cladosporium tenuissimum
   on watermelons, 273
Cleistothecium, 42, 43, 279, **65**
Clementines, 54, 287
   see also Citrus
Climacteric, 14-15, 279
Coatings, 20
Cochliobolus lunatus
   on Cape gooseberry, 166
Coelomycetes, 50-53
Colletotrichum acutatum, **412**
   on strawberries, 240-241, **410, 411**
Colletotrichum circinans, 283
Colletotrichum dematium
   on strawberries, 240-241
Colletotrichum fragariae
   on strawberries, 240-241
Colletotrichum gloeosporioides, 51, **325**
   see also Glomerella cingulata
Colletotrichum gloeosporioides f. sp.
   cucurbitae
   on melons, 272
   on watermelons, 272
Colletotrichum lagenarium
   syn. *Colletotrichum gloeosporioides* f. sp.
     cucurbitae
Colletotrichum musae, 51, **234**
   on bananas, 106-107, 110-111, **232, 233**
Colletotrichum orbiculare
   syn. *Colletotrichum gloeosporioides* f. sp
     cucurbitae
Colletotrichum sp., 51, **129, 135, 136**
   on blueberries, 253
   on kiwifruits, 269
   on persimmons, 154
   on tamarillos, 169
Compatibility, 25
Conidial state (anamorph), 47, 279
Coniella diplodiella
   on grapes, 264
Coniella granati
   on pomegranates, 163
Container transport, 24-26
Controlled atmosphere (CA), 279
   storage, 23
   transport, 25
Cooling, 20-21
Coprinus psychromorbidus
   on apples, 202
   on pears, 202
Core breakdown
   of pears, 205
Core flush
   of apples, 209, **374**
Core rot
   of apples, 182-183, **337**
Corky scab
   of bananas, 121, **256**
Corky spot
   of pears, 204, **370**
Corticium rolfsii, 45

on guavas, 125
on melons, 276
on pears, 202
on watermelons, 276
see also Sclerotium rolfsii
Corticium sp., 45, **79**
Corynespora cassiicola
   on papayas, 150
Corynespora sp., **119**
Cottony leak
   of melons, 273, **452**
   of watermelons, 273
Cottony rot
   of citrus, 66, **174**
   of strawberries, 250
Cracking
   in cherries, 236
   in grapes, 266
Cranberries, 253, 287, **429**
Creasing
   of citrus, 90
Crown mould
   of bananas, 110-111
Crown rot
   of bananas, 110-111, **239, 240**
Cryptosporiopsis malicorticis
   syn. *Gloeosporium perennans*
'Cukes', 103, **229**
Curing, 19
Currants, 253-254, 287
Curvularia sp., **118**
Curvularia tuberculata
   on guavas, 123
Custard apples, 164-165, 287
Cylindrocarpon mali
   on apples, 184-185, **342**
   on pears, 184-185, **343**
Cylindrocarpon sp., **98**
Cylindrocladium scoparium
   on guavas, 124, **260**
Cylindrosporium pomi
   on apples, 201
Cylindrosporium sp., **130**

Dates, 166-167, 287
Deep scald
   of apples, 213, **378**
   of pears, 213
Deformation
   of grapefruits, 90
Degreening of citrus, 20, 55
Deightoniella sp., **112**
Deightoniella torulosa, **253**
   on bananas, 116, 118
Dendrophoma obscurans
   syn. *Phomopsis obscurans*
Deterioration, 11-12, 16
Deuteromycetes, 44-53, 279
Diagnosis, 12, 29
Diaporthe actinidiae
   on kiwifruits, 269
Diaporthe citri, **188**
   on citrus, 74-75, 80-81, **186, 187, 196**
Diaporthe melonis
   on melons, 275
Diaporthe perniciosa
   on apples, 202
   on kiwifruits, 269
   on peaches, 234
   on plums, 234

Diaporthe phaseolorum
   on tamarillos, 169
Diaporthe sp., 43, **48**
Dichloran (Dicloran), 20
Didymella applanata
   on raspberries, 251
Didymella bryoniae
   on melons, 273, **451**
   on watermelons, 273
Didymella sp., **55**
Diphenyl (biphenyl), 20, 55
Diplodia natalensis
   syn. *Botryodiplodia theobromae*
Diplodia sp.
   on apples, 201
Diplodina persicae
   on peaches, 234
Discomycetes, 42
Disease, 16-17, 18, 29, 34-39, 279
Disorder, 17, 29, 30-33, 279
Dothiorella gregaria
   see Botryosphaeria ribis
Dothiorella sp., **139**
   on avocados, 99
   on mangoes, 135
Downy mildew, 41
   of grapes, 264
Durians, 167, 287

Elsinoë ampelina, **439**
   on grapes, 261
Elsinoë australis
   on citrus, 76
Elsinoë fawcettii
   on citrus, 76
Elsinoë mangiferae
   on mangoes, 137
Elsinoë sp., 43, **59**
Endogenous brown spot
   of pineapples, 162
Endosepsis
   of figs, 168
Enterobacter cloacae
   on papayas, 150
Enterobacter sp.
   on pineapples, 160-161
Erwinia ananas
   on melons, 272
   on pineapples, 158-159
   on watermelons, 272
Erwinia carotovora subsp. *carotovora*
   on avocados, 100
   on mangoes, 136
   on melons, 272
   on watermelons, 272
Erwinia chrysanthemi
   on pineapples, 160
Erwinia herbicola
   on mangoes, 136
   on melons, 272
   on papayas, 150
   on watermelons, 272
Erwinia sp.
   on guavas, 123
Erysiphe sp., 43, **42**
Etaconazole, 20
Ethylene, 280
   degreening of citrus, 20, 55
   disorders caused by, 33
   effect on climacteric fruits, 14

effect on non-climacteric fruits, 20
effect on vegetables, 15
production by extraneous sources, 15, 23
production by moulds, 17, 27, 55
production by plant tissues, 15, 16, 17
removal of, 23, 25, 105
sensitivity to, 21, 55, 267
see also under individual commodities
External carbon dioxide injury
of apples, 216, **382**

Feijoas, 167, 287
Figs, 167-168, 287
Film packaging, 21
Finger drop
of bananas, 121
Finger rot
of bananas, 112, **242**
Finger-stalk rot
of bananas, 110-111, **239, 240**
Flyspeck
of apples, 201
Forced-air cooling, 21
Forensic plant pathology, 26
Fosetyl aluminium, 20
Freckle
of bananas, 117
of citrus, 84
Freckle spot
of citrus, 62
Freezing injury, 31, 280
of apples, 210
of apricots, 236
of cherries, 236
of citrus, 88, **209**
of grapes, 266
of nectarines, 236
of peaches, 236, **406**
of pears, 210, **375**
of plums, 236
Friction discoloration
of apples, 207
of pears, 207
Fruitlet core rot
of pineapples, 158-159, **304, 305**
Fungi
as agents of disease, 39
classification, 7, 40-53
illustrated genera, 40-53
see also under individual names
Fungi Imperfecti (Deuteromycetes), 47-51
Fungicides, 20
Fusarium avenaceum
on apples, 201
on pears, 201
Fusarium culmorum
on apples, 201
on pears, 201
Fusarium equiseti, **296**
on Cape gooseberries, 166
on guavas, 124
on melons, 274
on papayas, 150
on watermelons, 274
Fusarium lateritium
on apples, 201
on pears, 201
Fusarium lunulosporum
on citrus, 67
Fusarium moniliforme

on bananas, 116
on citrus, 67
see also Gibberella fujikuroi
Fusarium moniliforme var. *fici*
on figs, 168
Fusarium moniliforme var. *subglutinans*
on pineapple, 158
Fusarium oxysporum
on citrus, 67
on guavas, 124
on melons, 274
on papayas, 150
on passion fruits, 153
on watermelons, 274
Fusarium pallidoroseum, **241**
on avocados, 100
on bananas, 110-111
on melons, 274
on watermelons, 274
Fusarium semitectum, **241**
syn. *Fusarium pallidoroseum*
Fusarium solani
on apples, 201
on citrus, 67
on guavas, 124
on melons, 274
on papayas, 150
on pears, 201
on watermelons, 274
Fusarium sp., **99, 105, 107**
on apples, 182, **368**
on avocados, 100, **226**
on bananas, 117
on watermelons, **453**
Fusicladium carpophilum
see Venturia carpophila
Fusicladium cerasi
see Venturia cerasi
Fusicladium pyrorum
see Venturia pirina
Fusicladium sp., **111**

Geotrichum candidum, **194**
on citrus, 78-79, **192, 193**
on litchis, 127
on melons, 277, **457**
on peaches, 235
on plums, 235
Geotrichum sp., **91**
Gibberella fujikuroi, **248**
on bananas, 116
Gibberella fujikuroi var. *subglutinans*, **306**
on pineapples, 158-159
Gibberella sp., **52**
Gibberellins, 20
Gliocephalotrichum bulbilium
on rambutans, 169
Gliocladium roseum
on sweet sops, 165
Gloeodes pomigena
on apples, 202
on citrus, 86
on mangoes, 137
on pears, 202
Gloeosporium album, **347**
on apples, 186-187
on pears, 186-187
Gloeosporium foliicolum
on citrus, 58
Gloeosporium fructigenum, 174

syn. *Colletotrichum gloeosporioides*
Gloeosporium perennans, **348**
on apples, 186-187
on pears, 186-187
Gloeosporium sp., **131, 137**
Glomerella cingulata, 51, **268**
on apples, 174-175
on avocados, 94
on bananas, 110-111
on cherries, 221
on citrus, 58
on grapes, 261
on guavas, 123
on mangoes, 130-131
on peaches, 221
on plums, 221
on sweet sops, 164
Glomerella sp., 43, **46**
Gnomonia comari, **415**
syn. *Gnomonia fructicola*
Gnomonia fructicola, **415**
on strawberries, 242-243
Gnomonia sp., **51**
Gooseberries, 238, 254, 287
see also Cape gooseberries
Granadillas, 287
see also Passion fruits
Granulation
in citrus, 90
Grapefruits, 27, 54-55, 90, 287
see also Citrus
Grapes, 255-266, 287
alternaria rot, 261, **437**
anthracnose, 261, **438**
aspergillus rot, 256, **431**
berry drop, 266, **445**
bitter rot, 262, **440**
black rot, 263
blue mould rot, 257, **432**
botryodiplodia rot, 263
cladosporium rot, 263, **443**
coniella rot, 264
cracking, 266
downy mildew, 264
grey mould rot, 258-259, **433, 434**
phomopsis rot, 264
physiological browning, 266
powdery mildew, 264
rhizopus rot, 260, **436**
ripe rot, 262
'slip-skin', 258
sour rot, 264
splitting, 266
sulphur dioxide injury, 265, **444**
Grease spot
of passion fruits, 153
Greasy spot
of papayas, 150
Greasy spot rind blotch
of citrus, 68-69, **176**
Green mould rot
of citrus, 70-72, **178, 181, 184**
Greeneria sp., **132**
Greeneria uvicola, **441**
on grapes, 262
Grey mould rot
of apples, 188-189, **349, 350**
of apricots, 227
of cherries, 227, **396**
of citrus, 73, **185**

of feijoas, 167
of figs, 168
of gooseberries, 254
of grapes, 258-259, **433, 434**
of kiwifruits, 268, **446**
of melons, 274, **454**
of nectarines, 227
of peaches, 227
of pears, 188-189, **351**
of plums, 227
of raspberries, 244-245, **417, 418**
of strawberries, 244-245, **416**
see also Botrytis cinerea
Growth-regulating chemicals, 18, 20
see also Ethylene
Guavas, 122-125, 287
anthracnose, 123
aspergillus rot, 123
bacterial soft rot, 123
botryodiplodia rot, 123
canker, 123, **258**
choanephora rot, 123
curvularia rot, 123
cylindrocladium rot, 124, **260**
dothiorella rot, 124
fusarium rot, 124
macrophoma rot, 124
macrophomina rot, 124
mucor rot, 124
phoma rot, 124
phomopsis rot, 124, **261**
phyllosticta rot, 124
phytophthora rot, 125
rhizopus rot, 125
sclerotium rot, 125
thielaviopsis rot, 125
Guazatine, 20
Guignardia bidwellii, **442**
on grapes, 263
Guignardia citricarpa, **169**
on citrus, 62-63, **167, 168**
Guignardia mangiferae
on mangoes, 136
Guignardia musae, **249**
on bananas, 117
Guignardia psidii
on guavas, 124-125
Guignardia sp., 43, **54**

Hainesia lythri, **428**
on strawberries, 252
Harvesting, 19
Helicobasidium sp., **77**
Helminthosporium sp., **123, 127**
Hemibasidiomycetes, 44
Hendersonia creberrima
on mangoes, 136, **274**
High temperature injury
of bananas, 121
Horned melons (kiwanos), 168, 270, 288
Hot water treatment, 20
Humidity
see Relative humidity
Hydro-cooling, 21
Hymenomycetes, 44
Hyphomycetes, 46-49
Hypobaric storage, 23

Identification procedures, 29
Imazalil, 20

Infection, 16-17, 280
Insect injury
of citrus, 91, **214**
of nectarines, 236, **407**
Insects, 20, 33
Integrated Pest Management (IPM), 18
Interfruitlet corking
in pineapples, 160
Internal blight
of papayas, 149
Internal breakdown
of apricots, 236
of mangoes, 139
of nectarines, 236
of peaches, 236
of pears, 205
of plums, 236
Internal browning
of apricots
of cherries, 236
of pineapples, 162, **309**
of plums, 236
Internal darkening
of avocados, 103, **230**
Internal yellowing
of papayas, 150
Iprodione, 20
Irradiation, 20
Itersonilia sp., **80**

Jackfruits, 168, 288
Jelly seed
of mangoes, 139, **278**
Johnston spot
of bananas, 113, **243**
Jonathan breakdown
of apples, 216
Jonathan spot
of apples, 211, **376**

Kaki, 154
Khuskia oryzae, **254**
on bananas, 119
Kiwanos (horned melons), 168, 270, 288
Kiwifruits, 267-269, 288
blue mould rot, 269, **447**
grey mould rot, 268, **446**
phomopsis rot, 269
premature ripening, 267
ripe rot, 269
Koch's postulates, 29

Lasiodiplodia theobromae
syn. *Botryodiplodia theobromae*
Leak
of raspberries, 246-247
of strawberries, 246-247, **419, 420**
Leather rot
of strawberries, 248-249, **422, 423**
Leathery pocket
of pineapples, 160
Lemons, 54-55, 288
see also Citrus
Lenticel spot
of apples, 211, **376**
Leptosphaeria sp., **58**
Limes, 54-55, 288
see also Citrus
Litchis, 126-127, 288
aspergillus rot, 127

chilling injury, 126
peronophythora rot, 127
pestalotiopsis rot, 127, **263, 264**
physiological browning, 126, **262**
sour rot, 127
yeasty rot, 127
Loculoascomycetes, 42
Longans, 168, 288
Loquats, 168, 288
Losses
see Post-harvest losses
Low temperature breakdown
of apples, 208, **373**
Low temperature injury
see Chilling injury
Lychees
see Litchis

McIntosh breakdown
of apples, 217
Macrophoma allahabadensis
on guavas, 124
Macrophoma mangiferae
on mangoes, 136
Macrophoma sp., **142**
Macrophomina phaseolina
on citrus, 84
Macrophomina sp.
on guavas, 124
on mangoes, 136
on melons, 273
on watermelons, 273
Magnaporthe grisea
on bananas, 113
Mandarins, 54-55, 288
see also Citrus
Mangoes, 128-139, 288, **265**
alternaria black spot, 136, **275**
anthracnose, 130-131, **266, 267**
bacterial black spot, 132, **269, 270**
bacterial rot, 132, 136
black mould rot, 133, **271**
black tip, 139
blue mould rot, 136
botryodiplodia rot, 134, **272**
charcoal rot, 136
chilling injury, 138, **277**
hendersonia rot, 136, **274**
internal breakdown, 139
jelly seed, 139, **278**
macrophoma rot, 136
mucor rot, 136
pestalotiopsis rot, 136
phyllosticta rot, 136
phytophthora rot, 137
powdery mildew, 137
rhizopus rot, 137, **276**
scab, 137
soft nose, 139
sooty blotch, 137
sooty mould, 137
spongy tissue, 139
stem-end breakdown, 139
stem-end rots, 135, **273**
Mangosteens, 168, 288
botryodiplodia rot, **317**
Melanconium fuligineum
syn. *Greeneria uvicola*
Melanose
of citrus, 74-75, **186, 187**

Meliola sp.
  on citrus, 86
  on mangoes, 137
Melons, 270-278, 288
  alternaria rot, 272, **450**
  anthracnose, 272
  bacterial rots, 272
  blossom-end rot, 278
  blue mould rot, 272
  botryodiplodia rot, 272
  charcoal rot, 273
  chilling injury, 278, **449**
  cladosporium rot, 273
  cottony leak, 273, **452**
  didymella black rot, 273, **451**
  fusarium rot, 274, **453**
  grey mould rot, 274, **454**
  myrothecium rot, 275
  phomopsis rot, 275
  phytophthora rot, 275
  pink mould rot, 275, **455**
  premature senescence, 278
  pythium rot, 273, **452**
  rhizopus rot, 276, **456**
  sclerotium rot, 276
  soil rot, 276
  solar injury, 278
  sour rot, 277, **457**
  vein-tract browning, 278
  watery soft rot, 277
Metalaxyl, 20
Microdochium sp., **100**
Mildew, 36
  see also Downy mildew, Powdery mildew
Minneolas, 54, 288
  see also Citrus
Modified atmosphere (MA), 280
  packaging, 21
  storage, 23
  transport, 24, 25
Moisture loss, 13-14, 16, 20, 22
Monilia implicata
  on peaches, 235
Monilia sp., **89**
  see also Monilinia spp.
Monilinia fructicola, **394**
  on apples, 180-181
  on apricots, 224-225
  on cherries, 224-225
  on nectarines, 224-225
  on peaches, 224-225
  on pears, 180-181
  on plums, 224-225
Monilinia fructigena, **336**
  on apples, 180-181, **333, 334**
  on apricots, 224-225
  on cherries, 224-225
  on nectarines, 224-225
  on peaches, 224-225, **392**
  on pears, 180-181, **335**
  on plums, 224-225, **393**
Monilinia laxa
  on apples, 180-181
  on apricots, 224-225
  on cherries, 224-225
  on nectarines, 224-225
  on peaches, 224-225
  on pears, 180-181
  on plums, 224-225
Monilinia sp., **63**

on blueberries, 253
Mould, 36, 280
Mouldy core
  of apples, 182-183, **338**
Mucor hiemalis
  on guavas, 124
Mucor piriformis
  on apples, 190, **352**
  on nectarines, 234
  on peaches, 234
  on pears, 190
Mucor sp., **35**
  on mangoes, 136
  on persimmons, 154
Muskmelons, 270
Mycocentrospora sp., **101**
Mycosphaerella caricae
  on papayas, 142-143, **281, 282**
Mycosphaerella citri, **177**
  on citrus, 68-69, **176**
Mycosphaerella fijiensis, **247**
  on bananas, 114-115
Mycosphaerella fijiensis var. *difformis*
  on bananas, 114-115
Mycosphaerella fragariae, **425**
  on strawberries, 250
Mycosphaerella musicola, **246**
  on bananas, 114-115
Mycosphaerella pomi
  on apples, 201
Mycosphaerella sp., **56**
  on avocados, 95
  on papayas, 147
Mycotoxins, 33, 280
  see also Toxins
Myrothecium roridum
  on melons, 275
  on watermelons, 275
Myrothecium sp., **94**

Naranjillas, 169, 288
Nashi (oriental pears), 170-171
Nectarines, 218-237, 288
  see also Stone fruits
Nectria galligena, **344**
  on apples, 184-185, **342**
  on pears, 184-185, **342, 343**
Nectria sp., **49, 66**
Nigrospora oryzae, 119, **254**
Nigrospora sp., **110**
Nigrospora sphaerica
  on apples, 201
  on bananas, 119
  see also Nigrospora oryzae
Nutritional disorders, 30

Oidium mangiferae
  on mangoes, 137
Oidium sp. **88**
  on apples, 196
  on gooseberries, 254
Oleocellosis
  in citrus, 89, **210**
Oomycetes, 40, 41
Oranges, 54-55, 288-289
  see also Citrus
Oriental pears (nashi), 170-171
Ortaniques, 54, 289
  see also Citrus

Packaging, 21, 28
Papayas, 140-151, 289
  alternaria rot, 148, **291, 297**
  anthracnose, 141, **279, 280**
  aspergillus rot, 148
  black rot, 142-143, **281, 282**
  black spot, 148
  blue mould rot, 149, **294**
  cercospora spot, 149
  chilling injury, 151, **297**
  cladosporium rot, 149, **295**
  fusarium rot, 150
  greasy spot, 150
  internal blight, 149
  internal yellowing, 150
  phytophthora rot, 144-145, **284, 285**
  pink mould rot, 150
  purple stain, 150
  rhizopus rot, 146, **287, 288**
  stem-end rot, 147, **289**
  stemphylium rot, 150
Passion fruits, 152-153, 289
  bacterial spot, 153
  brown spot, 152, **299**
  chilling injury, 152
  cladosporium rot, 153, **300**
  grease spot, 153
  septoria spot, 152
Patterns of damage, 27-28, 29
Patulin, 178, 257
Pawpaws
  see Papayas
Peaches, 218-237, 289, **384**
  see also Stone fruits
Pears, 170-217, 289
  alcohol formation, 216
  alternaria rot, 172-173, **320**
  aspergillus rot, 201
  bitter pit, 204, **370**
  bitter rot, 174-175
  black-end, 216
  black rot, 176-177
  blue mould rot, 178-179, **330**
  botryodiplodia rot, 201
  breakdown, 205, **371**
  brown core, **319**
  brown heart, 206
  brown rot, 180-181, **335**
  brown spot, 207
  bruising, 207
  calyx-end rot, 201
  carbon dioxide injury, 206
  chemical injury, 216
  cladosporium rot, 200, 201
  core breakdown, 205
  corky spot, 204, **370**
  cylindrocarpon rot, 184-185, **343**
  deep scald, 213
  freezing injury, 210, **375**
  friction discoloration, 207
  fusarium rot, 201
  gloeosporium rot, 186-187
  grey mould rot, 188-189, **351**
  internal breakdown, 205
  mucor rot, 190
  phacidiopycnis rot, 192-193, **354, 355**
  phytophthora rot, 194-195, **358**
  pink mould rot, 191
  powdery mildew, 196
  rhizopus rot, 197

297

ribbon scald, 213
russeting, 217
scab, 198-199
sclerotium rot, 202
senescent scald, 205
side rot, 200, **367**
snow-mould rot, 202
soft scald, 213
sooty blotch, 202
superficial scald, 214
trichoderma rot, 202
vascular browning, 205, **371**
vibration damage, 207
water core, 215
white rot, 176-177, **327**
Penicillium canescens
 on grapes, 257
Penicillium citrinum
 on grapes, 257
Penicillium crustosum
 on melons, 272
Penicillium cyclopium
 on grapes, 257
 on mangoes, 136
 on melons, 272
 on strawberries, 250
Penicillium digitatum, **183**
 on citrus, 70-72, **178, 181, 184**
Penicillium expansum, **331**
 on apples, 178-179, **329, 332**
 on apricots, 223
 on avocados, 100
 on cherries, 223
 on grapes, 257
 on melons, 272
 on nectarines, 223
 on passion fruits, 153
 on peaches, 223, **391**
 on pears, 178-179, **330**
 on plums, 223, **390**
 on strawberries, 250
Penicillium funiculosum
 on apples, 182
 on pineapples, 158-159, 160
Penicillium italicum, **180, 182**
 on citrus, 70-72, **178, 179**
Penicillium pallidum
 on melons, 272
Penicillium sp., **90, 103**
 on figs, 168
 on kiwifruits, 269, **447**
 on papayas, 149, **294**
 on persimmons, 154
 on pomegranates, 163
 on raspberries, 250
Penicillium viridicatum
 on melons, 272
Pepinos, 169, 289
Perfect state (teleomorph), 47, 281
Perithecium, 42, 43, 280, **66, 69**
Peronophythora litchii
 on litchis, 127
Peronospora sp., 41, **33, 39**
Persimmons, 154, 289
 alternaria rot, 154, **301**
 calyx separation, 154
 chilling injury, 154
Peruvian cherries
 see Cape gooseberries
Pestalotia longisetula

 on strawberries, 251
Pestalotiopsis mangiferae
 on mangoes, 136
Pestalotiopsis psidii, **259**
 on guavas, 123, **258**
Pestalotiopsis sp., **134, 138**
 on litchis, 127, **263, 264**
 on sapodillas, 169
Pestalotiopsis versicolor
 on avocados, 101
 on pomegranates, 163
Peteca
 of citrus, 90, **212**
Pezicula alba
 on apples, 186-187
 on pears, 186-187
Pezicula malicorticis
 on apples, 182, 186-187
 on pears, 186-187
Pezicula sp., **61**
Pezizella oenotherae, **428**
 on strawberries, 252
Phacidiopycnis malorum, **356**
 on apples, 192-193
 on pears, 192-193, **354, 355**
Phacidiopycnis pyri
 syn. *Phacidiopycnis malorum*
Phacidiopycnis sp., **146, 150**
Phaeoisariopsis sp., **114**
Phaeoramularia angolensis, **202**
 on citrus, 84
Phialophora malorum
 on apples, 200
 on pears, 200, **367**
Phlyctaena vagabunda
 syn. *Gloeosporium album*
Phoma caricae, **283**
 syn. *Phoma caricae-papayae*
Phoma caricae-papayae, **283**
 on papayas, 142-143
Phoma cucurbitacearum
 on melons, 273
 on watermelons, 273
Phoma exigua
 on tamarillos, 169
Phoma psidii
 on guavas, 124
Phoma sp., 51, **140, 147, 148**
 on kiwifruits, 269
 on persimmons, 154
 on pomegranates, 163
 on raspberries, 251
Phomopsis actinidiae
 on kiwifruits, 269
Phomopsis anonacearum
 on sweet sops, 164, **311**
Phomopsis caricae-papayae, **290**
 on papayas, 147
Phomopsis citri, **188**
 on citrus, 74-75, 80-81, **186, 187, 196**
 on mangoes, 135
Phomopsis cucurbitae
 on watermelons, 275
Phomopsis destructum
 on guavas, 124
Phomopsis mali
 on apples, 182, 202
 on peaches, 234,
 on plums, 234, **404**
Phomopsis obscurans, **427**

 on strawberries, 251
Phomopsis perseae
 on avocados, 99
Phomopsis psidii
 on guavas, 124
Phomopsis sp., 51, **144, 149**
 on apples, 182
 on kiwifruits, 269
 on papayas, 147
 on pomegranates, 163
 on sapodillas, 169
Phomopsis vaccinii
 on blueberries, 253
Phomopsis viticola
 on grapes, 264
Phragmidium rubi-idaei
 on raspberries, 252
Phragmidium sp., **75**
Phycomycetes, 41
Phyllosticta ampelicida
 on grapes, 263
Phyllosticta citricarpa
 on citrus, 62
Phyllosticta mangiferae
 on mangoes, 136
Phyllosticta musarum
 on bananas, 117
Phyllosticta psidiicola
 on guavas, 124-125
Phyllosticta sp., **283, 141**
Physalospora rhodina
 on citrus, 80-81
 on mangoes, 134
Physiological browning
 of grapes, 266
 of litchis, 126, **262**
Phytophthora cactorum, **424**
 on apples, 194-195
 on feijoas, 167
 on melons, 275
 on pears, 194-195
 on strawberries, 248-249, **422, 423**
 on watermelons, 275
Phytophthora capsici
 on melons, 275
 on watermelons, 275
Phytophthora citricola, **172**
 on avocados, 101
 on citrus, 64-65
 on feijoas, 167
 on guavas, 125
 on watermelons, 275
Phytophthora citrophthora, **173**
 on citrus, 64-65
 on strawberries, 248-249
Phytophthora drechsleri
 on melons, 275
 on watermelons, 275
Phytophthora hibernalis
 on citrus, 64-65
Phytophthora nicotianae var. *parasitica*
 on bananas, 117
 on citrus, 64-65
 on guavas, 125
 on mangoes, 137
 on melons, 275
 on passion fruits, 153
 on strawberries, 248-249
 on watermelons, 275
Phytophthora palmivora, **286**
 on breadfruits, 166

on durians, 167
on papayas, 144-145, 147, **284, 285**
on sapodillas, 169
on sweet sops, 165
Phytophthora sp., 41, **30, 38**
Phytophthora syringae, **359**
on apples, 194-195, **357**
on citrus, 64-65
on pears, 194-195, **358**
Pineapples, 155-162, 289
aspergillus rot, 160
bacterial soft rot, 160
black heart, 162
black rot, 156-157, **302**
botryodiplodia rot, 160
chilling injury, 162, **309**
endogenous brown spot, 162
fruitlet core rot, 158-159, **304, 305**
interfruitlet corking, 160
internal browning, 162, **309**
leathery pocket, 160
pink disease, 160-161
rhizopus rot, 161
yeasty rot, 161, **308**
Pink disease
of pineapples, 160-161
Pink mould rot
of apples, 191, **353**
of avocados, 101
of bananas, 117
of citrus, 84, **203**
of melons, 275, **455**
of papayas, 150
of pears, 191
of plums, 234
of sweet sops, 165
of watermelons, 275
Pitting
of bananas, 113, **243**
of citrus, 87, **7, 208**
Plantains, 104-105, 290
see also Bananas
Plasmodiophoromycetes, 40, 41
Plasmopara sp., 41, **32**
Plasmopara viticola
on grapes, 264
Plectomycetes, 42
Pleospora herbarum **206**
on apples, 182
on citrus, 85, **204**
Pleospora sp., 43, **60, 67**
Plums, 218-237, 290
see also Stone fruits
Podosphaera leucotricha, **361**
on apples, 196, **360**
on pears, 196
Podosphaera sp., 43, **44**
on apricots, 234
on cherries, 234
on plums, 234
Pome fruits, 170-217
see also Apples, Pears
Pomegranates, 163, 290
anthracnose, 163
aspergillus rot, 163, **310**
chilling injury, 163
coniella rot, 163
pestalotiopsis rot, 163
scald, 163
Pomelos, 54

see also Citrus
Post-harvest
handling, 19-22
losses, 11-12, 28
physiology, 13-16
treatments, 20
Potebniamyces pyri
on apples, 192-193
on pears, 192-193
Powdery mildew, 43
of apples, 196, **360**
of apricots, 234
of cherries, 234
of gooseberries, 254
of grapes, 264
of mangoes, 137
of peaches, 234
of pears, 196
of plums, 234
of raspberries, 251
of strawberries, 251
Pre-cooling, 20-21
Pre-harvest factors, 18
Premature ripening
of bananas, 27, 104-105, **231**
of kiwifruits, 267
Premature senescence
of watermelons, 278
Pressure-cooling, 21
Prickly pears, 169, 290
Prochloraz, 20
Propiconazole, 20
Pseudocercospora purpurea
on avocados, 95, **218, 219**
Pseudocercospora sp.
on bananas, 114-115
Pseudomonas ananas
on pineapples, 158-159
Pseudomonas syringae pv. *lachrymans*
on melons, 272
on watermelons, 272
Pseudomonas syringae pv. *papulans*
on apples, 201
Pseudomonas syringae pv. *passiflorae*
on passion fruits, 153
Pseudomonas syringae pv. *syringae*
on avocados, 100
Pseudothecium, 42, 43, **67**
Puccinia sp., 45, **73, 83**
Puffiness
of citrus, 90
Pummelos, 54, 290
see also Citrus
Purple stain
of papayas, 150
Pustular spot
of apricots, 228-229
of cherries, 228-229
of nectarines, 228-229
of peaches, 228-229, **397**
of plums, 228-229
Pycnidium, 51-53, 281
Pycnium, 44, 281
Pyrenochaeta furfuracea
syn. *Phacidiopycnis malorum*
Pyrenochaeta mali
on apples, 202
Pyrenochaeta sp., **145**
Pyrenomycetes, 42
Pyrenopeziza sp., 43, **62**

Pyricularia angulata
on bananas, 113
Pyricularia grisea
on bananas, 113, **243**
Pyricularia sp., **97**
Pythium aphanidermatum
on melons, 273
on watermelons, 273
Pythium butleri
on melons, 273
on watermelons, 273
Pythium myriotylum
on melons, 273
on watermelons, 273
Pythium sp., 41, **29**
Pythium ultimum
on melons, 273
on strawberries, 251
on watermelons, 273

Rambutans, 169, 290, **215**
Ramularia tulasnei
on strawberries, 250
Raspberries, 238-252, 290, **409**
alternaria rot, 250
anthracnose, 240
blue mould rot, 250
cladosporium rot, 250
grey mould rot, 244-245, **417, 418**
leak, 246-247
phoma rot, 251
powdery mildew, 251
rust, 252
Recommended storage conditions, 286-290
Red currants, 254
Refrigerated storage, 22-23
Refrigerated transport, 24-25
Relative humidity, 281
during storage, 22-23
during transport, 24-25
recommended conditions, 286-290
Respiration, 13, 14-15, 16, 17, 20, 30, 281
Rhizoctonia solani
on melons, 276
on strawberries, 251
on watermelons, 276
Rhizoctonia sp. **82, 86**
Rhizopus arrhizus, 230
Rhizopus oryzae
on apples, 197
on apricots, 230-231
on cherries, 230-231
on grapes, 260
on guavas, 125
on litchis, 127
on mangoes, 137
on melons, 276
on nectarines, 230-231
on peaches, 230-231
on pineapples, 161
on plums, 230-231
Rhizopus sexualis, **421**
on raspberries, 246-247
on strawberries, 246-247, **420**
Rhizopus sp., 41, **36, 40, 41**
on figs, 168
on persimmons, 154
on pomegranates, 163
Rhizopus stolonifer, **401**
on apples, 197

on apricots, 230-231
on avocados, 101, **227**
on bananas, 117, **250**
on cherries, 230-231
on citrus, 85, **205**
on grapes, 260, **436**
on guavas, 125
on mangoes, 137, **276**
on melons, 276, **456**
on nectarines, 230-231
on papayas, 146, **287, 288**
on passion fruits, 153
on peaches, 230-231, **399, 400**
on pears, 197
on pineapples, 161
on plums, 230-231
on raspberries, 246-247
on strawberries, 246-247
Rhubarb, 13, 14
Ribbon scald
  of apples, 213, **378**
  of pears, 213
Rind discoloration
  of avocados, 103
  of citrus, 87, 90
Ripe rot
  of grapes, 264
  of kiwifruits, 269
Ripening, 15, 16, 17, 281
Rosellinia sp. **47**
Russeting
  of apples, 217
  of citrus, 90
Rust, 37, 45, 281
  of apricots, 232-233
  of nectarines, 232-233
  of peaches, 232-233, **402**
  of plums, 232-233
  of raspberries, 252

Saccharomyces spp.
  on strawberries, 252
Sapodillas, 169, 290
Satsumas, 54, 290
  see also Citrus
Scab, 37
  of apples, 198-199, **363, 364**
  of apricots, 234
  of avocados, 98, **223**
  of cherries, 234
  of citrus, 76, **19, 189**
  of loquats, 168
  of mangoes, 137
  of nectarines, 234
  of peaches, 234
  of pears, 198-199
  of plums, 234
Scald
  of apples, 213, 214
  of pears, 213, 214
Scale insects
  on citrus, 91
  on nectarines, 236, **407**
  on peaches, 236
Sclerotinia fructicola, **394**
  syn. *Monilinia fructicola*
Sclerotinia fructigena, **336**
  syn. *Monilinia fructigena*
Sclerotinia fuckeliana, **435**
  syn. *Botryotinia fuckeliana*

Sclerotinia minor
  on citrus, 66
Sclerotinia sclerotiorum
  on apples, 201
  on bananas, 119
  on citrus, 66, **174**
  on melons, 277
  on pears, 201
  on strawberries, 250
Sclerotinia sp., 43, **64, 68, 70**
Sclerotium rolfsii, 45, **87**
  on guavas, 125
  on melons, 276
  on watermelons, 276
  see also Corticium rolfsii
Sclerotium sp., **81, 85**
Seal-packaging, 20, 55
sec-butylamine, 20
Senescence, 14-15, 33
Senescent breakdown
  of apples, 212
Senescent scald
  of pears, 205
Septoria citri
  on citrus, 77
Septoria depressa
  on citrus, 77
Septoria limonum
  on citrus, 77
Septoria passiflorae
  on passion fruit, 152
Septoria sp., 51, **152, 156**
Ships, 24-25
Side rot
  of apples, 200
  of pears, 200, **367**
Sigatoka disease
  of bananas, 114-115
Skin discoloration
  of peaches, 237
Smut fungi, 37, 45
Snow mould rot
  of apples, 202
  of pears, 202
Sodium *ortho*-phenylphenate (SOPP), 20
Soft fruits, 238-252
  see also Raspberries, Strawberries
Soft nose
  of mangoes, 139
Soft scald
  of apples, 213, **378**
  of pears, 213
Soiling
  of citrus, 20, 55, 71, **181**
Solar injury
  of apples, 217
  of melons, 278
Solarization, 18
Sooty blotch
  of apples, 202
  of avocados, 101
  of citrus, 86
  of mangoes, 137
  of pears, 202
Sooty mould, 281
  on bananas, 117-118, **251, 252**
  on citrus, 86, **207**
  on mangoes, 137
SOPP (sodium *ortho*-phenylphenate), 20
Sour rot, 20

  of citrus, 78-79, **192, 193**
  of figs, 168
  of grapes, 264,
  of litchis, 127
  of melons, 277, **457**
  of peaches, 235
  of plums, 235
Sour sops, 164-165, 290
Spartan breakdown
  of apples, 217
Sphaceloma ampelinum
  on grapes, 261
Sphaceloma fawcettii var. *scabiosa*, **190**
  on citrus, 76
Sphaceloma perseae
  on avocados, 98, **223**
Sphaceloma punicae
  on pomegranates, 163
Sphaceloma sp., **128**
Sphaeropsis malorum
  on apples, 176, **326**
  see also Botryosphaeria obtusa
Sphaeropsis sp., **153, 157**
Sphaerotheca macularis
  on raspberries, 251
  on strawberries, 251
Sphaerotheca mors-uvae, **430**
  on gooseberries, 254
Sphaerotheca pannosa
  on apricots, 234
  on peaches, 234
Sphaerotheca sp., 43, **43**
Spilocaea eriobotryae
  on loquats, 168
Spilocaea pomi
  on apples, 198-199
Split pits
  in peaches, 237
  in plums, 237
Spongospora sp., **27, 37**
Spongy tissue
  of mango, 139
Star fruits
  see Carambolas
Stem-end breakdown
  of mangoes, 139
Stem-end rind breakdown
  of citrus, 91, **213**
Stem-end rots
  of avocados, 99, **224**
  of citrus, 80-81, **195, 196**
  of mangoes, 135, **273**
  of papayas, 147, **289**
Stemphylium botryosum
  on apples, 202
  on papayas, 150
  see also Pleospora herbarum
Stemphylium lycopersici
  on papayas, 150
Stemphylium sp., **116**
  on melons, 272
Stemphylium vesicarium
  on apples, 202
Stenella citri-grisea
  on citrus, 68
Stigmina carpophila, **398**
  on apricots, 228-229
  on cherries, 228-229
  on nectarines, 228-229
  on peaches, 228-229, **397**

on plums, 228-229
Stigmina sp., 51, **113**, **133**
Stomiopeltis citri
   on citrus, 86
Stone fruits, 218-237
   alternaria rot, 220, **386**
   anthracnose, 221, **387**
   bacterial spot, 222, **388**, **389**
   bitter rot, 221, **387**
   black mould rot, 234
   blue mould rot, 223, **390**, **391**
   botryodiplodia rot, 234
   botryosphaeria rot, 234
   brown rot, 224-225, **392**, **393**
   bruising, 236
   cladosporium rot, 226, **395**
   coryneum blight, 228-229, **397**
   cracking, 236
   diplodina rot, 234
   freezing injury, 236
   grey mould rot, 227, **396**
   insect injury, 236
   internal breakdown, 236
   internal browning, 236
   mucor rot, 234
   phomopsis rot, 234, **404**
   pink mould rot, 234
   pitting, 237
   powdery mildew, 234
   pustular spot, 228-229, **397**
   rhizopus rot, 230-231, **399**, **400**
   rust, 232-233, **402**
   scab, 234
   shrivelling, 236-237
   skin discoloration, 237
   sour rot, 235
   split pits, 237
   surface pitting, 237
   'woolliness', 237
Storage, 22-23
   recommended conditions, 286-290
Stowage, 24-25, 28
Strasseria oxycocci
   on cranberries, 253
Strawberries, 238-252, 290, **408**
   alternaria rot, 250
   anthracnose, 240-241, **410**, **411**
   black seed disease, 250
   black spot, 240-241, **410**, **411**
   blue mould rot, 250
   calyx blotch, 242-243, **413**, **414**
   cottony rot, 250
   grey mould rot, 244-245, **416**
   leak, 246-247, **419**, **420**
   leather rot, 248-249, **422**, **423**
   pestalotia rot, 251
   phomopsis rot, 251
   powdery mildew, 251
   pythium rot, 251
   rhizoctonia brown rot, 251
   stem-end rot, 242-243, **413**, **414**
   tan brown rot, 252
   yeasty rot, 252, **426**
Stylar-end breakdown
   of citrus, 91
Sugar apples
   see Sweet sops
Sulphur dioxide injury
   of grapes, 16, 255, 265, **444**
Sunburn

of apples, 217
Sunscald
   of apples, 217, 383
Superficial scald
   of apples, 214, 379
   of pears, 214
Surface pitting
   of cherries, 237
'Sweating', 237
Sweet sops, 164-165, 290
   anthracnose, 164
   black canker, 164, **311**
   botryodiplodia rot, 165, **312**
   chilling injury, 164
   gliocladium rot, 165
   phytophthora rot, 165
   pink mould rot, 165
Systemic fungicides, 20

Tamarillos, 169, 290
Tan brown rot
   of strawberries, 252
Tangelos, 54, 290
Tangerines, 54, 290
Tangors, 54, 290
Teleomorph (perfect state), 47, 281
Temperature
   during storage, 22-23
   during transport, 24-27
   recommended conditions, 286-290
   recorders, 26
Thanatephorus cucumeris, 45, 283
   on melons, 276
   on watermelons, 276
Thanatephorus sp., **78**
Thiabendazole (TBZ), 20
Thielaviopsis paradoxa
   on guavas, 125
   on pineapples, 156-157, **302**
   see also Ceratocystis paradoxa
Thielaviopsis sp., **109**
Thrips damage
   on bananas, 121, **256**
   on citrus, 91, **214**
Thyronectria pseudotrichia
   on avocados, 99
Toxins, 33, 123, 178, 257
   see also Mycotoxins
Transport, 24-26
Trachysphaera fructigena, **237**
   on bananas, 108-109
Tranzschelia pruni-spinosae var. *discolor*, **403**
   on apricots, 232-233
   on nectarines, 232-233
   on peaches, 232-233, **402**
   on plums, 232-233
Tranzschelia sp., **72**
Tree tomatoes
   see Tamarillos
Trichoderma harzianum
   on apples, 202
   on pears, 202
Trichoderma sp., **93**
Trichoderma spp.
   as antagonists, 189, 276
   see also Biological control
Trichoderma viride, **200**, **201**
   on apples, 202
   on citrus, 82-83, **198**, **199**

on pears, 202
Trichothecium roseum, **341**
   on apples, 182, 191, **353**
   on avocados, 101
   on bananas, 117
   on citrus, 84, **203**
   on melons, 275, **455**
   on papayas, 150
   on pears, 191
   on plums, 234
   on sweet sops, 165
   on watermelons, 275
Trichothecium sp., **102**, **106**
Tropical fruits, 92-169

Ugli fruits, 54, 290
Ulocladium chartarum
   on melons, 272
Ulocladium sp., **115**
Uncinula necator
   on grapes, 264
Uncinula sp., 43, **45**, **65**
Urocystis sp., 45, **71**
Uromyces sp., 45, **76**, **84**

Vacuum cooling, 21
Vascular browning
   of pears, 205, **371**
Vascular rot, 38, **22**
Vegetables, 14-15
Vein-tract browning
   of melons, 278
Ventilation
   of containers, 24, 25
   of packages, 21
   of ships' holds, 24, 25, 28
Venturia carpophila, **405**
   on apricots, 234,
   on nectarines, 234
   on peaches, 234
   on plums, 234
Venturia cerasi
   on cherries, 234
Venturia inaequalis, **366**
   on apples, 198-199
Venturia pirina, **365**
   on pears, 198-199
Venturia sp., 43, **57**
Verticillium sp., **95**
Verticillium theobromae, **238**
   on bananas, 108-109, 110-111, 116, **235**, **236**
Vibration damage, 16, 27
Viruses, 29, 39, **23**, **24**

Watercore
   of apples, 215, **380**
   of pears, 215
Watermelons, 270-278, 290, **448**
Watery breakdown
   of citrus, 87
Watery soft rot
   of bananas, 119
   of melons, 277
Waxing, 20
White currants, 254
White rot
   of apples, 176-177
   of pears, 176-177, **327**
Wilting, 13-14, 21, 22, 33

Wind-scarring
 of citrus, 91
'Woolliness'
 of nectarines, 237
 of peaches, 237
Wound healing, 17

Xanthomonas campestris pv. *mangiferaeindicae*
 on mangoes, 132, **269, 270**
Xanthomonas campestris pv. *melonis*
 on melons, 272
 on watermelons, 272
Xanthomonas campestris pv. *passiflorae*
 on passion fruits, 153
Xanthomonas campestris pv. *pruni*
 on apricots, 222
 on cherries, 222
 on nectarines, 222
 on peaches, 222, **388, 389**
 on plums, 222

Yeasty rot
 of citrus, 86
 of figs, 168
 of litchis, 127
 of pineapples, 161, **308**
 of strawberries, 252, **426**

Zygomycetes, 40, 41
Zygophiala jamaicensis
 on apples, 201
Zythia fragariae
 on strawberries, 242, **413, 414**
Zythia sp., **143**